MPS

Market-focused Production Systems

Manufacturing Practitioner Series

Stop Wasting Time: Computer-Aided Planning and Control, *Norman Sanders*
Just-in-Time Manufacturing in Perspective, *Alan Harrison*
Managing for Total Quality, *Nickolas Logothetis*
Beyond Partnership, *Richard Lamming*

Market-focused Production Systems: Design and Implementation

David J. Bennett
and
Paul L. Forrester

Prentice Hall

New York London Toronto Sydney Tokyo Singapore

First published 1993 by
Prentice Hall International (UK) Ltd
Campus 400, Maylands Avenue
Hemel Hempstead
Hertfordshire, HP2 7EZ
A division of
Simon & Schuster International Group

Typeset in 10/12 Times by
Mathematical Composition Setters Ltd, Salisbury, Wiltshire

Printed and bound in Great Britain by
T. J. Press, Padstow, Cornwall

Library of Congress Cataloging-in-Publication Data

Library of Congress Cataloging-in-Publication Data is available from the publisher.

British Library Cataloguing in Publication Data

A catalogue record for this book is available from the British Library.

ISBN 0–13–065129–X (hbk)

1 2 3 4 5 97 96 95 94 93

Contents

Foreword

I am delighted to have been invited to write the foreword to this book because it deals with what I believe has become one of the most important challenges facing manufacturing industry.

In recent years manufacturing has witnessed the most dramatic changes. Under relentless competitive pressure, companies like ours have needed to raise their efficiency levels, and to become more flexible. At ICL this has enabled us quickly and reliably to deliver competitively priced products that are manufactured in wider varieties and to improved levels of quality. It also means that our new products can be introduced into the marketplace faster, and in greater numbers than previously would have been considered possible.

Achieving these changes is, of course, no easy task. At ICL it has meant refocusing our business and developing new production systems to provide a competitive advantage and meet the exacting demands of the market. As a consequence we are one of the few computer manufacturers in the West to have been consistently profitable over the last decade. David, Paul and the other members of their research team spent considerable periods of time within ICL during the introduction of our new production systems, and the results of their study form the basis for much of this book. They have documented the experiences of other successful companies as well as evaluating existing design methodologies. The outcome is a book which balances theory with practice thus appealing to industrialists and researchers alike.

This is a timely book which answers the need for a comprehensive guide to the design and implementation of market-focused production systems. I am sure it will be well received by all who are concerned both with improving customer service and increasing productive efficiency.

Alastair Kelly
Director of Manufacturing and Supply
International Computers Ltd, UK

Preface

During the 1980s and 1990s the growing challenge for manufacturing companies has been to satisfy greater customer demands within a climate of increasingly intense competition. A new philosophy towards production systems design has, therefore, been required in which the emphasis is placed on improving product quality as well as performance, and where maximizing customer service is as important as minimizing cost.

This book considers the question of how companies can address this challenge by developing 'market-focused production systems'. Such a concept does not mean that production simply becomes subservient to marketing; market-focused production is a strategic perspective which requires a wholly different approach to material control, production control, quality and human resource management based on adopting proactive, market-oriented, strategies in each of these areas of management. Also, the demand for organizations to have greater flexibility and responsiveness means that creating market focus invariably involves restructuring the physical system.

This book serves two main purposes. First, it offers a thorough account of the principal features of market-focused production systems and describes the type of structured approaches that can be used in their design. Second, it provides a detailed description of a methodology and framework, called DRAMA, which takes the form of a set of guiding principles that can be used to assist the implementation and analysis of production systems within a market-focused context.

Market-focused Production Systems: Design and Implementation does not replace existing books on manufacturing strategy and system design, rather it is intended to supplement them. Its aim is to synthesize current thinking and it deliberately avoids taking the type of rigidly prescriptive approach often found in other texts. The ten components of DRAMA take the reader through the strategic, tactical and operational activities that are common to implementation projects. The starting point of DRAMA is the market and environment, being the foundation on which the other components are built. The DRAMA *methodology* allows industrial users to identify practical solutions within their own specific conditions. The DRAMA *framework* provides the student or academic researcher with a vehicle for understanding, describing or analyzing the total production system design process and its various elements.

Acknowledgements

This book is largely based on research work supported by the United Kingdom Science and Engineering Research Council under its ACME (Application of Computers to Manufacturing Engineering) Directorate, for which the authors would like to record their appreciation. David Bennett was Principal Investigator on this project and Paul Forrester a Research Fellow. Paul Forrester's subsequent work in developing the ideas emerging from this research have formed the basis for the bulk of this book; he also conducted the majority of work in preparing the text for publication.

It is impossible to acknowledge all the people and organizations that were involved in the original project but some of them must merit special mention.

For allowing access to their operations and providing industrial collaboration, the authors are particularly indebted to International Computers Limited. Many people at all levels within ICL provided help and much of their valuable time but particular thanks should be extended to Peter Wright, Manufacturing Engineering Manager, who acted as the Company's project co-ordinator and provided invaluable advice as the work proceeded. Thanks are also due to the numerous people in many other companies from which additional research data were gathered.

For their role in helping to manage the project the authors would also like to acknowledge the efforts of Mark Oakley of Aston University and John Hassard of Keele University, each of whom provided their own expertise to guide the work of the researchers.

Finally, it is imperative to mention the work of Sushil Rajput. As the other Research Fellow on the project 'Raj' conducted a substantial part of the investigative work and was heavily involved in developing the DRAMA methodology. His own input to the work underlying this book has been considerable and must not be under-estimated.

List of Industry Applications

The spanner symbol at the head of the page identifies an industry application on that page.

Introduction

Market-focused Production Systems is divided into three distinct parts, and the way it is used will depend on the background, prior knowledge, needs and interests of its readers.

Part I (Chapters 1 and 2) is essentially the background material necessary to provide an understanding of the principles and features of market-focused production. It will be of relevance to students and other readers with only a limited knowledge of the challenges faced by manufacturing companies in today's competitive environment and the responses that are necessary to meet them. More experienced readers who are familiar with modern industry could either disregard this material or use it as a means of relating their experience to the theoretical precepts and descriptive framework that has been used to articulate the important concepts contained.

Part II of the book (Chapters 3 and 4) is about production systems design models and methodologies. It links the concepts described in Part I with the specific methodology subsequently contained in Part III. It surveys a number of established models and methodologies before going on to describe the research basis and practical development of a specific methodology and framework for designing, implementing and evaluating market-focused production systems (DRAMA).

This part of the book will be of interest to researchers and practitioners who would like to supplement their existing knowledge and wish to understand the background to the DRAMA methodology and framework. The discussion on models and methodologies will also be of benefit to students by supplementing the more detailed descriptions provided elsewhere. It should be stressed, however, that the brief descriptions and evaluation of existing models and methodologies are by no means intended to provide a substitute for consulting the original sources. The extensive list of references and bibliography will allow readers to explore beyond this book if they wish to consider this material in greater detail.

Part III (Chapters 5 to 14) constitutes the bulk of the book. The ten chapters in this part each deal with one component of the DRAMA production system design methodology and framework. A particular feature of all the chapters in this section is that they follow a common format throughout, namely:

1. Theory and current practice.
2. Key parameters.
3. Selection and configuration of design options.
4. Generalized methodology.
5. Conclusions.

DRAMA is unique in that its scope is more extensive than that of most methodologies, which normally deal with a limited aspect of the production systems design process, such as manufacturing strategy, the planning of facilities or control system design. The ten components of DRAMA cut across the responsibilities and activities of a large number of business functions, a feature which in turn means that the book's potential readership goes beyond just those concerned with manufacturing engineering. The structure and content of Part III allows an organization to consult and apply the methodology in different ways. For example, a multidisciplinary approach could be taken in which an individual or small project group examines every component. Alternatively, the independent and distinct nature of the DRAMA components can enable separate functional specialists to refer only to those components that are of particular interest to themselves, possibly with a co-ordinator ensuring that a common corporate perspective is maintained. DRAMA's hierarchy embraces the strategic, tactical and operational domains of decision making, so it will be relevant to personnel at all levels of an organization. At the top level, senior managers and directors can gain an appreciation of the implications for the production function of adopting a market-driven strategy. At the lower organizational levels, managers and engineers can use the methodology to guide their production systems design decisions.

Although the methodology in Part III is of immediate practical relevance to industries who are developing a market-focused system, it can also provide value to readers other than practitioners. Once again, students and researchers will find benefit in much of the material, particularly the methodological framework, the approach taken towards linking strategy with production systems design and the discussion about theory and current practice.

Within each chapter are a number of Applications containing illustrations of some of the concepts and practices discussed. These include a series of applications relating to ICL, the company in which the original research work was conducted. In particular, the tactical and operational domain components of DRAMA are illustrated via examples based on the development of new production systems at ICL's plant at Ashton under Lyne, in the UK. Readers of the book will find that these Applications provide supplementary information on the practical approaches adopted in industry.

PART I

Market-focus and the Design of Production Systems

1 The Nature and Importance of Market-focused Production

Objectives of the chapter

On completion of this chapter you should:

1. Understand the meaning of 'market-focused' production and the ways in which it contrasts with a 'technology-led' approach.
2. Recognize the importance of 'downstream' production activities, particularly the assembly function, within market-focused organizations.
3. Appreciate how a market-focus demands a unique approach to the design of production systems.
4. Understand how product introduction, material planning, production control and quality management are affected by the need to adopt a market focus.

■ Chapter outline

Market-focused production is the name given to a philosophy based on the response required to meet the ever increasing demands of customers and the constant threat of competition. Traditional 'technology-led' production is oriented towards maximizing product performance and minimizing cost. Market-focused production highlights the importance of quality and customer service. In market-focused production, volumes are lower and 'economy of scope' replaces economy of scale. There is also a tendency towards greater customization.

To facilitate market-focus the emphasis is generally placed on 'downstream' production activities. By doing so greater flexibility and responsiveness can be achieved. Thus, greater attention is paid to assembly and final processing of products where materials represent a high proportion of manufacturing costs and added value is relatively low.

The management of market-focused production differs from the technology-led approach in a number of different ways. Materials management is usually of the 'make to order' type and material ordering is facilitated using the 'pull' rather than 'push' approach. Production control in market-focused production is based on the 'just in time' principle where the supply of materials and parts at each stage of production takes place at precisely the time and in exactly the quantities required. As well as minimizing inventory costs this enables the production systems to operate with maximum flexibility. Finally, a new approach to quality management is required based on creating a culture which is customer-oriented and sees quality as an asset rather than simply in terms of costs which must be minimized.

1.1 The emergence of market-focused production

The 1980s and 1990s has seen the emergence of a new philosophy towards production systems design and management. This new philosophy is generally based on the responses required to meet the ever increasing demands of customers and the constant threat of competition. Different names have been given to the type of organization that has emerged but the term 'market-focused' is the one we have adopted in this book, as it embraces the twin concepts of customer orientation and competitiveness.

What, then, is a market-focused organization? The idea of firms responding to market requirements is not new. Business strategists have long advocated the need to be more customer oriented and the term 'market economy' is widely used to describe the business environment of capitalist countries, as opposed to the alternative 'planned' or 'command' economy.

From the perspective of production management, however, 'market-focus' means more than simply having a manufacturing function that is subservient to marketing. Market-focused production means restructuring the physical system and the associated planning and control systems in order to equip production with the means of deriving a greater customer orientation and competitive edge, thereby having a *proactive* rather than a *reactive* manufacturing function. To understand the way the philosophy achieves its objectives it is necessary to compare 'market-focused' production with the more traditional 'technology-led' approach.

Technology-led production systems have, for many years, predominated in the established industrialized economies. A technology-led system is one in which product and process technology determines manufacturing strategy. It is based on a presumption that customer choice is influenced by two main factors, namely product performance and price. The result of such an approach is that production is of highly standardized products, which are manufactured in large volumes to achieve economies of scale and reduce cost. Such products then 'look' for markets, that is to say they try to attract potential customers by using their price advantage and technical performance as the main competitive factors.

The technology-led approach was appropriate when markets were less discriminating, when customers had less buying experience, when disposable income was lower and when financial credit not so widely available. The main objective, therefore, was to provide 'value for money' expressed as a function of technical performance and price.

Market-focused production systems, on the other hand, have been developed in response to more sophisticated markets. Here the emphasis is on quality as well as performance, and on customer service as well as cost. There is a tendency towards greater product customization and lower volume production with 'economy of scope' (the ability to produce, economically, a wide product range) replacing economy of scale.

A good illustration of market-focused production can be provided by the case of the electronics industry. In a relatively young industry which originally developed in the West, most of the major players relied on technological advances and production volumes to achieve and retain competitiveness. Their main focus was therefore on process innovations (such as the development of 'surface mount technology', automated testing, etc.) and product innovations (such as the development of 'large scale' integrated circuits and fiber optic data communications).

As the industry grew and matured, however, the major Western manufacturers became subject to increasing competition from, predominantly, the Far East. They found that 'clones' of their products could be produced quickly by companies at greatly reduced cost and, therefore, the basis on which they were competing was rapidly being eroded.

Faced with a situation which could eventually threaten their existence, the response among the traditional Western manufacturers has been dramatic. There have been efforts to reduce costs and to keep ahead technologically; however, the most significant response has been to change the basis of competition. The following company examples exemplify this change:

EXAMPLE

Xerox:

What Kearns (the Xerox CEO) sees as vital to his company's survival is to transform it from a seller of boxes to a purveyor of systems. The boxes are the copiers, electronic typewriters and laser printers that Xerox sells as stand-alone items. The systems are the integrated office products that enable companies to use their computers, printers and local area networks to create, manipulate and print documents entirely on their own premises ... To change the thinking of Xerox executives raised on boxes will be a mammoth task ... (Skapinker, 1986)

DEC:

One major stumble was its misguided attempt to launch personal computers onto the market, a wrong-footed effort to 'sell Rolls-Royces to people who want Volkswagens', in the estimation of one insider ... (Piper, 1985)

Hewlett-Packard:

... scrapped their structure of scores of divisions, organized by products, and regrouped the company around five 'business sectors' – organized by markets. The shift was a countercultural activity ... from product- to market-orientation ... Now more overt processes are needed to make sure the company understands its customers. (Knobel, 1988)

ICL:

... the first major strand of ICL's strategy is concentration on a few vertical markets rather than selling boxes across the board ... industry-specific business centres get close to customers, tailor the systems to specific applications and anticipate their needs. We've had to shift the emphasis from excellence in technology to excellence in markets. (Caulkin, 1987)

ICL used to 'deliver to the kerbside' but now it monitors the situation right through to the point where the system is up and running in the computer room. (Farish, 1989)

Philips:

... is trying multidisciplinary business units that cut straight across existing divisional lines ... (Hill, 1986)

Although different in their specific detail, all these statements represent an articulation of market-focus rather than technological dominance. They illustrate a desire to bring production closer to the customer and to 'provide solutions' or 'satisfy needs' rather than simply 'selling boxes of electronics'. In manufacturing terms it places an emphasis on the organization's interface with the customer which implies a shift in concentration from 'upstream' to 'downstream' activities.

The driving force behind market-focused production is, therefore, customer requirements. Of course these are likely to be diverse. They may be product-related requirements associated with performance, quality and reliability, variety, maintainability, etc., or they may be systems-related requirements, such as delivery times, delivery reliability, responsiveness to special customer requests, etc. Satisfying customers' requirements within a market-driven strategy should essentially be based on the process of creating superior value:

> Superior value is created when the benefits customers derive from superior performance is worth more than the price premium they have to pay. There are a myriad of ways to favourably distinguish a business or its products. Providing superior service, utilizing a strong brand name, offering innovative features or providing superior product quality are some of the favoured routes. (Day, 1990).

This emphasis on satisfying customers' requirements means that a company's marketing department must play a key role in determining the objectives to be met in designing and implementing a market-focused production system. The marketing department's role must cease to be one of simply selling products (or boxes) and should become one of primarily determining and serving customers' needs. Production's role then ceases to become one of simply making products for sale; instead it becomes one of supporting marketing's endeavours to satisfy customers' requirements. It is important that this should be seen as a logistics, or supply, role rather than one involving just manufacturing, in this way customers will be assured of their requirements being met from whatever source. A possible way of facilitating this new relationship between marketing and production can even be to separate them organizationally and allow them to operate as independent functions under a customer–supplier relationship.

Application 1.1 'Sell one, make one' in the factory of the future

It's all change on the production line. Manufacturing enterprises, at least those that are world leaders, are undergoing big changes. By the end of the decade almost all the yardsticks of performance they use today will have been replaced by simpler and tougher measures. Production as we know it will not exist in 10 years' time, nor will the accepted functions of companies involved in areas such as marketing, personnel and information management.

These forecasts are made by Dr Michael Oliff, of the International Institute for Management Development in Lausanne, Switzerland, in the context of a 10-year research programme he heads called Manufacturing 2000. Backed by 13 top manufacturing companies – including IBM, Sony, BP Chemicals, Exxon Chemical, Nestlé, Du Pont and others such as Andersen Consulting – the project aims to define and create the standards against which leading companies in the next century will measure themselves.

Dr Oliff explained his project, which is in its second year, on Wednesday at the Computers In Manufacturing (CIM) conference and exhibition at Birmingham's National Exhibition Centre tomorrow. He points out that he is dealing with the 'manufacturing enterprise' rather than with the technology or processes involved. Peter Dempsey, an associate of Andersen Consulting sees the factory of the future as an office-type environment in which mini-factories or work cells are enclosed within a plant, with the operators – managers, really – controlling them from outside.

Gone will be the pungent smell of cutting oil; instead each mini-factory will produce components for an end product to meet a current order. Flow lines will be short – and to the point. While each cell is completing one order its machines will already be being reprogrammed to handle the next.

These manufacturing plants will probably not be owned or controlled by its users. The latter will have had their design teams produce all the instructions – programs – for the factory to make what they want. The materials will be delivered to the point of use just before they are needed.

In the case of a complex product such as a car, many of the components can be pre-produced and delivered to the factory as 'materials'.

Perhaps curiously, in this age, Mr Dempsey predicts that the dependence on computers will be restricted to key aspects of manufacturing where commands can be issued in simple terms. Like Dr Oliff, he believes that people will become more important in the factory of the future, and that companies will invest more in humans than in capital goods or technology. But neither believes that industry will need more people – just better trained ones.

Central to the philosophy of Manufacturing 2000 are Dr Oliff's 'Ten Commandments for the Nineties'. The first is 'dominate or die'. He argues that the rate of innovation will be faster in the next 10 years than ever before. And, given the change in industrial leadership in world terms in the past 30 years, it is no longer enough for a manufacturer to be good in world terms, or even top in national terms; the necessity is to be the world leader. Parameters such as quality, competitive prices and fast delivery times will no longer set the best apart, because these will be natural to all companies.

The best will be those firms that 'create value with, not for, their customers', Dr Oliff says. The top companies will know their customer's business as well as the customer does himself. 'They will develop, produce and deliver products' he says. For example, in all its key markets today, Nissan takes 48 hours from the time an order is received to assemble and deliver a car to the dealer. It is introducing a concept called 'Sell One, Make One' (Somo); no car is assembled until it is ordered, but when it is, it becomes a 'special' model, with all the variations the customer wants built in. To achieve this, 90 per cent of the components are standard and are either on-hand or produced quickly; the remaining 10 per cent make it the special car, add the real value and govern the final price.

Dr Oliff argues that whatever the business – from steel to microelectronics – by the year 2000 all the top companies will have similar programmes – only they will be able to cut the lead time to 24 hours or less. Today's manufacturers use a mass production plant; those of the future will work through 'mass customization'. These manufacturers will also be responsible for the product's performance during its life, as computer systems makers increasingly are now, and the makers of products such as white goods are not.

For this to happen the people, the career structures, the design and fabrication processes will have to change. Companies will no longer have many career levels; sideways promotion will become the norm as people are repeatedly retrained to tackle new tasks. 'Restructuring will be a continual process – not a one-time activity' he says.

The other commandments elaborate on elements of the first. They are 'tear down the walls' (abandon conventional structures); 'create values' (by giving customers what they want); 'invest environmentally' (by recycling as well as avoiding costly and harmful pollution); 'eliminate waste' (of time, money and materials); 'whatever you do, do it quickly' (act rapidly, and if you're wrong, change it fast); 'get wisdom, not gold' (make sure your information is correct and up-to-date and is not being corrupted by out-of-date information technology professionals).

None of these commandments will be easy to meet – especially changing people's attitudes to work, promotion, status and the like. But Dr Oliff says every aspect of his current vision of the manufacturing enterprise of 2000 is visible in one form or another in leading firms today.

Source: John Stansell, *The Independent*, Monday 7 October, 1992, p. 16.

1.2 Downstream production activities and the assembly function

To understand the influence a market-focused strategy exerts on the design of a production system it is necessary to understand the distinction between the 'upstream' and 'downstream' activities in production. This is illustrated in Figure 1.1.

Upstream activities are those which take place early on in the chain of manufacturing. They involve production of component products and parts or they represent the early stages in a series of manufacturing processes. Production volumes for the upstream activities are generally large and varieties are small, the output being highly standardized. Material cost is also low compared with labour and overhead costs and, as a result, the added value element is high. Technology-led companies tend to concentrate on improvements in their upstream activities, as it is here that the greatest technological and cost benefits can be achieved. This would apply in the case of electronic manufacturing, as mentioned above, where the production of circuit boards and other component parts constitutes an area for technological and cost improvement. The concept also applies to other industries. For example, the engineering and

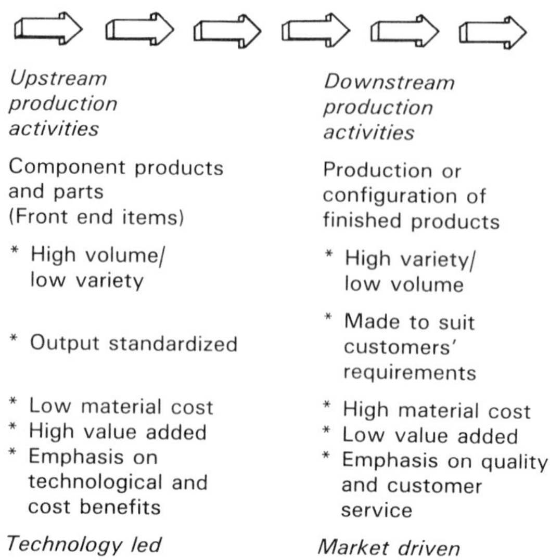

Figure 1.1 Upstream and downstream production activities.

automotive industries have upstream activities such as component and subassembly manufacture where technology-led companies will concentrate their efforts.

Downstream activities, on the other hand, take place later in the manufacturing chain. They occur where finished products are being produced or configured. This usually involves bringing together components and subassemblies in such a way as to suit customer requirements. Volumes for these products are often low and a wide variety of end-products are made. Materials usually represent a very high proportion of manufacturing cost so added value is relatively low. Market-focused companies tend to concentrate their efforts in improving the management of their downstream activities, as it is here that the greatest impact on quality and customer service lies. Market-focused companies demand greater responsiveness and flexibility from their production systems and try to develop an organizational culture which promotes an orientation towards product and process quality.

Downstream activities are usually assembly operations. However, they could also be final material processing operations or other customization procedures, such as software programming. In the electronics industry the final assembly, test and end-product configuration operations are all examples of the type of downstream activities that will affect the level of customer service and product quality. Other industries have similar types of downstream activities. For instance, the automotive industry provides many good examples of companies that have improved their quality and service by concentrating their efforts on improving their final assembly operations.

As well as moving the emphasis from the upstream to the downstream activities, market-focused production is achieved by adopting a number of philosophies towards the design and control of production systems. These philosophies often run contrary to more conventional, accepted, practice. They are now described under the areas of manufacturing to which they relate, namely material planning, production control and quality management.

■ 1.3 Product introduction in market-focused production systems: the increasing importance of 'time to market'

A significant feature of modern production systems is that the technology for manufacturing products has become increasingly available to companies on a world-wide basis. This extensive availability of production equipment not only relates to the products of independent suppliers, it is becoming more common for companies who develop their own processes in-house to then sell this technology to third parties, who may even include their competitors. The result of this situation has been a convergence in the technical capabilities of manufacturers, which is evident in the similarity of competitors' products.

As a result of this phenomenon, the way in which companies derive competitive edge has shifted to factors other than product and process technology. One of the more important of the alternative factors in which market-focused companies can compete is the speed with which they are able to introduce new products. This particular aspect of competition is often referred to as 'time to market'.

Typical product life-cycle curves for a company using 'time to market' as a means to compete are shown in Figure 1.2, where they are compared with life-cycle curves found in more traditional companies, which tend to take longer to introduce new products. By introducing products in advance of its competitors a company can generally expect a measure of market dominance until competing products from other manufacturers come on-stream. This market dominance can be exploited in two ways: a price premium can be added or extra sales can be 'creamed-off' while the product still has novelty value and no competition. Either way this will be reflected in increased revenue for the company, as illustrated in the figure.

Of course, it is easy to state that by reducing the 'time to market' a competitive advantage can be gained; it is more difficult to say how the reduction can be achieved. There are numerous complex 'process planning' activities that must take place from the time a product is originally conceived until it is eventually launched into the marketplace. Some of these are shown in Figure 1.3.

There may be scope for rationalization where there is overlap or duplication and, in certain cases, redundant activities could be eliminated. This, however, only partly addresses the problem; there will still remain a huge technical and organizational task, which can be approached in a number of ways:

1. **Product and parts standardization.** The tendency in many companies when introducing a new product is to start with a 'clean sheet of paper'

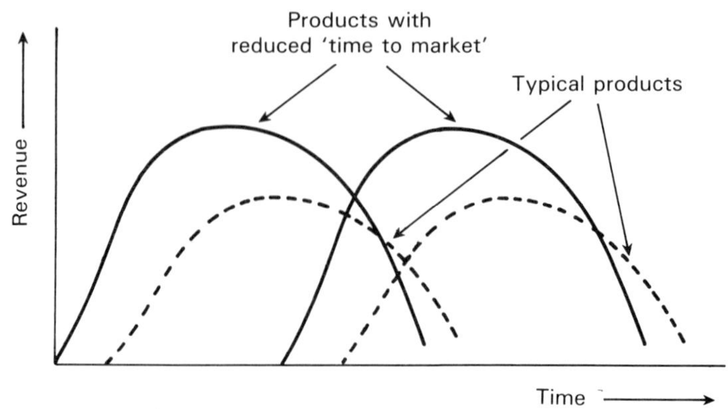

Figure 1.2 Product life-cycle curves and the 'time to market' concept.

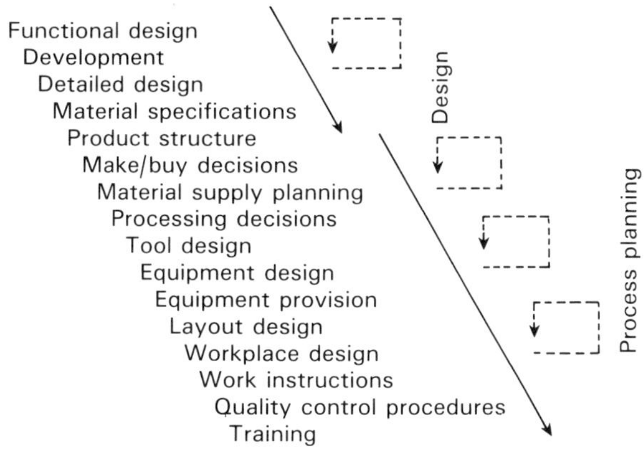

Figure 1.3 Typical activities required to bring a product to market.

and generate designs that are as original as possible. From a marketing point of view this may have the advantage of resulting in unique products but it is a time-consuming process and entirely new products often prove functionally unreliable.

Standardization may at first seem to work against the idea of creating new and innovative products. However, if product performance is not adversely affected and it is used within areas of the product which do not impact on the customer, standardization can prove effective in enabling products to be brought to market more quickly.

2. **Applying technology to product introduction activities.** Bringing a new product onto the market is traditionally a complex, time-consuming and labour-intensive process involving many separate functions of an organization. Technology, particularly in the form of computer-aided tools and techniques, can help the process to be carried out more quickly and efficiently. Computer-aided design (CAD), computer-aided engineering (CAE) and computer-aided process planning (CAPP) are all examples of the type of technology available. A further feature of modern computer-aided techniques is their use of common databases to avoid costly duplication of data and allow easier access to all the information necessary to design, develop and manufacture products.

3. **Using project management and other related techniques to reduce the process planning time-scale.** The introduction of a new product (and the implementation of a production system) is a 'project', and project management is an important means of minimizing the time taken and resources used in process planning.

Project management will be considered in detail in Chapter 9 but it is appropriate here to provide a brief description of two important techniques, network analysis and simultaneous engineering.

Network analysis is one of the established project management techniques that can be used to good effect. It identifies which activities are 'critical' to ensuring that a project is completed on time and provides the means of avoiding delays to these activities. More recently, simultaneous or concurrent engineering has emerged as another effective means of shortening the process planning period. Quite simply this is an approach to process planning which addresses the problems associated with carrying out the necessary product introduction activities sequentially. Instead of waiting for each activity to be completed before starting the next, information is passed between functions as soon as possible in order to allow commencement of the next planning stage and in this way the overall time scale is reduced.

Given the benefits of bringing products to market more quickly what are going to be the costs? While some of the methods described do not involve the organization in any significant extra expenditure, others, such as using computer-aided tools may require significant capital investment. These costs would generally be justified by the increased revenue generated after the product's introduction. In terms of cash flow this argument is illustrated in Figure 1.4.

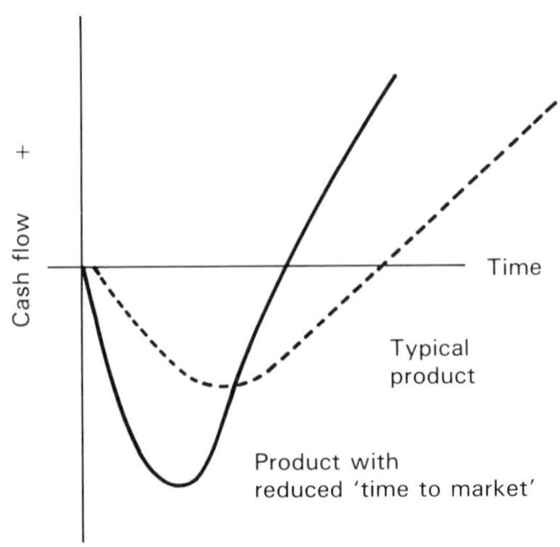

Figure 1.4 The costs and benefits of shortening the 'time to market'.

1.4 Material planning in market-focused production systems

A further feature which distinguishes technology-led from market-focused production systems is the approach taken towards material planning. There are two relevant aspects of material planning: (1) The concept of 'make-to-order' versus 'make-to-stock'; and (2) 'Pull' versus 'push' material ordering systems. Generally, market-focused production is based on make-to-order and pull material ordering while technology-led production is concerned with make-to-stock and push ordering.

Make-to-order versus make-to-stock

Make-to-order means that finished products are made only when they are subject to a particular demand, and then they are only manufactured in the exact quantities required to meet the customer's order.

Make-to-stock means that finished products are held in stock in sufficient quantities to meet anticipated demand for a period. Orders are generated by an inventory control system to replenish the stock. These orders are placed either when stocks reach a prescribed level or at regular review periods.

Make-to-stock systems are typically used in technology-led organizations because they are ideally suited to situations where products are highly standardized, where volumes are large and varieties low. One of the reasons for this is that safety stocks need to be kept to allow for possible increases in customer demand and the expense associated with keeping safety stocks can be kept at manageable levels where the range of items is narrow.

Traditionally, make-to-order was limited to items manufactured to customer specifications, such as heavy engineering products. Make-to-order, however, is now one of the principal features of market-focused production because it facilitates many of the objectives associated with achieving higher levels of customer service. An additional benefit, of course, is that it avoids the cost associated with holding finished goods stocks.

Pull versus push material ordering systems

A further philosophical distinction within material planning relates to the idea of 'pull' and 'push' ordering systems. In some respects this distinction is similar to the idea of making finished products to order or to stock but it is applied to the various component parts, subassemblies and raw materials from which the product is manufactured.

'Pull' material ordering systems have acquired a recent popularity in connection with market-focused production systems, although the principle is actually very old – it is the idea behind conventional inventory-based systems. In an inventory-based material ordering system a stock of materials, components, etc. is held at each stage of production and used as required in the manufacturing process. When the stock of parts needs to be replenished a demand, often an 'economic order quantity', is placed on the immediately preceding process. Parts are thereby 'pulled' from one stage of production to the next as shown in Figure 1.5. An example of an inventory-based approach is the simple 'two bin' system where the maximum stock level is when both containers of parts are full. Parts are taken from one container only, until it is empty. At this point an order is placed to replenish the stock in this container and parts are now taken from the second container.

A 'push' material ordering system is one where orders for finished products are 'broken down' into orders for the appropriate parts, which are then produced according to a schedule which determines when they are to be made and in what quantities. In this way they are 'pushed' through the various stages of production, as shown in Figure 1.6. The best known push system is material requirements planning (MRP). Inventory control-based systems and MRP are described more fully in Chapter 11. Push systems can be readily applied in technology-led companies because, again, the larger volumes and lower varieties of standardized products enable the necessary material requirements and order schedule calculations to be computed more readily and it is easier to balance load and capacity.

The renewed popularity of push systems in market-focused production is largely due to the development of the up-to-date equivalent of the

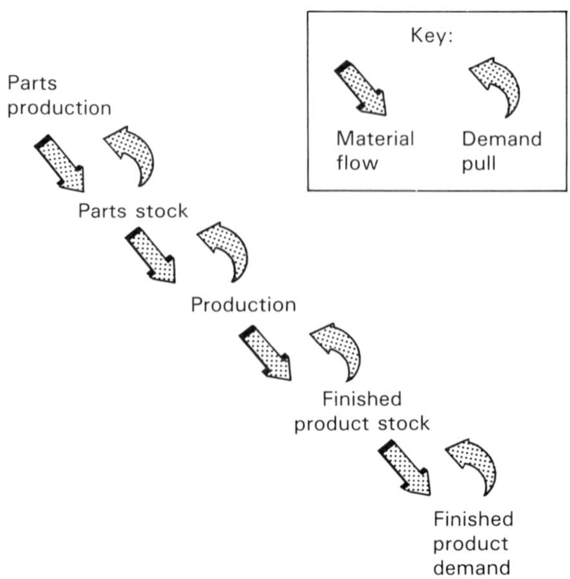

Figure 1.5 Inventory-based (pull) material planning systems.

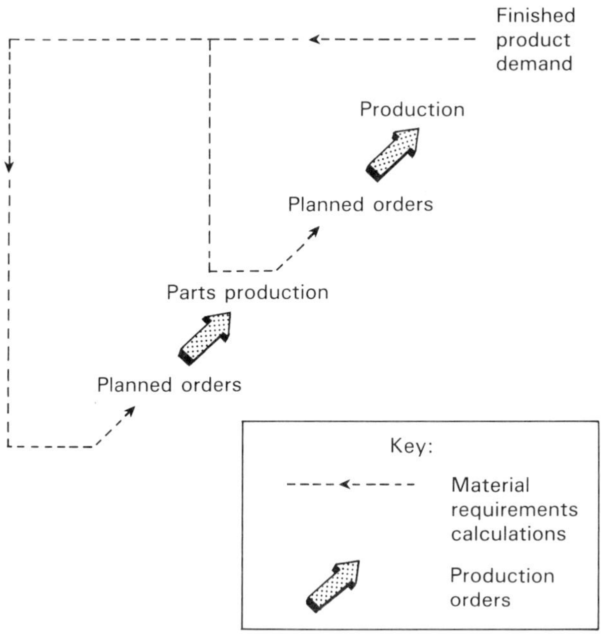

Figure 1.6 Requirements-based (push) material planning systems.

conventional inventory-based approach in the form of the well-known 'kanban' system. Kanban was originally developed by the Toyota Motor Company in Japan. In Japanese the word 'kanban' simply means 'card' and the kanban system is based on the idea of using cards to signal and authorize the production of parts or move parts between processes. The cards or kanbans remain within the physical production system and must accompany their associated parts or materials when they are moved. Thus the number of kanbans in circulation at any one time, and the quantity of parts they authorize to be processed or moved, places strict control over the amount of inventory that is allowed to be held within the system. The operation of a typical kanban system is shown in Figure 1.7.

The kanban system is ideally suited to market-focused production because experience has shown that it works well with low volumes and high varieties of parts, provided their demand is clearly understood. It also ignores the idea of economic order quantities, being based on the 'improvement approach' rather than the 'optimization approach', as described later in Chapter 2. An additional benefit is that it does not necessarily require the use of sophisticated computer software for it to function effectively. Being a 'pull' system it is also sensitive to short-term customer demands. It is superior to more conventional inventory-based systems because of its ability to keep stock to a minimum, as well as ordering parts when they are needed. It is, therefore, usually used as the means of ordering within the 'just-in-time' type of production control

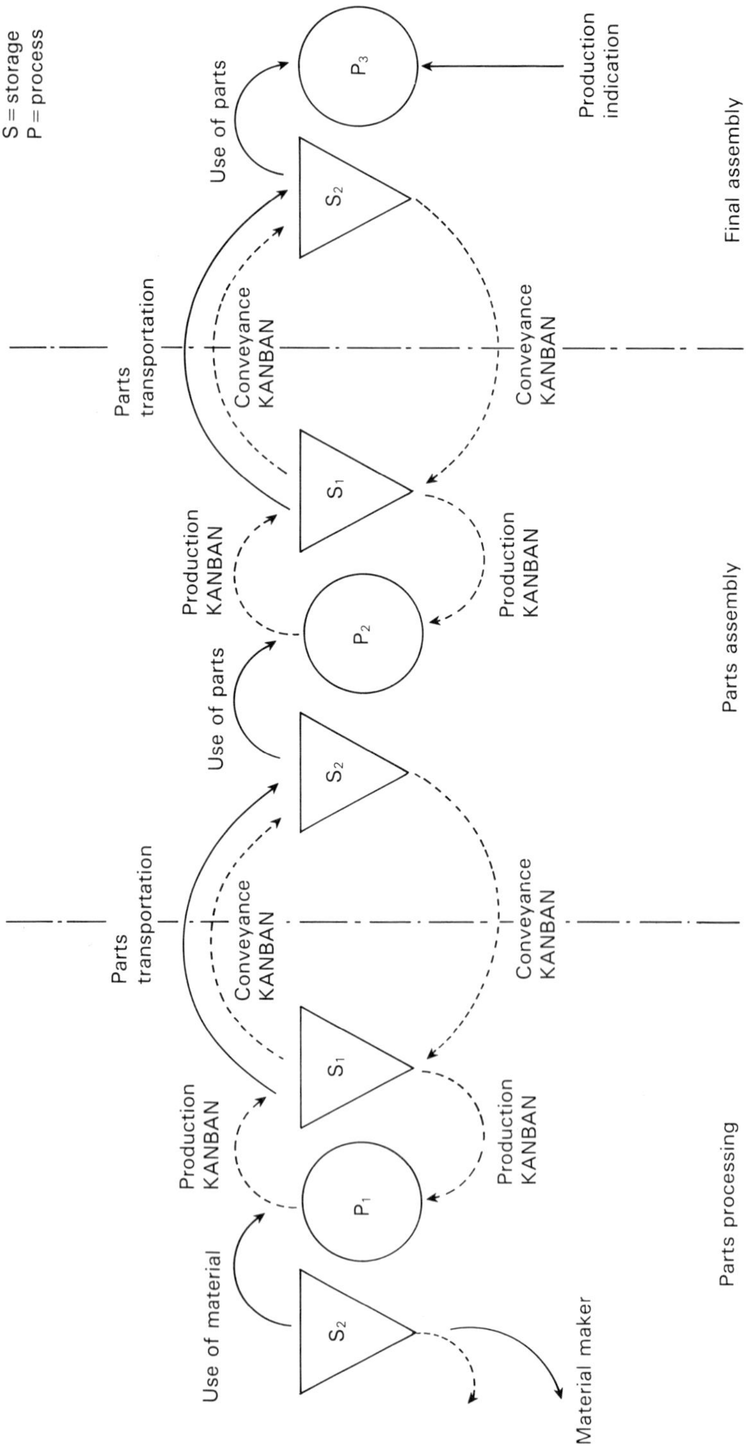

Figure 1.7 Circulation of kanbans within a just-in-time production system.

system, which is described in the next section and will be considered in greater detail in Chapter 11.

In practice the kanban ordering system has been developed and refined in a number of different ways. There are in particular several variations on the original idea of using cards to requisition and accompany materials while still maintaining the principle of keeping work in progress to a minimum. One variation is to use containers for this purpose, with each container being exclusively assigned a quantity of a specific part. Another approach is to assign parts to a marked space of a certain size on the factory floor (sometimes called a 'kanban square'). Of course this can only be used for requisitioning purposes and another means must be used to identify parts moving between processes.

1.5 Production control in market-focused production systems

> Production control has been defined as: The function of management that plans, directs and controls the material supply and processing activities of an enterprise, so that specified products are produced by specified methods to meet an approved sales programme, these activities being carried out in such a manner that the labour, plant and capital available are used to the best advantage. (Burbidge, 1978)

Once again there are different methods of achieving the broad objectives of production control depending on the philosophy underlying the particular approach being adopted. The approach that is most appropriate to market-focused production is generally referred to as 'just-in-time' production control.

In its simplest sense 'just-in-time' means, literally, that the supply of materials and parts at each stage of production takes place at precisely the time it is required. It also implies that they are supplied in the exact quantities necessary to satisfy immediate needs (Monden, 1986). The 'kanban' system, described previously, is a mechanism whereby 'just-in-time' supply can be achieved. The 'just-in-time' approach therefore contrasts with the opposite approach, which could be described as 'just-in-case' production control (although it is not formally referred to in this way). The idea of supplying materials and parts in advance of when they are required and in larger quantities than are actually needed is justified 'just-in-case' an unforeseen circumstance arises. 'Just-in-time' production control requires that the company either removes the unforeseen circumstances or protects against them in order that costly safety measures (in terms of inventory holding) can be avoided.

With the background to 'just-in-time' production control explained it can be seen that a further dimension to the approach is the development of the ways and means of removing, or protecting against, the unforeseen circumstances referred to. Simply to implement a 'kanban' type ordering system is not enough. A company's ability to effectively tackle the reasons for ordering materials 'just-in-case' is the key to successful implementation of 'just-in-time'.

There are a number of areas of manufacturing that must usually be improved before 'just-in-time' production control can be adopted successfully. These include quality management, ordering procedures, process set-up times, maintenance policies, work in progress control and purchasing and supply chain management. Quality management will be considered in detail in the next section and the other areas receive attention elsewhere in this book. The importance of purchasing and supply chain management to the success of just-in-time production control merits more discussion here because, in the past, its role has often been under-estimated.

If required, just-in-time production control can be implemented only within the production operations of a single site but will be more effective if extended to the organization's series of suppliers (i.e. along the supply chain). This, however, requires profound changes in the policies of both customers and suppliers (Lamming, 1987). A particular aspect of change is the move towards new supplier relationships and, in particular, the creation of partnerships, where suppliers can be viewed as extended factories (Suzaki, 1987). In such a partnership customers and suppliers work together to develop an integrated manufacturing system by eliminating the wastes that are typically accumulated between the organizational boundaries. For example, the kanban system would be applied between customer and supplier to enable frequent, small-batch, on-time delivery of materials.

Other features of an effective partnership include co-operation on design, quality and cost improvement. Furthermore, the relationship can be developed along the supply chain and into an extended family grouping where concern and support is freely given and received (Macbeth and Ferguson, 1991).

■ 1.6 Quality management in market-focused production systems

It is unusual nowadays to find a manufacturing company that does not recognize the importance of quality management. Quality management, however, is a subject which can be approached in different ways depending on the view taken by the organization towards the concept of quality. The conventional view of quality is to regard it as a burden that is seen only in terms of costs; as a result the company's objective will be to achieve a minimum level of product quality that will be accepted by customers for a given price. This can be described as a *reactive* view of quality and typifies the attitude towards quality management that is often taken by technology-led companies. They will see quality costs as comprising two components, the cost of defectives and the cost of prevention. As the amount of quality control activity increases it is assumed that the cost of defectives is reduced but the cost of prevention becomes greater. Thus it is argued that there is an optimum amount of quality control activity that will minimize the total cost.

The alternative view of quality is to regard it as an asset. In this case it is seen as a competitive weapon which can be used to increase sales or justify a price premium. This view of quality can be described as *proactive* and is the key to quality management in market-led companies. As well as appreciating the use of quality as a means of competing they also adopt a wider perspective towards quality costs, recognizing the consequential and legal implications of producing defective products as well as the direct costs.

Market-led companies usually try to create a 'quality culture', which goes far beyond simply trying to improve the quality of products. They understand that customers' perceptions of quality embrace their overall impression of an organization. Quality management is therefore implemented as a company-wide philosophy based on the idea of creating an awareness of everybody's contribution to quality, no matter what their function. The expression total quality control (TQC) or total quality management (TQM) is often used to embrace this concept. The idea is in marked contrast to the more traditional approach taken by technology-led, functionally oriented, companies where quality management is regarded as a separate activity and inspection and quality control techniques are carried out by specialist staff.

The creation of a quality culture in an organization cannot be achieved through instant solutions or a 'quick fix'. People's attitudes, values and behaviour must be changed and this can only be accomplished through leadership, management style and example.

Within this general approach to changing culture, however, there are a number of specific techniques and procedures which can be used to enable and assist the 'quality process'. Included among these are a number of well-known approaches that have been developed in the United States and, more recently, in Japan. 'Zero defects programmes', 'quality circles' and 'kaizen' are three such approaches which merit discussion.

Zero defects

The idea that everyone in an organization can make a positive contribution to quality improvement is embodied in the 'zero defects' philosophy. The zero defects idea resulted from a desire within the US defence industry to reduce the number of system defects, almost literally, to zero. The Martin Company, contractors for the Pershing missile system, developed the concept at its plant at Orlando, Florida (Halpin, 1966). The need for such an approach arose due to the complex nature of the Pershing system. Experience showed that with 25,000 parts the system was likely to suffer from a high number of mission failures unless a concerted effort was made to detect and rectify each and every quality defect. The company's solution was to devise a programme which encouraged a contribution from every employee towards the total quality effort.

Success of a zero defects programme depends on planning, organization and co-ordination. It would normally run under the direction of a Zero Defects Administrator, who has an open channel of communication with all functions likely to be in a position to influence quality in any way. Participation in a zero defects programme should be voluntary, although full commitment is expected from those who do participate. To this extent a 'pledge' is often signed by participating company employees. The introduction of a zero defects programme is usually heralded by a launch or 'kick off'. This will often take the form of a rally attended by all employees and addressed by senior executives. Awareness is also raised via a campaign of posters, badges, stickers, etc.

Of course the promotional and motivational aspects of a zero defects programme cannot alone ensure success, although they are essential ingredients. Equally important is the underlying organization, which ensures that each individual contribution produces maximum yield. Defects can only be reduced towards zero by preventing their occurrence rather than relying on detecting them. Therefore, important features are 'error cause identification' (ECI) and 'error cause removal' (ECR), whereby an employee can make suggestions regarding elimination of the reasons for generating defective items.

Financial rewards are not normally given to employees in return for their contribution to quality improvement, but certificates, prizes, badges, etc. are commonly used to recognize special achievement.

In the Martin Company an audited saving of nearly $1.7 million was recorded over the first two years after the introduction of its zero defects programme and there was a 54 per cent drop in defects during the same period. With respect to individual efforts, a forklift truck driver moved around 7 million kilograms of delicate electronic hardware without a single incident of damage and a solderer made almost half a million connections without error. Such results demonstrate that workers are capable of delivering high quality and high productivity at the same time if the job is properly designed and management creates the climate for producing minimal defects.

Quality circles

Quality circles go under various other names such as Quality Control Circles, Quality Control Groups, Improvement Teams, etc. They were established in Japan in 1962 at about the time zero defects programmes were being introduced in the United States.

The quality circles idea was initiated through the Union of Japanese Scientists and Engineers (JUSE) following publication of a magazine *Quality*

Control for Foremen. As their name suggests, quality circles are based on the creation of groups rather than placing the focus on individual contributions to quality improvement. Circles are usually led by a departmental foreman, supervisor or other qualified person, and consist of between five and ten employees. They are usually drawn from the same department, although cross-functional groups can also be created. As with zero defects programmes, participation should be voluntary. By comparison, however, their introduction is usually gradual so that successes can be built upon progressively.

In many companies the overall quality circle programme is co-ordinated by a 'facilitator', who organizes the training and acts as catalyst to enable the groups to maximize their contribution to the overall goals of the organization. Apart from their focus on groups the other distinctive feature of quality circles is the training aspect. Training of group leaders is fundamental to the success of the idea. A leadership course is usually provided, aimed at teaching basic problem-solving techniques which the leader will then pass on to group members. This training passed on by the group leader is usually supplemented by additional courses in quality control methods.

Ideally every member of the group should be equipped to use the so-called quality circle 'tools', i.e.:

1. Pareto analysis.
2. Cause and effect (or 'fishbone') diagrams.
3. Stratification.
4. Check sheets.
5. Histograms.
6. Scatter diagrams.
7. Control charts.
8. 'Brainstorming' techniques.

Despite their proven success in Japan it was often claimed that quality circles could not be applied in the West because they were only suited to the Japanese industrial environment. However, evidence now appears to prove otherwise and there are many cases of their successful introduction in the United States and Europe, as well as many other countries around the world. There are frequently quoted quality circle 'failures', but these often prove to be because they have been introduced involuntarily or too many groups have been created to allow them to receive sufficient attention from management.

'Kaizen'

Kaizen is a Japanese term meaning *continuous improvement*. It is not an alternative approach to zero defects or quality circles or any other similar programme, rather it is intended to supplement them with the aim of providing added impetus. A problem with any quality improvement programme is that its effect is often short lived and the initial enthusiasm of participants is diminished or lost. Kaizen addresses this problem by emphasizing the fact that quality improvement is a never-ending mission and not an end in itself. It often involves the establishment of management-initiated corrective action teams, which are quality- (and in particular process-) focused and concentrate on identifying causes of errors or problems and proposing solutions.

According to Imai, kaizen should be a part of the working ethos of all members of the organization (Imai, 1986). It involves continuously exploring for improvement via:

1. Achievement of customer orientation.
2. Total quality control.
3. Technology and automation.
4. Suggestion schemes.
5. Workplace discipline.
6. Just-in-time.
7. Zero defects.
8. Co-operative labour and management relations.
9. Productivity.
10. Innovation.

The distinctive features of kaizen are, therefore, that it is a *dynamic process* rather than relating to *static problem solving*. It is also applied to all areas of production rather than just quality.

Application 1.2 The application of kaizen within British Alcan

Kaizen was introduced to British Alcan in 1990 through contact with Alcan's Japanese associate company, Nippon Light Metal, who have achieved significant benefits in using the technique. The literal meaning of the Japanese word 'kaizen' is 'continuous improvement', but it has acquired a rather special meaning in British Alcan.

It describes a simple, structured method of driving performance improvement in a specific part of the business using a small, full-time team of employees over a period of a few days. These people are not predominantly improvement professionals but may be drawn from many different parts of the organization.

The team is set a challenging target, to be achieved by the end of the project, and their efforts are concentrated on identifying and eliminating waste (or 'muda' in Japanese), defined as any activity which does not add value to the business. Through careful selection of kaizen projects by management, the overall aim is to increase the productivity of the whole organization and to provide the customer with a more responsive and flexible service.

British Alcan originally used kaizen to tackle problems in their manufacturing operations but, as experience of the technique has grown, they have started to apply it in administrative and service functions also.

The emphasis of kaizen is on implementing change immediately rather than making recommendations for subsequent consideration and approval. In this way it harnesses the potential of the individual, often under-utilized by traditional management practice, and is a powerful vehicle for achieving results quickly. Kaizen is particularly exhilarating for employees working in the project area who have often lived with the problems for a long period, despite having raised them many times.

This style of generating improvement requires a good understanding of the process and strong commitment to change from management. From defining the project scope and targets to supporting the teams' work, their role as leaders and facilitators is crucial.

Before a project starts, the kaizen team receives a basic, intensive training in the methods of kaizen. The team then works on the project for a fixed duration to achieve a pre-defined target. The approach follows a systematic problem-solving procedure summarized by the '10 Steps':

1. Prepare
2. Observe and understand
3. Measure
4. Chart
5. Analyze
6. Improve and implement
7. Check the results
8. Document and train
9. Communicate the results
10. Standardize and improve

The outcome of the project will be an improved working method which becomes the standard until further improvements are developed and incorporated. The safety and quality requirements of the job will always be identified and incorporated in this standard. Each project is distinct but the overall process is continuous. The steps of a kaizen project define a cycle of involvement, analysis, action and monitoring which is mirrored in the organization as a whole. Kaizen projects can only be effective when the organization has a culture of continuous improvement which supports the identification and elimination of waste.

1.7 Conclusions

This chapter has highlighted the emergence of market-focused production as a response to the need to satisfy the ever increasing demands of customers and the threat of competition. Market-focus means more than simply having a production function that is subservient to marketing, it requires a different approach to product introduction, material planning, production control and quality management. This generally means adopting proactive strategies based on decreasing the 'time to market', using 'make to order' production and 'pull' type ordering, adopting 'just-in-time' control and the creation of a 'quality culture'.

Market-focused production also involves restructuring the physical system. As well as facilitating the flexibility and responsiveness required to achieve greater customer orientation, physical system redesign also removes or reduces many of the constraints which limit the possible improvements in materials efficiency and quality. The next chapter takes up this theme by examining the structure and features of conventional production systems and the ways in which alternative systems can be developed from which improved performances can be achieved.

■ 1.8 Further reading

Gaynor, G. H. (1991) *Achieving The Competitive Edge Through Integrated Technology Management*, McGraw Hill, New York.

Harrison, A. (1992) *Just-in-time Manufacturing in Perspective*, Prentice Hall, Hemel Hempstead.

Lamming, R. (1993) *Beyond Partnership: Strategies for Innovation and Lean Supply*, Prentice Hall, Hemel Hempstead.

Puttick, J. (1987) Marketing pull – manufacturing push, in R. Toone and D. Jackson (ed.) *The Management of Manufacturing*, IFS Publications, Bedford, UK.

Stephanou, S. and F. Spiegl (1992) *The Manufacturing Challenge*, Van Nostrand Reinhold, New York.

Steudel, H. and P. Desruelle (1992) *Manufacturing in the Nineties*, Van Nostrand Reinhold, New York.

2 Concepts and Principles in Production System Design

Objectives of the chapter

On completion of this chapter you should:

1. Recognize the various elements which comprise a total production system and understand the way work organization and facility layout determines the nature of the system.
2. Be aware of the features, benefits and deficiencies associated with conventional systems.
3. Appreciate how alternative production systems can overcome many of the deficiencies associated with conventional approaches and enable production to become more market-focused.

Chapter outline

This chapter outlines the importance of the design of production systems to the overall performance of an organization and explains how the drawbacks associated with conventional production systems can lead to the need for 'trade-offs' between business objectives. It then outlines how alternative production systems can enable the reduction and removal of the constraints within which optimal solutions would otherwise need to be sought.

The main premise is that a system's transformation process (i.e. its combination of physical facilities and human resources) provides the means of extending the limits of performance. Planning and control systems are important but what they can achieve is restricted by the system's inherent characteristics.

In order to describe the various benefits and drawbacks of conventional production systems they will firstly be defined according to the type of work organization and facility arrangement they represent. This same framework will then be used to explain how alternative systems can be based on designs that seek to adopt the favourable features of the conventional systems while at the same time avoiding many of the associated drawbacks.

The chapter includes a brief account of cellular organization as an effective alternative to conventional approaches to both batch and flow production. Two forms of cellular organization, Group Technology and Autonomous Working, are described and their implications for production control, product design and payment systems are considered.

The chapter concludes by stressing the importance of implementing a detailed design of production systems that will satisfy the market, strategic and organizational objectives of the enterprise. It finally justifies the need for a detailed and thorough methodology to facilitate new production system design and implementation.

2.1 The elements of a production system

The production system is that part of an organization comprising the facilities, labour and managerial functions required to manufacture physical products and other outputs from materials and other types of input. Materials may be in many forms including raw material, component parts, subassemblies, etc. The other inputs to the production system are often numerous and can be both tangible and intangible. These may include (to mention only a few) consumable materials and supplies, financial resources, skill and knowledge (of products and processes), outside services, etc. (Bennett, 1986).

The outputs from a production system also can be both tangible and intangible. For example, apart from the physical product other outputs may include instruction in the product's use, after sales service and (in the case of computer equipment) software.

Most production systems are extremely complex and considerable time and effort is usually expended on their design and implementation. One of the major problems when designing and operating a production system is the difficulty of being able to regard the system as anything other than an amorphous whole, or the proverbial 'black box', which can be viewed only in its totality.

Engineers, of course, can see the physical facilities in terms of individual machines and other items of equipment. However, production system designers and operations managers also need a means of identifying the various elements which together comprise the total production system.

At the most basic level it is useful to think of the production system as actually comprising two constituent parts (or subsystems), as shown in Figure 2.1.

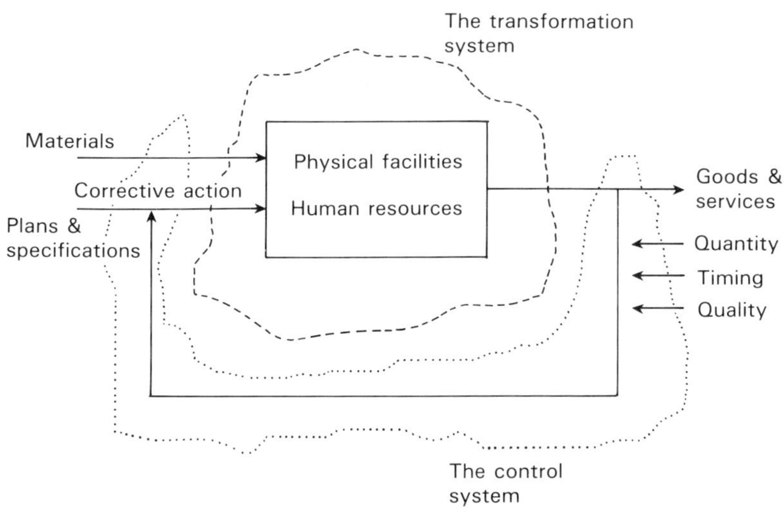

Figure 2.1 The components of a production system.

The first of these systems is the transformation system. This is essentially that part of the total production system which is necessary to physically transform materials into finished products. It therefore comprises all the machines, tools and equipment needed to cut, mould, assemble, inspect, move and store all the direct materials and consumables used in the production process. We can call all these items the physical facilities. Apart from the physical facilities there are, of course, all the people needed to set, operate and maintain the machines, move the material, inspect the products, etc. These we will call the Human Resources.

Thus, production systems are characterized by:

1. Organization of work (the way the human resources are organized).
2. The choice and arrangement of the physical facilities.

Much of the effort to enhance the performance of production systems has been directed towards planning and control. The result has been the development of a plethora of techniques and software packages aimed at improving capacity planning, work scheduling, inventory control, etc. However, many of the weaknesses associated with a production system are a result of the underlying transformation system and any benefit derived from implementing better control systems will be limited by the inherent constraints on material flow efficiency imposed by the choice of work organization or arrangement of physical facilities. This chapter therefore concentrates not so much on the planning and control aspects of production system design, rather it considers the less fashionable, but perhaps more important, question of design of the transformation system. It is here where many of the constraints on improving planning and control can be removed and where systems can be developed with greater market focus.

■ 2.2 Work organization and facility layout options

It was stated earlier that the two main parameters shaping the design of a production system relate to: (1) the human resources; and (2) the physical facilities. Let us now examine the choices open to the designers of production systems within these two broad parameters.

Work organization options

Taking first the human resources. Work organization can take one of three forms (based on, and extending, the ideas of Hayes and Schmenner, 1978). It can be:

1. Product oriented.

2. Process oriented.
3. Task oriented.

Product-oriented work organization

This is based on the idea of a worker or team performing all the work elements required to manufacture a product. The work elements are performed as a set, though not necessarily together nor in a particular order. It is represented conceptually in Figure 2.2.

It might be speculated that this was the original way in which workers organized themselves. Its attraction lies in the fact that it does not necessarily require the preparation of formal production plans. Individuals or groups can work independently and different products can be manufactured in parallel. However it does not make the most efficient use of resources and workers do not have the opportunity of becoming specialized.

Process-oriented organization

This is a means of enabling specialization to be achieved, which is done by enabling similar operations to be performed repeatedly on a whole range of components and products. The process thereby exploits the principle of division of labour. However, it requires more planning and co-ordination. Figure 2.3 illustrates the general concept.

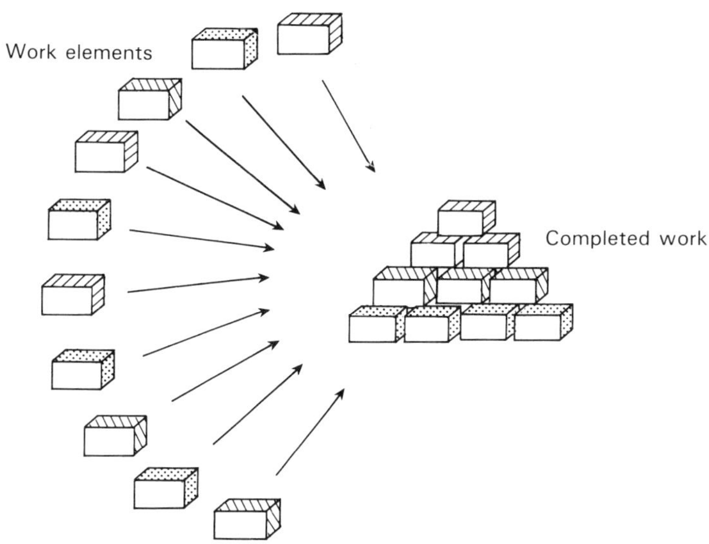

Figure 2.2 Product-oriented work organization.

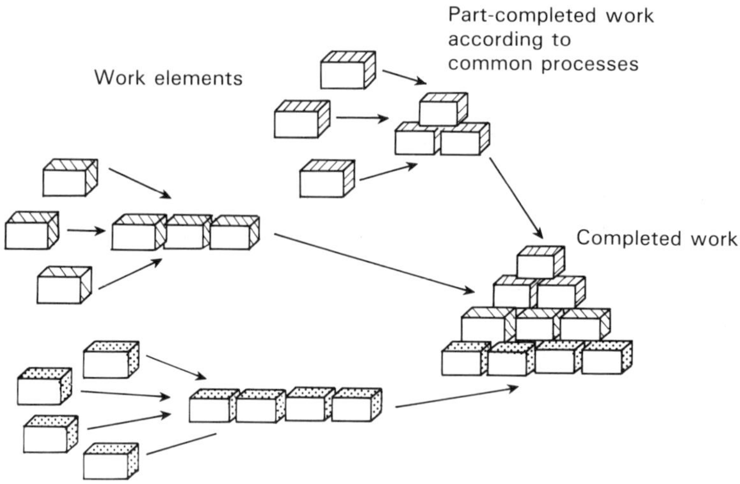

Figure 2.3 Process-oriented work organization.

Task-oriented organization

This takes the idea of specialization to its logical conclusion. It involves repeatedly carrying out small work tasks on part-completed components and products which are manufactured continuously, as represented in Figure 2.4. Task-oriented work organization requires a considerable amount of planning but the economies of scale can be considerable.

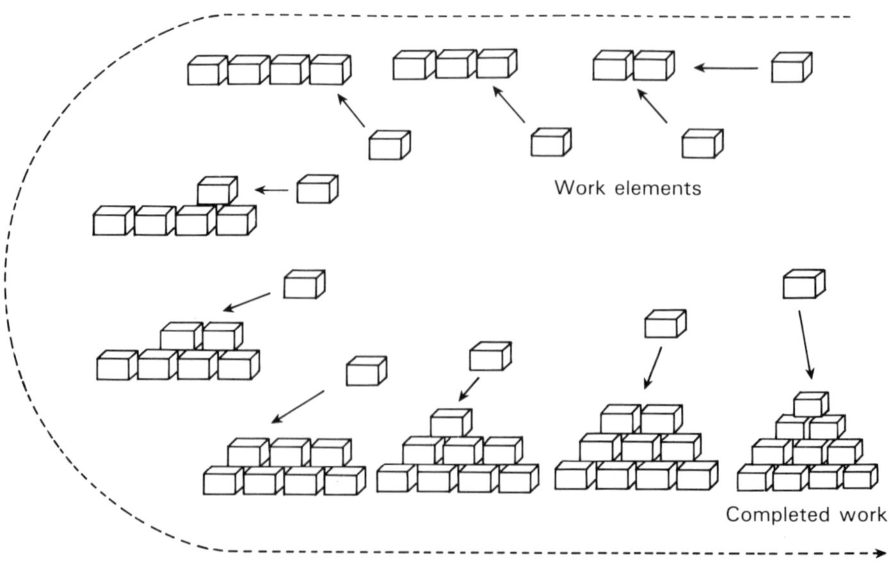

Figure 2.4 Task-oriented work organization.

Facility layout options

There are three options for the arrangement of facilities (based on Wild, 1990):

1. By fixed position.
2. By function.
3. By operation sequence.

Fixed position arrangements

These occur where the product remains stationary and all the productive resources (materials, equipment, labour, etc.) are brought to the place of work, as shown in Figure 2.5. By their nature, some products naturally need to be manufactured in a fixed position, but for others this arrangement represents just one of the available options.

Arrangements by function

These occur where machines, processes and equipment of the same type are grouped together in the same department or area, as in Figure 2.6. During manufacture products move from one department to another according to the operations required. In the industrial revolution there was probably a need to adopt this type of arrangement when machines were driven from a single

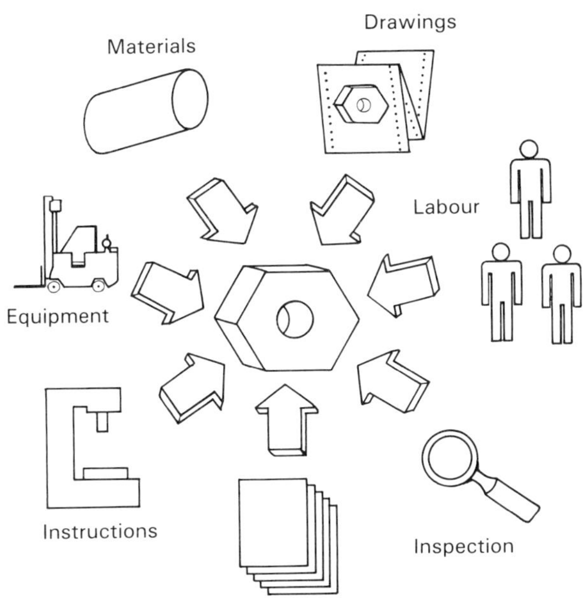

Figure 2.5 Fixed position arrangement.

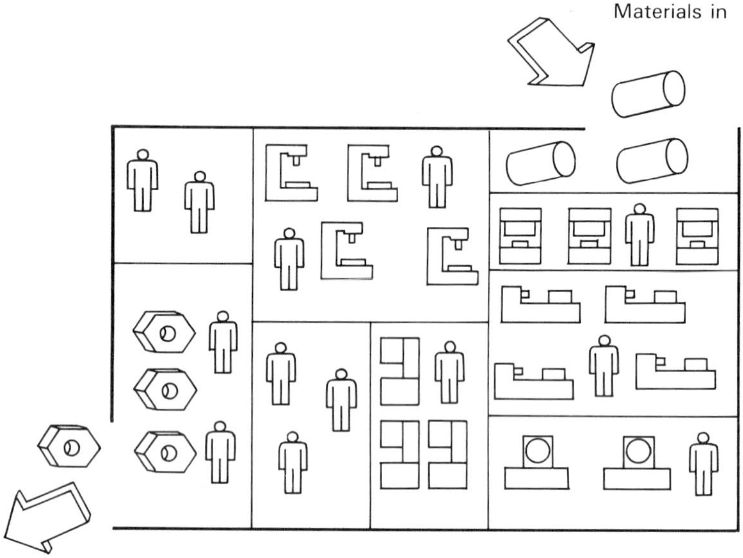

Figure 2.6 Arrangement by function.

power source. Today it is still a widely used method of arranging physical facilities but it gives rise to numerous problems of production planning and control.

Operation sequence arrangements

These occur where machines, equipment and workplaces are arranged according to the order in which operations need to be carried out to produce a complete component, product or subassembly. An example of this type of arrangement is given in Figure 2.7. They usually take the form of unidirectional lines, although there are other means of arranging facilities according to sequence. In some ways the planning of production is relatively uncomplicated because product flow is straightforward. When used for assembly,

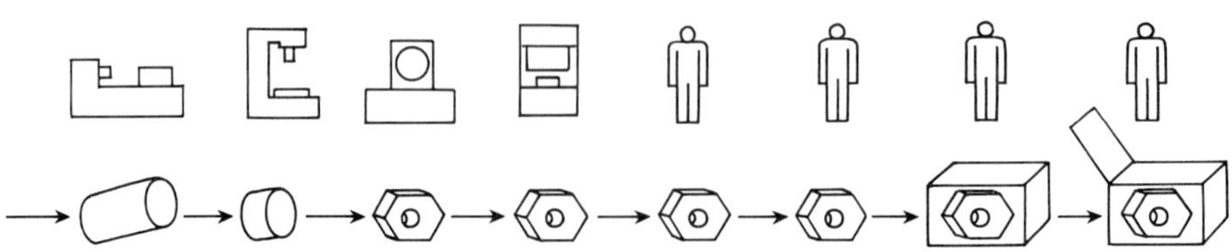

Figure 2.7 Operation sequence arrangement.

however, the scheduling of material supplies to each stage of production can still be a complex problem.

2.3 Conventional production systems: features, benefits and drawbacks

Given the fact that there are three alternative types of work organization and three ways of arranging the physical facilities for production, it further follows that a matrix can be drawn to contain the nine theoretical combinations of these two physical system parameters.

When we talk about a *conventional* production system we usually refer to either job, batch or flow, which in fact can be identified as particular combinations of parameters (Figure 2.8):

1. The 'job' system = product-oriented organization combined with fixed position arrangement
2. The 'batch' system = process-oriented organization combined with arrangement by function
3. The 'flow' system = task-oriented organization combined with operation sequence arrangement

There is, in fact, another factor to be considered within the context of the design of production systems, which must also be considered alongside the

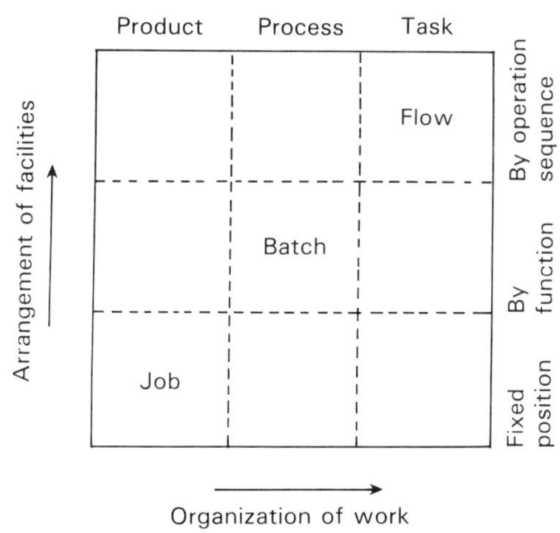

Figure 2.8 The conventional production system types.

combinations of productive resources. This factor is demand (in terms of quantity or pattern), which again could be viewed as being divided into three types:

Demand quantity		*Demand pattern*
low	or	unique
medium	or	intermittent
high	or	continuous

If demand is also considered, therefore, certain conventions also prevail:

1. The job system = product-oriented organization
   + fixed position arrangement
   + low or unique demand
2. The batch system = process-oriented organization
   + arrangement by function
   + medium or intermittent demand
3. The flow system = task-oriented organization
   + operation sequence arrangement
   + high or continuous demand

Figure 2.9 shows these three conventional production systems positioned within a 'production system cube'. The edges of this cube are defined by the three characterizing factors (work organization, facility arrangement and demand). As each factor was categorized into three types, then the total number of combinations (i.e. spaces within the cube) is 27, of which the conventional systems represent just three. Later in this chapter some alternative production systems will be described, which occupy more novel positions on the production system cube. To understand the rationale behind these alternative systems, however, it is first necessary to examine the characteristics, benefits and drawbacks of the conventional systems.

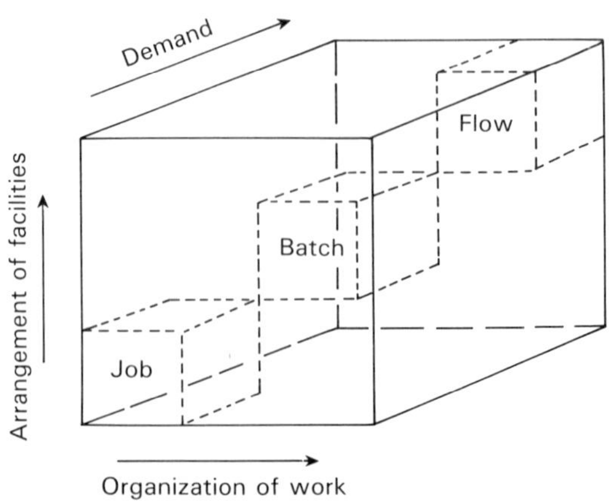

Figure 2.9 The production system cube.

	Characteristic / System type	Job production	Batch production	Flow production
Physical facilities	Type of layout	Fixed position	By function	By operation sequence
	Type of plant and equipment	Mainly general purpose	Mixture of general purpose and adapted	Mainly special purpose
Human resources	Proportion of managers	Relatively few managers	Relatively more managers	Most managers
	Number of management levels	Few levels (i.e. shallow hierarchy)	Relatively more levels	Most levels (i.e. deep hierarchy)
	Skills, number and type	Large numbers of 'direct' skilled operators	Moderate number and range of skills	Skills confined to 'indirect' functions (setting, maintenance, etc.)
	Training requirements	Extensive training of most direct operators	Moderate training needs	Most training aimed at indirect workers
Materials control	Amount of product movement	Negligible movement of product	Extensive movement of part-finished products	Minimum movement
	Queuing and work in progress	Dependent on resource availability & allocation	Long queues at facilities High work in progress	Limited queuing and lower work in progress levels
	Throughput time	Dependent on resource availability & allocation	Long throughput time compared with total operation time	Short throughput time
System capability and cost implications	Product range and quantities	High variety/low volume		Low variety/high volume
	Production costs	Low fixed/high variable costs		High fixed/low variable costs

Figure 2.10 Characteristics of the conventional production systems.

Features of the conventional production systems

The three conventional production systems differ in many respects and each has its distinctive advantages and disadvantages, which derive from their characteristics. These characteristics can be itemized according to their relationship with the three resources (physical facilities; human resources; materials). They are as follows:

1. Characteristics relating to the physical facilities, i.e. types of layout, plant and equipment.
2. Characteristics relating to the human resources, i.e. the proportion of managers, the number of levels of management, the number and type of skills and the training requirements.
3. Characteristics relating to materials control, i.e. the amount of product movement, the queuing and work in progress and the throughput time.

The nature of each of the characteristics is shown in Figure 2.10. This figure also indicates the capability of the system types in terms of product range and volumes, together with the implications for fixed and variable production costs.

Benefits and drawbacks of conventional systems

Now let us look at the benefits and drawbacks which are derived from these characteristics. They are shown in Table 2.1.

Table 2.1 The benefits and drawbacks of conventional systems

Benefits	Drawbacks
Job production	
flexibility (product)	low resource utilization
low fixed costs	duplication of facilities
job enlargement	high training costs
planning can be simplified	no skill specialization benefits
Batch production	
flexibility (process)	high working capital costs
specialization (of operational tasks and supervision)	frequent set ups
	high material handling costs
isolation of process where necessary	long delivery lead times
ability to make priority changes	
Flow production	
few set ups	'human' problems, i.e. recruitment difficulties, high absenteeism, high labour turnover
low working capital costs	
low material handling costs	
low direct training costs	'physical' problems, i.e. high capital costs, interdependency can cause unreliability, inflexibility (product and process)
specialization enables reduction in task times and application of direct incentives easily automated	

To overcome the drawbacks associated with the conventional systems it may be necessary to adopt alternative approaches using different combinations of work organization and facility arrangement. If such production systems can extend the limits on the efficiency of material flow there will be clear strategic benefits and the control system will not be constrained by the previous restrictions imposed. The ideas behind how such alternatives can be developed are discussed in the next section.

2.4 Development of alternative systems for market-focused production

To explain the reasoning behind the development of alternative production systems it is first necessary to understand that the design of conventional production systems is usually influenced by a number of underlying assumptions. The main such assumption is that 'trade-offs' must be reached both within and between the customer service and resource efficiency objectives of a production system which in turn influence its design parameters. (For a seminal discussion of trade-off decisions in manufacturing see Skinner, 1969.) The existence of trade-offs means that optimum solutions must be sought within the inherent limits (or constraints) of the production system. These optimum solutions in turn keep the system in 'balance' as shown in Figure 2.11.

To explain this idea further it is necessary to understand that the resource efficiency of the system determines the cost of producing a product, while customer service will influence its selling price. The profitability of the organization is a function of both total production cost and sales revenues and the elements that go to make up each are represented in Figure 2.12.

The need for trade-offs can be demonstrated within this model. It is assumed that to gain an improvement on one side of the profitability model there will be a consequent penalty to be paid on the other. Alternatively, the improvement and the penalty may both lie on the same side of the model. However, this makes no difference to the general idea which may be illustrated with the following examples:

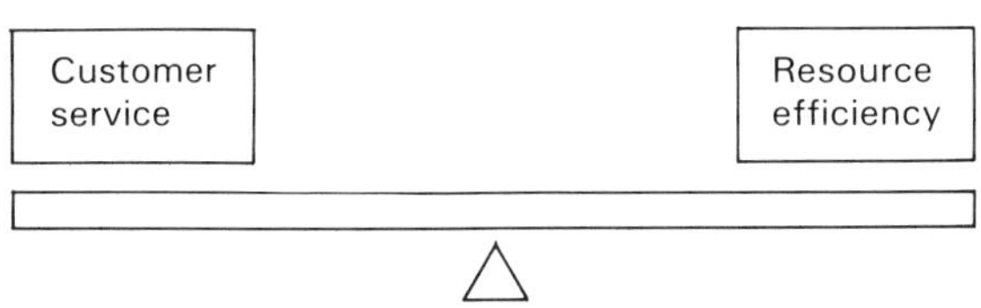

Figure 2.11 The balance or 'trade-off' between customer service and resource efficiency in a production system.

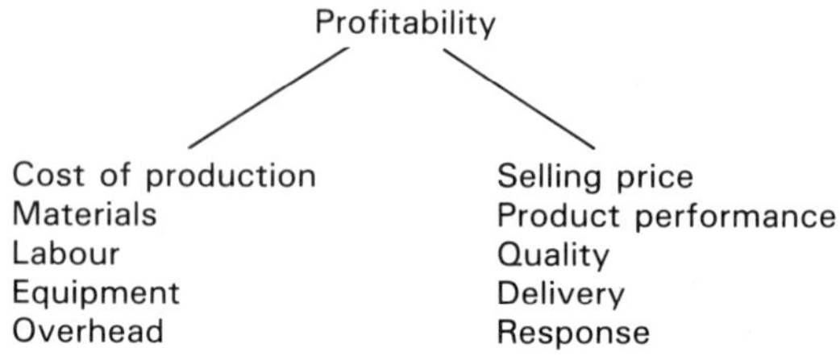

Figure 2.12 The determinants of profitability leading to the trade-off assumptions.

EXAMPLE

Trade-off 1:

Delivery performance versus *Material cost*

ASSUMPTION UNDERLYING CONVENTIONAL PRODUCTION SYSTEMS DESIGN: Improved delivery is achieved by holding high stocks of work in progress . . . low stocks make delivery performance worse.

Trade-off 2:

Quality versus *Overhead cost*

ASSUMPTION UNDERLYING CONVENTIONAL PRODUCTION SYSTEMS DESIGN: Better quality is achieved by investing in inspection equipment and procedures . . . reducing the inspection overhead has an adverse effect on quality.

Trade-off 3:

Response versus *Equipment cost*

ASSUMPTION UNDERLYING CONVENTIONAL PRODUCTION SYSTEM DESIGN: Greater customer response can be obtained by investing in more flexible machinery . . . reduced capital investment leads to poorer response.

Trade-off 4:

Overhead versus *Materials cost*

ASSUMPTION UNDERLYING CONVENTIONAL PRODUCTION SYSTEM DESIGN: Machine overheads can be reduced by increasing their utilization and building stocks of materials . . . reducing material stocks adversely affects utilization and increases overheads.

This idea of trade-offs has in turn led to a number of approaches designed to optimize objectives within the constraints imposed by the constituent influencing factors.

Examples of the optimization approach include the 'economic batch quantity', 'optimum maintenance policies' and 'the cost of quality concept'. They are illustrated in Figures 2.13, 2.14 and 2.15, respectively.

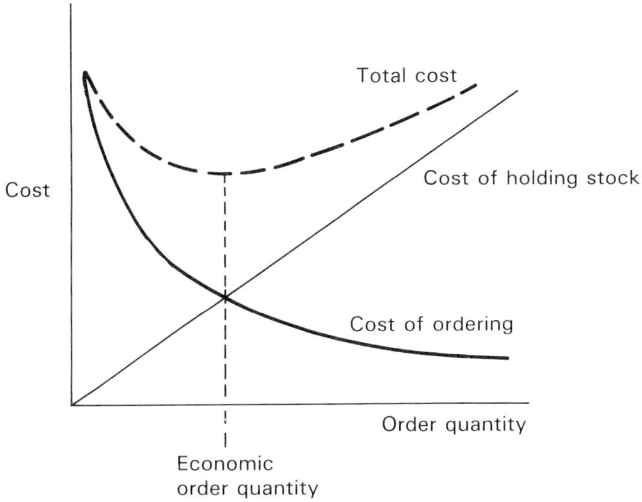

Figure 2.13 Optimization applied to stock control.

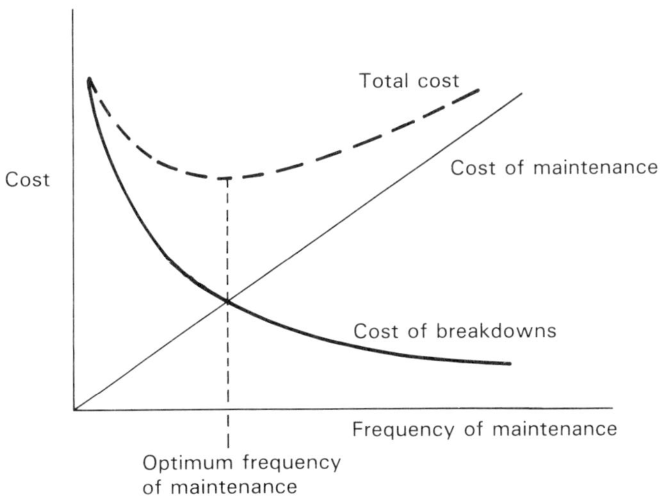

Figure 2.14 Optimization applied to maintenance.

The idea behind any optimization approach is that the underlying constraints are identified and accepted. No attempt is made to reduce or remove the constraints, a balance is simply struck which minimizes (or maximizes) the total effect. A limit to the solution is thereby imposed by the constraints. If the constraints could be reduced or even removed then solutions might be achieved which are many orders of magnitude better than any of those that are feasible under the optimization approach. Thus we can think

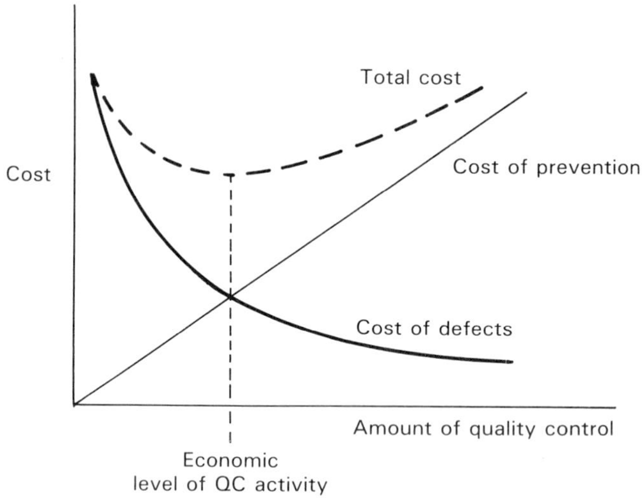

Figure 2.15 Optimization applied to quality management.

of a different philosophy as being necessary when designing production systems that overcome the deficiencies of conventional systems. This can be termed 'the improvement approach', which is based on the idea of deliberately exposing constraints and finding ways of reducing and removing them rather than simply optimizing within them, as illustrated in Figure 2.16.

Redesigning the transformation system

An essential part of the improvement approach is the development of new types of transformation system which allow the limits on system performance to be extended. The objective here is to design transformation systems based on combinations of work organization and facility arrangement which will exhibit the beneficial features of conventional systems while having fewer of their drawbacks. In this way many of the intrinsic constraints of conventional production systems can be removed and the improvement approach can thereby be facilitated.

The optimization approach v	The improvement approach
* Identifies and accepts the constraints and uses control techniques to minimize or maximize objective functions	* Deliberately exposes constraints and finds ways of reducing or removing them

Figure 2.16 The essential differences between the optimization approach and the improvement approach towards production systems design.

Within the context of market-focused production there is an additional and complementary objective of improving the system's customer orientation. This is usually achieved by developing production systems with a greater product orientation, normally only associated with 'job' production. The answer often lies in the use of cellular or modular organization. There are basically two types of cellular organization:

1. For batch production of components.
2. For continuous production (via assembly) of end products.

They are based, respectively, on the established principles of group technology and autonomous working. Detailed descriptions are provided in Appendix I but their context within the overall framework of production system design theory will be briefly explained here.

Cellular organization for component production

Cellular organization for the production of intermittent batches of components is based on the principles of 'group technology' (GT), (Snead, 1989). Group technology has three aspects:

1. Grouping *parts* into *families*.
2. Grouping *processes or machines* into *cells*, with the aim of reducing material movement.
3. Creating groups of *multiskilled operators*.

Grouping parts into families

Grouping parts into families has the objective of reducing the amount of setting time between batches and thereby avoids the need to produce in large batches to minimize the unit ordering and setting cost. Parts family formation is based on the idea of grouping parts together according to geometrical or processing similarity. Sometimes a 'composite component' is conceived, which has all, or most, of the features of the constituent parts. The composite component may be 'real', although more often than not it will probably be hypothetical. If the process is set up to produce the composite component the setting time between all its constituent parts can, by definition, be minimized.

Grouping into cells

Processes or machines are grouped into cells with the intention of reducing material movement. Whereas a functional layout involves extensive movement of material between departments with common processes, a cell comprises all the processes or machines required to manufacture a family of components.

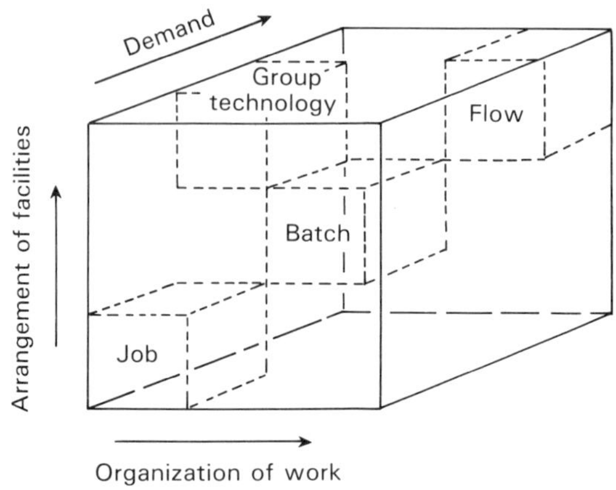

Figure 2.17 Group technology within the production system cube.

Material movement is therefore restricted to within the cell and throughput times are thereby reduced.

Multiskilled operators

Creating multiskilled groups enables increased autonomy and flexibility on the part of operators. This enables easier changeovers from one part to another and increases the job enrichment of members of the group. This in turn can improve motivation and have a beneficial effect on quality.

An additional feature of group technology is that cells can be a single planning point rather than individual machines. Production control need therefore only plan down to level of the cells and becomes more simplified. Planning within the cell is carried out by the group members themselves, which enables them to apply their detailed knowledge of process capacity, capability, etc.

The position of group technology within the 'production system cube' is illustrated in Figure 2.17. Work organization is product oriented, in common with job production. Facility arrangement is by operation sequence, so in this way it is similar to flow production. Group technology is therefore able to derive the advantages of these two types of production system while still being suited to intermittent batch production.

Application 2.1 Cellular manufacturing at GPT Limited

The Business Systems Group of GPT Limited is located at Beeston, near Nottingham, U.K. On this site GPT develop and manufacture telecommunication equipment, including exchange and switching systems, local area network (LAN) products and computer-supported telephone systems. The manufacturing processes include automatic and manual component insertion and onsertion, flow soldering, automatic circuit testing of assembled printed circuit boards (PCBs) and other functional and systems testing.

Within the Beeston factory there has, since the late 1980s, been a recognized need for local ownership of processes whereby the production personnel take increased responsibility for ensuring the product is manufactured under controlled conditions, process and product inspection, maintenance and housekeeping. In terms of production systems design this has manifested itself with the conversion of conventional 'herring-bone' or 'spine and spur' assembly to 'U-cells' in the manufacture of telephone products. The former herring-bone layout involved operators performing an allocated task before moving the product down the line. During 1991/2 these assembly operations were being rearranged into U-shaped facilities. Utilizing existing benching and tooling, the conversion of individual operations into a cell took as little as 48 hours to complete, with a minimum of capital investment.

To communicate the details of the planned changes, a series of 1-hour presentations were given by the Manufacturing Engineering Team, starting with top management but then cascaded throughout the organization. Presentations to the workforce were made by their own Production Managers, but supported by Manufacturing Engineering personnel. Many questions were asked of this new configuration of production facilities and the consequences of it. A great deal of interest and enthusiasm for the change in manufacturing approach was thus generated from the outset.

GPT call their new production system 'U-shaped (people-based) flexible JIT cells'. The emphasis is on enabling people to have the flexibility and scope to take more responsibility for their own work, and for cells to manage their own operations. Specifically, the company, with the introduction of cells, is keen to maximize team-working, product ownership and customer focus. The key feature of U-cells is that the work layout should facilitate the introduction of just-in-time (JIT) to ensure work is pulled through the cells in a continuous flow and production is levelled with a minimum of work-in-progress (WIP) between individual operations. The requirement for JIT and smoothed production demands the development of multiskilled people within cells to facilitate operator mobility between tasks so that workloads can be balanced.

Managers at GPT express surprise at the enthusiasm with which U-cells have been greeted by the workforce. Operators within cells now freely admit they have assumed greater pride in their work and their teams. Specific benefits arising from the introduction of U-cells include:

1. Product quality has improved to the extent that faulty work shipments are virtually a thing of the past.
2. WIP has been dramatically reduced, both within and between cells. Cells pack the assembled products and ship directly from there. Moves are now afoot to introduce supplier deliveries direct to point of use.
3. Savings in labour content: most cells comprise four operators and these

are typically producing the same output as seven people in the old systems. Overall a labour cost saving of 30 per cent has been reported.

4. Operators can now work at their own pace instead of this being governed by preceding operations. When an operator finishes a task they will now look around the cell and move to those tasks where they will be more gainfully employed instead of waiting for more work as they did in the old system.

The aim of GPT at Beeston is now to incorporate all processes associated with an assembly into its production cell. Manufacturing engineers are now working to include flash, acoustic and burn-in testing in cells. In some cases this may require the reverse of increased automation, where semiautomated processes with their reduced production rates are more suited to the output rates of U-cells. GPT now consider that they have taken the U-cell manufacturing principle to a point of refinement comparable with the best in the world.

Source: by kind permission of GPT Limited (a GEC-Siemens Company)

Cellular organization for the assembly of end products

The general approach to cellular organization for end-product assembly is based on the principle of autonomous working (Sandberg, 1982). This is an alternative to conventional flow production systems and, like flow production, it is associated with high (continuous) demand. Autonomous working addresses both the 'human' and 'physical' problems associated with conventional flowlines.

To implement autonomous working in a practical environment the broad options available are:

1. Parallel assembly cells.
2. Cells based on line assembly.

Each type can be based on assembly being carried out by individual operators or groups.

Both approaches change the work organization from being task oriented to being product oriented. Parallel assembly, however, uses a fixed position facility arrangement, while line assembly retains the operation sequence arrangement usually associated with flowline working (Karlsson and Bennett, 1991). The two types of autonomous working are thus positioned on the 'production system cube' as shown in Figure 2.18.

The benefits of autonomous working can be summarized as follows:

1. The balance and system losses associated with conventional flowlines are reduced or eliminated with the longer cycle times of autonomous working.
2. Different products can be produced at the same time in different cells or modules.

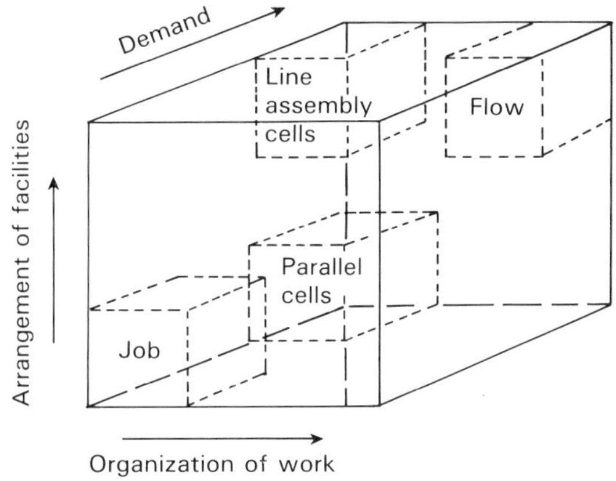

Figure 2.18 Autonomous working within the production system cube.

3. Response and volume flexibility is greater than with flowlines.
4. Model changes can be effected more easily.
5. The autonomous nature of assembly cells means there is no work station interdependency.
6. Job enrichment and autonomy lessens the human problems associated with flowline working.
7. Quality tends to improve because responsibilities are more clearly assigned.

Production control implications of cellular organization

Some of the implications for production control of adopting cellular organization have already been addressed in the discussion on the benefits of group technology. The greater complexity of autonomous work systems for assembly necessitates a control system that can monitor products and subassemblies, as well as guide them to the correct work places at the correct time.

The reduced setting times, more efficient material flow and modular nature of cellular organization facilitates 'just-in-time' production using 'kanbans'. Material requirements planning (MRP) can be used for longer-term material purchasing where procurement lead times are too long for 'just-in-time' ordering.

Product design implications of cellular organization

One of the most important requirements for successful implementation of cellular organization is that the design of the product lends itself to the 'completeness of task' principle. With complex products this principle may require the design of modularized products, which enable the complete production of a module to be assigned to an individual or group. Product design is one of the keys to successful implementation of cellular organization. The potential benefits of both group technology and autonomous working are often restricted when products have been designed to be manufactured on conventional systems.

Payment systems and cellular organization

Conventional payment systems, particularly those based on an individual financial incentive, are usually inappropriate for cellular organization. A more relevant payment system would probably comprise several components depending on the nature of work organization and the motivational and reward factors being emphasized. Some of the features of a payment scheme for autonomous work groups might be as follows:

1. An individual element based on job evaluation, e.g. education, training, range of skills, extra responsibilities, timekeeping record, etc.
2. A group element based on delegated responsibilities, e.g. production planning, quality responsibility, cost accounting, administration, social responsibility (no rejection of members).
3. A result related element (possibly plant-wide), e.g. cost of factory workers, cost of staff, rejects and adjustment, operational supplies and tools, added material, store value, quality index.

Application 2.2 Technology makes craftsmanship possible at Volvo

Volvo strives to manufacture cars of a very high quality. That has been the intention ever since the first car was built in 1927. Even then, the goal was to give customers as much value as possible for the money they invest in their new car. At the same time, Volvo aims for a high level of efficiency and long-term profitability.

Achieving these goals requires a skilled and dedicated workforce. That is why the Volvo Car Uddevalla Corporation, a subsidiary wholly owned by Volvo Car Corporation, has built a very unusual car factory, which will match up to Volvo's high demands well into the twenty-first century.

When Volvo first began building cars, it was natural to use what was then the most technically advanced and efficient method of production – the conveyor belt. In those days there was still a large measure of craftsmanship involved in car building. This disappeared successively in the post-war years as the different stages in the assembly process were made progressively shorter.

But in the early 1970s the pendulum began to swing back. Volvo opened Kalmarverken, a plant which created echoes throughout the automotive industry. The mechanical conveyor belt was replaced by self-propelled trolleys – carriers – on which the car bodies were moved. The workteam became a factor in the endeavour to break down the impersonality of a big factory into sections which were easier to take in. These changes are now also being introduced into Volvo's biggest plant, Torslandaverken in Göteborg.

The Volvo Car Uddevalla Corporation is a natural development of Volvo's concern with high quality production in a sound working environment. Modern and flexible production systems make big demands, and the know-how, involvement and responsibility of the people who work with them is becoming increasingly important.

The Uddevalla plant's new production methods are based on small workteams. As each workteam has a number of different tasks to perform, and thanks to the greatly lengthened work cycles which the new method gives, the individual team member's work can be more varied. Each workteam also handles a large part of the administration necessary in the building of cars. In this way every employee is brought close to decision-making, and the number of levels of decision from plant manager to car builder are relatively few.

Much emphasis is also placed on the composition of the workteams. A blend of men and women of different ages complement one another in each team, and this has led to an increase in involvement and job satisfaction. The Uddevalla plant's attention to the overall picture is also evident in the healthy working environment. It is an environment where the development of competence, business strategies and modern technology are put into practice.

Material from hundreds of different subcontractors, both in Sweden and abroad, is received at the Uddevalla plant's centrally-positioned materials and parts depot. As the various components are delivered as they are needed, stocks can be maintained at a relatively low level. In the materials and parts depot, the deliveries are spot-checked before they are used in production.

Work with the over 3000 parts which make up each car is made considerably easier by the use of a powerful computer-based information system, which contains information about, for example, availability of various parts and the specifications of the cars which are to be built. This advanced information system takes care of a large part of the vital flow of information. It keeps track of what parts are used in each car and what parts are in stock – the stores inventory. It also contains a weekly plan of which cars are to be produced. By means of the information system, every car is followed all the way through production to ensure that it is built to the right specification. However, this information system is simply an aid – employees still supervise and make decisions.

In the materials and parts depot, components which on assembly will be in the same area of the car are made up into kits. At the Uddevalla plant, part of this work is carried out by automatic sorting and packing machines. These have been developed, chiefly by Volvo, especially for the Uddevalla plant, and can pack up to 18,000 kits a day. A monotonous job which the employees no longer have to do, thanks to technology's help.

The kits also contain larger components which have been preassembled in the depot, such as axles and exhaust systems.

At the depot there are also several multilevel car part stores where computer-controlled cranes steered by bar codes and other systems move parts and materials when necessary.

Car bodies are not made at Uddevalla. They are delivered, newly painted, from the Torslanda plant in Göteborg and are kept in the multilevel store before inspection, preparation for and inclusion in production.

The depot also receives larger components, such as gearboxes, seats and wheels, from Volvo's other plants in Sweden. All materials and parts which are distributed from the depot, including bodies and engines, are transported on the automatic self-propelled trolleys – the carriers. Volvo originally developed this system for the Kalmar assembly plant. These carriers transport everything necessary for the building of a complete car to each of the work-teams in the six product-assembly units. This is the chief and basic difference between this and earlier production methods. So, instead of the bodies moving along on a conveyor belt through a large warehouse where the parts are added as they pass, in the Uddevalla plant the parts are moved through the factory to where the cars are being assembled. The advantages of this are that the car is stationary during assembly, and that several cars can be assembled simultaneously by the same workteam, besides which the time-span for assembly tasks can be increased to allow a broader work content. In this way the work is made more efficient and the flexibility of the workteam increased, which in turn leads to greater job satisfaction and a better quality product.

From the outset, the bodies are placed in tilts, which are adjustable so that much of the assembly work can be done standing upright. A lot of thought has gone into efforts to avoid the risks of various kinds of occupational injuries, and ergonomics is a regular topic on the agenda at workteam meetings. At Volvo's Uddevalla plant there is everything from advanced robots to ordinary screwdrivers and other simple tools. Many of the tools have been specially developed to be as ergonomically correct as possible.

When the car has been fully assembled, it is again transported by carrier to the roll-test station. There are two of these test stations – located centrally between three product-assembly units. Any mechanical faults are corrected by members of the work-team that assembled the car – in this way, the teams quickly get to know about anything being wrong.

The workteams have direct responsibility for maintaining the level of quality throughout the entire assembly process. Quality consciousness runs through the whole organization of the plant. By means of the unique assembly procedure, in which each workteam assembles complete cars, the team does not release a car until all quality requirements have been fulfilled. Any faults found are rectified by members of the team which assembled the car. In this way, the employees are in control of the entire process. This is a necessary condition for the warmth, humanity and job satisfaction which must play a vital role in the production of real quality cars.

A number of cars are also taken aside every day for quality assessment. This involves detailed appraisal of the cars, where any faults are graded according to a points system.

In the plant complex there is an administration building which also houses the canteen for all the staff.

The work-teams have a large amount of freedom to decide how their work is to be done. This cuts out unnecessary stress and gives each person the opportunity to take responsibility for his or her own work effort.

Good lighting and low noise levels enhance job satisfaction, and the production areas all have large skylights to admit natural daylight. In addition there is artificial lighting, which, combined with a bright colour scheme, gives an impression of working in daylight. The noise level is low – building a Volvo shouldn't be any noisier than driving one.

The new assembly method, with carriers which transport material to the stationary cars, gives team members the opportunity to develop production according to their own ideas, to have longer and more complex assignments and to make the work more stimulating and varied.

'Master', 'journeyman' and 'apprentice' are words which became rare in daily use a long time ago. However, at the Uddevalla plant, the masters at least have returned. A master is someone who, in principle, can build a whole car alone. Every car is divided up into seven work zones or "balances". Generally, people can manage two to three of these after the 16-month training period. Much of the training is informal and takes place when colleagues in a work-team share their know-how and skills with each other.

Source: by permission of Volvo Cars Limited.

2.5 Conclusions

Conventional production systems are based on combinations of work organization and facility arrangement, which give rise to numerous drawbacks as well as benefits. These in turn have necessitated that 'trade-offs' must be achieved both between and within the system's customer service and resource efficiency objectives. An optimization approach has therefore become established where objective functions are minimized or maximized within identified and accepted constraints.

The key to achieving greater competitiveness within the context of a market-focused production system lies in the idea of rejecting the concept of optimization. Instead an improvement approach should be adopted based on deliberately exposing constraints and reducing or removing them.

The development of alternative production systems can enable the reduction or removal of constraints by seeking a combination of design factors that will yield many of the advantages of the conventional systems while avoiding their drawbacks. For market-focused production, a suitable alternative may be found within the general scope of cellular organization based on group technology for batch production of components and autonomous working for product assembly.

Despite the obvious attractiveness of cellular organization, however, the issue still remains of how to actually design and implement a system that will be compatible with the overall market, strategic and organizational objectives of the firm. The remaining two parts of this book will therefore address this issue. In Part II it will examine the use of design models and methodologies before introducing a specific methodology (DRAMA) for market-focused production. In Part III the methodology and its practical application will be explained in detail.

2.6 Further reading

Bennett, D. J., C. D. Lewis and M. H. Oakley (1988) *Operations Management*, Philip Allan, Hemel Hempstead.

Gallagher, C. and W. Knight (1986) *Group Technology Production Methods in Manufacture*, Ellis Horwood, Chichester.

Gerwin, D. and H. Kolodny (1992) *Management of Advanced Manufacturing Technology*, Wiley, New York.

PART II

Production System Design: Models and Methodologies

3 The Use of Structured Approaches in Designing Production Systems

Objectives of the chapter

On completion of this chapter you should:

1. Understand the need for systems modelling and the adoption of methodologies in the design and analysis of production systems.
2. Be aware of the various techniques that are available to the practitioner in process design activity.
3. Be able to put these approaches into practice.

■ Chapter outline

The need to structure production system design decisions is becoming increasingly recognized as a prerequisite for successful implementation of effective systems. To do this, systems designers are turning increasingly to established models and methodologies – to appreciate the process of design, to understand the operation of proposed and existing systems, and for prescription on 'best practice' for design. This chapter introduces some of the most widely used models and approaches, which together comprise the current concepts and techniques for production systems design.

3.1 The use of models to analyze production system design activities

Introduction

This chapter focuses on the process of production system design and offers a number of structured models and methodologies that have been developed to assist in carrying out this process. A logical and rational approach to the design process should emerge from the use of one or more of these models. Likewise, the models also assist in the analysis of existing facilities by offering a framework for evaluation. Production system design has primarily focused in the past on consideration of physical systems and job design. However, recent advances in process and control technology now demand that the information and decision-making activities in the day-to-day running of operations be structured and conducted in a format that lends itself to ease of automation (i.e. common shared databases, knowledge-based systems, etc.). In response to this demand a number of conceptual models and methodologies have been developed to assist in the analysis of the decision and information activities within advanced production systems. These design tools, however, need not be limited merely to the development of advanced systems: they offer considerable assistance for the development of new systems and the evaluation of existing facilities within manufacturing organizations generally.

This chapter will first introduce a number of established models, which are now widely accepted and used in design projects. The majority of these derive from what is commonly known as the 'systems approach', although systems theory in itself covers a wide diversity of models and methodologies. The types of systems approaches will be introduced through an initial categorization into so called 'open' and 'closed' models. The distinction between 'hard' and 'soft' systems will then be drawn, showing how a 'soft systems methodology' can prove useful in guiding attempts to structure decision making in the design of production systems (Checkland, 1981; Checkland and Scholes 1990).

Following this introduction to systems theory, a number of models are presented which seek, primarily, to formulate the structure of decision processes within organizations including those pertaining to production systems design. Systems theory tends to be 'problem oriented', whereas decision process models from organization behaviour do not necessarily consider the starting point for a sequence of decisions as being a 'problem'. The decision process models reviewed in this chapter include the structure developed by Mintzberg and colleagues in Canada (Mintzberg *et al.*, 1976) and the conceptual model of production systems design developed at the Graphe a Resultats et Activites Interrelies (GRAI) Laboratory at the University of Bordeaux, France (Doumeingts *et al.*, 1986; Doumeingts and Chen, 1992).

The model developed by Mintzberg, Raisinghani and Theoret (hereafter called the Mintzberg model) provides a longitudinal 'through-time' insight into decision sequences (Mintzberg *et al.*, 1976). Thus it is useful for tracking and understanding the chronology of decisions from the recognition of the need for a decision to be made through to the final solution. The GRAI decision analysis model provides a structural vehicle for decision-making, where the decision is set with the context of an organizational hierarchy and is related to the physical system activity and the corporate information system. The authors have found that the Mintzberg and GRAI models can be used in conjunction and are complementary to one another in providing an appreciation of both the longitude and hierarchy of decision-making within the process of production system design.

A number of contemporary modelling tools and methodologies have also been developed for particular application in the design and development of advanced manufacturing systems. In many cases, however, they are applicable to the general systems design work, whether the systems are defined as 'advanced' or not. These include SADT (Structured Analysis and Design Technique; Ross, 1977), IDEF (ICAM Definition; Yeomans, 1985) and CIM-OSA (Computer Integrated Manufacturing Open Systems Architecture; AMICE, 1989 (European Computer Integrated Manufacturing Architecture, AMICE, 1989)). It has been found that these methodologies offer guiding principles for the design of systems. They are particularly applicable for developing architectures for computer-based manufacturing control systems. Herein, however, lies some of the limitations for advanced systems methodologies, in that they offer limited assistance in terms of linking systems design activities with the wider business and organizational contexts of production systems.

To address the issue of developing appropriate systems two methodologies for evolving manufacturing strategy and its links to production systems design are presented. These are the Department of Trade and Industry (DTI)/Cambridge Group's methodology for developing manufacturing strategy based primarily upon the work of Platts and Gregory at Cambridge University (DTI, 1988) and, second, Hill's framework for reflecting production and operations choices within the context of corporate objectives (Hill, 1985). These are described in detail in the later part of the chapter.

The modelling approach

Let us start by asking the question 'Why use models at all?' The reasons can be categorized into two areas. First, the extensive, complex and integrative nature of technology in modern production systems usually means that no one system designer can fully grasp all the fine details relating to the overall system, so there is a need for greater powers of conceptualization on the part of the system designer and, consequently, by greater use of models. Second, increasing demands for more frequent product introduction and the need to offer of a wide range of variants has called not only for the shortening of product design lead-times, but also for greater flexibility from production systems. This requires reduced time-scales for the introduction of new facilities, as new product designs could not be produced if the corresponding physical production facility were not in place at the right time. Time pressure, and the resulting requirement for effective project management, calls for efficient methodologies to be adopted. They should be developed on the basis of experiences in manufacturing companies generally, so that the time-consuming delays and problems that frequently occur during the systems design process can be identified and, as far as is possible, avoided.

Understanding decision and design processes

Recent manufacturing trends show accelerating rates of innovation in product, process and work organization design (Clark, 1987). This trend emerged in the downstream assembly operations of companies, but is now passing back through the preceding activities (Bennett *et al.*, 1992). The turbulent nature of contemporary markets has meant that industrial firms must develop production systems which are capable of satisfying more sophisticated and ever-changing customer requirements. In such circumstances designers of production systems should possess a clear and precise understanding of the process of designing and implementing new production systems in order to control and manage such projects effectively.

A large number of manufacturing companies, however, find difficulty in transferring the lessons learned from past experience into comprehension of the various activities which together comprise the decision and design processes for new operations systems. A lack of effective organizational learning hinders progression along the technological 'learning curve' and, to some degree, this explains why the performance of assembly systems, after implementation, often fall short of original expectations. Practising managers and engineers often have a myopic view of design, which ignores the sequence of, often confusing, events that occur within a decision-making process. As a result many tend to 'postrationalize' their descriptions of a production system design project and reconstruct the decision sequence as though it were a single, simple, chain of events (Bennett *et al.*, 1990). Given these problems, certain theoretical models and conceptual frameworks have been developed which help to explain how such decision processes are enacted. These enable the practitioner to better understand the dynamics and complexities of systems design.

The current view, therefore, is that systems designers should adopt meaningful models to assist in conceptualizing decision-making activities and systems configurations. They should also make use of appropriate methodologies to enable the process of translating designs into operating systems. This chapter separates its discussions on some of the more widely used approaches by looking first at the modelling of decision processes and, second, by considering a range of design methodologies. Moreover, it will be suggested where each approach may be considered appropriate, and what are the strengths and limitations of each.

Models and methodologies: the overview

Following an introduction to a group of approaches commonly categorized under the title of 'systems theory', an explanation is given here of how the 'soft systems methodology' can be used by the systems designer and how it was employed in structuring of the DRAMA methodology. The decision framework provided by Mintzberg and colleagues and the decision analysis model developed by the GRAI Laboratory at the University of Bordeaux are also outlined. These provide, respectively, processual and structural models of design making. Additionally, a model is described which was developed by the authors and other members of the IDOM (Innovation, Design and Operations Management) Research Team at Aston University. This considers the extent to which systems and organizational design parameters are fixed and variable in relation to time and the level of decision making.

Contemporary approaches and methodologies to facilitate effective system and organization design are reviewed in the second part of the chapter. The methodology developed at Cambridge University for developing manufacturing strategies is covered, as are the IDEF and CIM-OSA methodologies for the conceptual design of systems architectures. Finally, some current ideas of production systems design are summarized with an emphasis on alternative approaches for organizing production activities.

These models may well be usefully consulted in understanding the process of systems design within an organization, and may provide an invaluable insight for the project manager into the dynamics of the design, development and implementation phases of a new system.

■ 3.2 Modelling decision processes

What is a model?

In a general sense, a model can be defined as something which 'represents reality'. Obviously there are many forms of model ranging from physical models of different scales to more abstract types such as mathematical and process models. This chapter will concentrate on a few of the many models that have been developed to assist the systems designer in the design, implementation and management of new manufacturing facilities. Therefore, a more specific definition for the types of models that we shall consider is that 'a model is an abstract representation of the manufacturing systems and/or of the activities and decision processes involved in their design and development.'

Models can be used to test the robustness of an eventual system design and to map the course of actions and decisions taken in their design. Models emerging from systems theory and organizational behaviour can also be used to map out a process model of the decision sequences over time and so help the systems designer to develop a greater understanding of the design process itself. This section will now consider models from systems theory and organization theory and it will be shown how these can be adapted and used to assist the systems designer. Modelling aids developed by the field of systems analysis will then be covered and their value to production systems design described.

Systems theory

Systems theory forms an important part of the theory of organizational decision making and constitutes the basis of many of the models representing these processes. The systems approach has derived from a desire to structure processes in an easily understandable and relatively standardized format. The basic system is represented as a simple input/output model, and the other systems models described below build on this fundamental concept. Despite its somewhat 'mechanical' appearance, systems thinking has become an important part of organizational theory, with behaviouralists realizing the benefits and flexibility of the approach. Some of the applications of systems modelling are now explored.

Open and closed systems

In the analysis of business organizations a distinction is frequently drawn between conceptualizing the firm as an 'open' or 'closed' system. Whereas the open system perspective analyzes an organization in terms of the way it maintains a steady state through interactions with the environment, the closed system perspective accounts solely for intraorganizational factors. Developed upon the pioneering work of Ludwig von Bertalanffy (von Bertalanffy, 1968), contemporary wisdom in organizational analysis suggests that we should, primarily, conceptualize the activities of work organizations in open systems terms. The dominant perspective in organizational analysis has long been that of the open systems approach (for example Katz and Kahn, 1975) and many writers talk of an open systems 'orthodoxy' in research and conceptualization (Silverman, 1970; Reed, 1985).

The analysis of organizational decisions, including those in production systems design, has led to the development of models which suggest that decision-making processes can usefully be researched through reference to partially open or, in some cases, closed system perspectives. These are approaches that focus firmly upon the interaction of decision processes and events within the organization. Advocates of this approach to decision modelling, whilst recognizing the limitations of closed systems approaches, suggest that such perspectives are fruitful for directing attention to the existence of structural similarities and in the identification of key internal relationships in decision processes (Silverman, 1970). The decision to apply an open, partially open or closed systems methodology should, it appears, always be based upon the level of conceptualization required by the system designer.

Checkland's soft systems model

Checkland's work at Lancaster University (Checkland, 1981; Checkland and Scholes, 1990) has drawn a distinction between what he calls 'hard' and 'soft' systems. He argues that orthodox systems thinking concentrates on the solving of 'well-defined' problems by the use of structured methodologies. He feels, however, that a 'softer', more generic approach is necessary for the analysis of 'purposeful human activity systems' such as those unstructured problems most commonly requiring solution in industry. What is needed, he argues, is a systems view which accounts for the solution of such unstructured problems. This he terms an 'action approach'.

The soft systems model involves a series of stages and transfer of the thought processes of the system designer between the 'real world' and the 'abstract world of systems thinking' (Figure 3.1). Thus problems are expressed and structured in the real world prior to the identification of 'root definitions' of those systems that appear to be applicable through systems thinking. Existing methodologies and systems thinking are used to develop and test conceptual models, which are then compared with the expressed situation in the real world. Having determined whether changes are both 'feasible and desirable', they are then implemented and their effect in the real world monitored with a view to further calibration if necessary.

Summary of systems theory

Systems approaches provide techniques for analyzing the design and decision processes within manufacturing and other organizations. Although specific situations can be tackled using closed or partially open systems analysis, for higher levels of analysis it would probably be more appropriate to utilize soft and open systems models.

Despite their attractions in terms of structuring the problem and the decision-making process, there remain many drawbacks associated with the systems approaches. The more general and open the model, the more conceptual is the form of analysis produced. In general, open models would appear somewhat detached and less applicable in terms of practical production

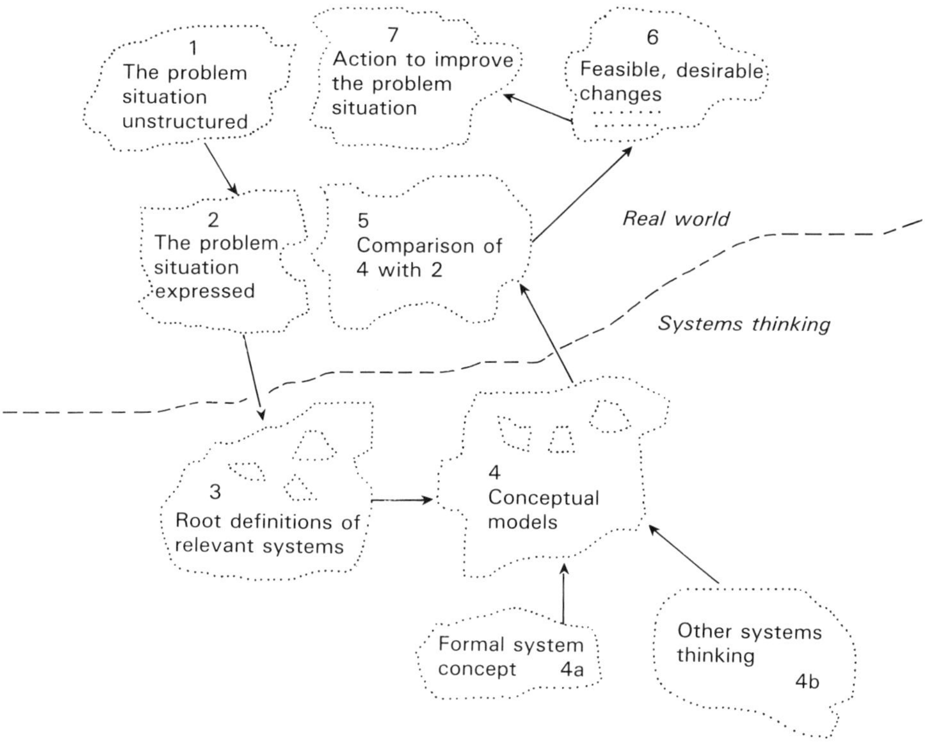

Figure 3.1 Checkland's soft systems model.

systems design. Closed systems, however, tend to be limited in their scope, whereas soft systems are limited to conceptualization and do not readily provide the systems designer with the techniques for more detailed analysis. Additionally, methodologies derived from systems theory tend to be more problem oriented in their rationale. Regarding all stimuli to the decision-making process as problems tends to neglect wider influences upon design, such as strategic opportunity for the introduction of new systems. It is with such problems in mind that other forms of decision modelling have also been considered in developing the DRAMA methodology for production systems design.

The Mintzberg model

The decision process model developed by Mintzberg, Raisinghani and Theoret (Mintzberg *et al.*, 1976) is regarded by a significant proportion of social scientists as the best currently available technique for comprehending complex strategic decisions (for example, see Hickson *et al.*, 1985). The model overcomes the problem orientation criticism of systems modelling by categorizing stimuli to the decision process as 'opportunity, problem and crisis'. The Mintzberg model emphasizes decisions made within organizations as opposed to the decision-making behaviour of individuals (described by Newell and Simon, 1972). It is superior to previous instruments for modelling organizational and group decision making (as developed for example by Cyert and March, 1963), in that it considers a greater number of decisions across a wider range of contexts.

In essence, the Mintzberg model separates the overall decision process into seven steps or 'routines'. Despite a largely phase-wise progression, it is recognized that one routine does not necessarily follow another in adherence to a strict sequence (Figure 3.2). In reality, decision processes are liable to move forwards and backwards between routines in a sporadic manner due to the incidence of 'interrupts', 'speed-ups' and 'delays' in the decision process, which together cause a delineation of the basic steps.

Structuring the decision process in this way is an invaluable aid for leading the practitioner or researcher through the stages of decision making in a concise and comprehensible fashion. The model's applicability to manufacturing system design has been demonstrated and the model has been used in a number of situations as a tool for analyzing the decision processes inherent in a large-scale project (Figure 3.3; Bennett *et al.*, 1990). If organizations could conceptualize their decision-making activities in this way, it would enable

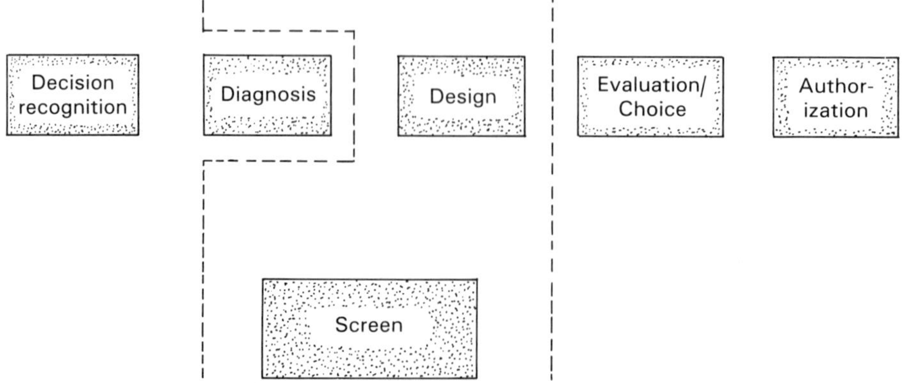

Figure 3.2 The Mintzberg *et al.* decision process model.

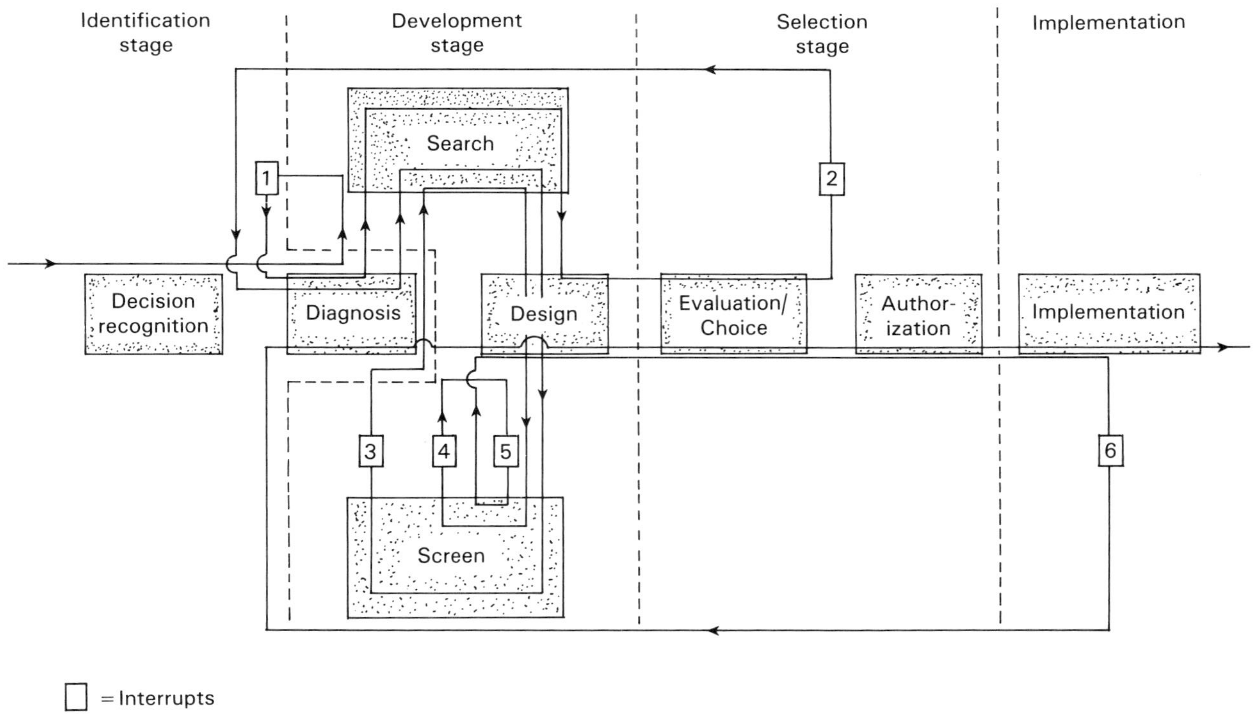

Figure 3.3 Mintzberg's decision process model applied to the analysis of a system design case.

greater understanding of the rationale behind design projects and, therefore, increase the effectiveness of future systems design projects.

The Mintzberg model, however, has a major shortcoming in that implementation is not recognized as a distinct phase. It is argued here that the addition of an implementation phase to follow the seven previous routines is necessary to capture fully the entire decision process (Bennett *et al.*, 1990). The implementation phase has been added in Figure 3.3. The omission of implementation is a criticism that also applies to other decision models, as few offer a detailed analysis of how solutions are developed from conception through to systems installation. Considerable scope exists for future research which seeks to analyze the implementation of complex design projects and how this process may be better understood and improved.

The GRAI model

The GRAI Laboratory at the University of Bordeaux has developed a conceptual model for total system design in tandem with a more specific technique for designing production control systems (Doumeingts *et al.*, 1986). In terms of practical application it is the production control system design methodology that has been widely adopted and used as a control aid, particularly under the funding of the European Community's ESPRIT programme (described in the next section on Design Methodologies). However, the widely cited conceptual model of hierarchical decision making could offer the greatest potential for the future development of generic, higher level methodologies.

This hierarchical decision model is illustrated in Figure 3.4. More commonly known as the GRAI conceptual model, it disaggregates manufacturing organization into three components, the 'physical system', the 'decision system' and the 'information system'. The decision and information systems, which together comprise the 'production control system', are then split into levels of decomposition to propose an organizational model with a hierarchical structure of decision centres. The interaction of the information system with the external environment initiates the open systems nature of the model.

The strength of the GRAI conceptual model is the way it provides a visual representation of the relationship between the decision centres at different levels within an organization. At the operational level it is a simple input–output model of the manufacturing system, converting raw materials and other resources into goods and services. Superimposed on this is the organization structure, each level having a set of decision centres which collectively comprise a single decision centre at the next level up. Most commonly this is dictated by the management reporting system, a number of subordinates reporting to a single manager. The information system is the element which links the various levels of organization with each other.

In the dynamic working situation one can visualize an organization operating according to this model, with information and communication

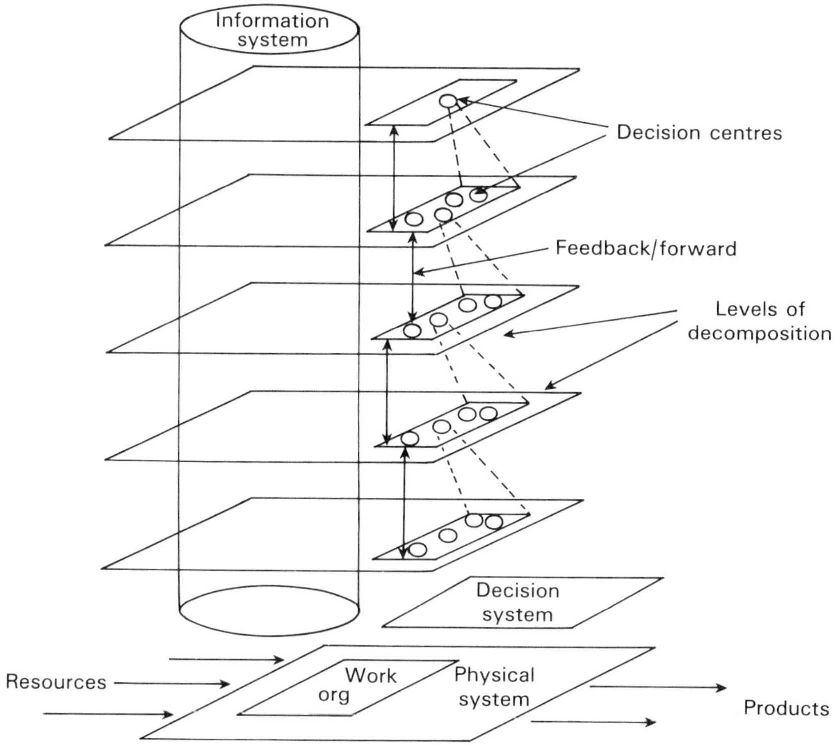

Figure 3.4 The GRAI conceptual model of decision making.

moving up and down the structure in the form of 'feedforward' and 'feedback' interaction. It is possible to envisage decision centres at the various levels as not just being passive receivers and senders of information: they frequently act as 'filters' in both a positive and negative way. Positively, filtering of information will draw out the main points from the complexity and mass of information whereas, negatively, managers at decision centres can manipulate information and pass on only what they want to. Thus it has a 'political' dimension. Figure 3.5 illustrates how the GRAI model can be applied to represent the structure of organizational decision making with a manufacturing company.

A major criticism of this basic GRAI model is that it only concerns itself with structure, or the static situation, and little with the process of design and its dynamics through time. Thus it is more useful as a means for analyzing a proposed or existing decision making structure than for analyzing decision and design processes.

The Design Attribute Time Analysis model

In addition to its use of existing models and theories to analyze decision

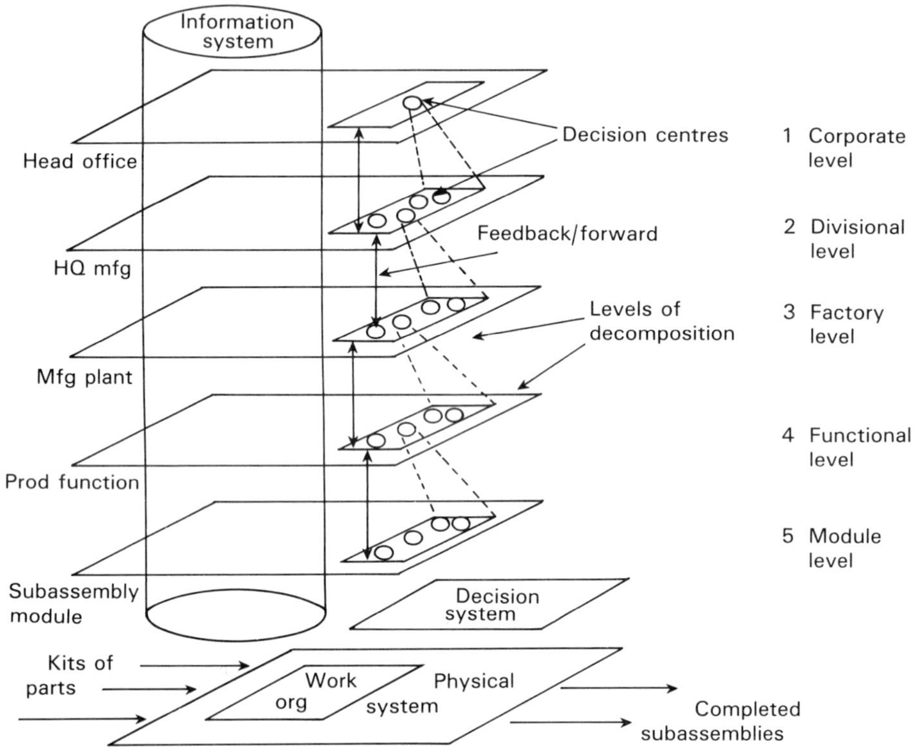

Figure 3.5 The GRAI conceptual model of decision making applied to ICL.

processes and information systems the IDOM Research Unit at Aston Business School has, during its research, developed a number of analytical frameworks to enable its researchers to structure their observations and findings. Of most relevance to the current discussion is a model which supplements the GRAI notion of hierarchical decision centres with a time frame. The 'Design Attribute Time Analysis' (DATA) model expresses the concept of time and the degree of variability of design attributes for each level of organization. The model is illustrated in Figure 3.6.

A three-level structural disaggregation of organization has been adopted for the DATA model with decision making and design activity categorized into three hierarchical 'domains'. In common with other analyses these have been ascribed the terms of the 'strategic', 'tactical' and 'operational' domains. The degree to which specific design attributes are considered long-term or short-term depends upon the domain within which the attributes decision process is contained. Decisions regarding the configuration of design attributes can, consequently, be positioned in the relevant sector of the model depending

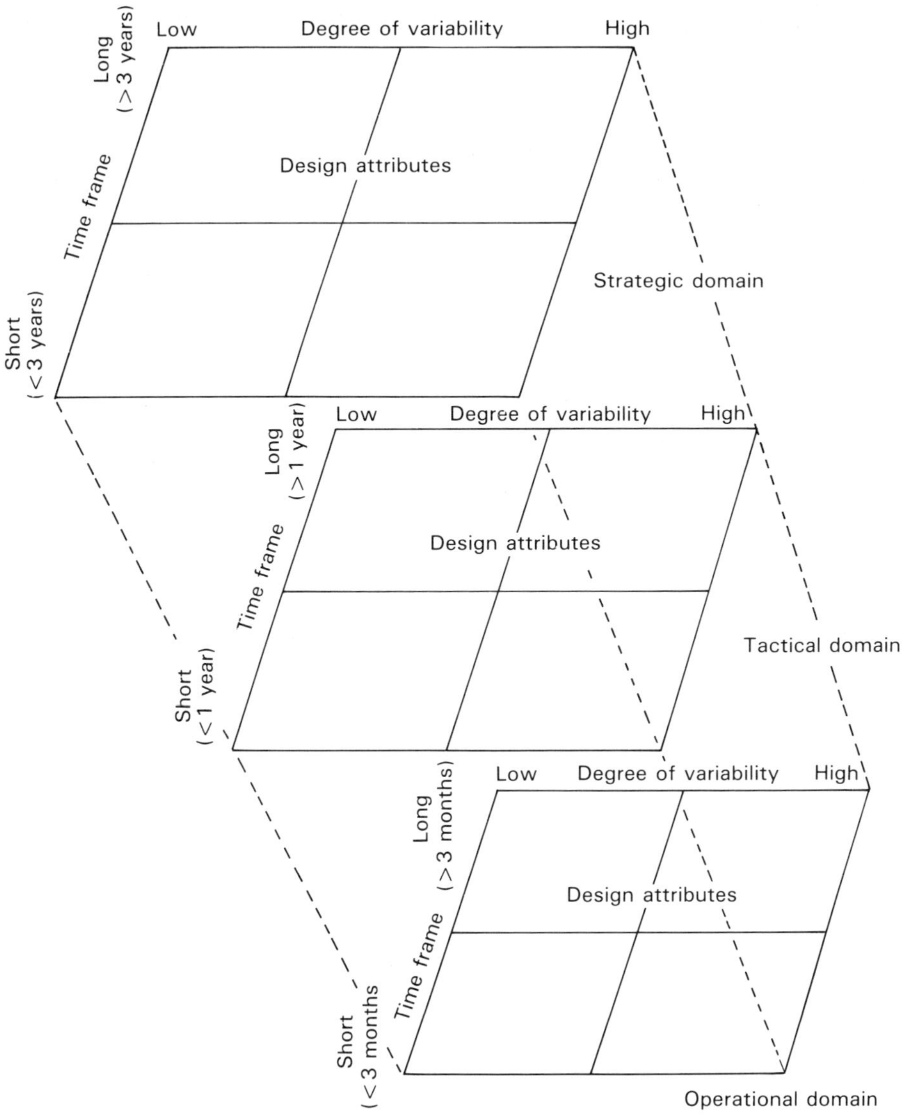

Figure 3.6 The IDOM design attribute time analysis (DATA) model.

upon whether they tend towards having low to high variability in relation to the short-term and long-term poles on the time axis. This method of categorizing decisions can be adopted in a structured way to analyze elements in the design components of the DRAMA methodology to test whether individual decisions on design are of long-term or short-term type, and whether they are relatively fixed or have degrees of variability over time.

Application 3.1 Modelling tools for process planning at Lucas Rists

Lucas Rists Wiring Systems is a medium-sized UK manufacturer of electrical components in the automotive supply industry. The company has traditionally used manual methods for the breakdown of assemblies into operations and sequential build instructions, and little progress has been made in specifying tooling requirements. The nature of this assembly task in the context of CAPP system development has been divided into three phases. Figure 1 shows the three phase development of this CAPP system. Development of these phases is being undertaken in a serial manner as each phase needs to be implemented successfully before the next stage commences. This will allow the company to gain maximum benefits from each stage and will also allow available data to be input at later stages.

The research team from Keele University is participating in the development of a sophisticated CAPP system within the host company. The main

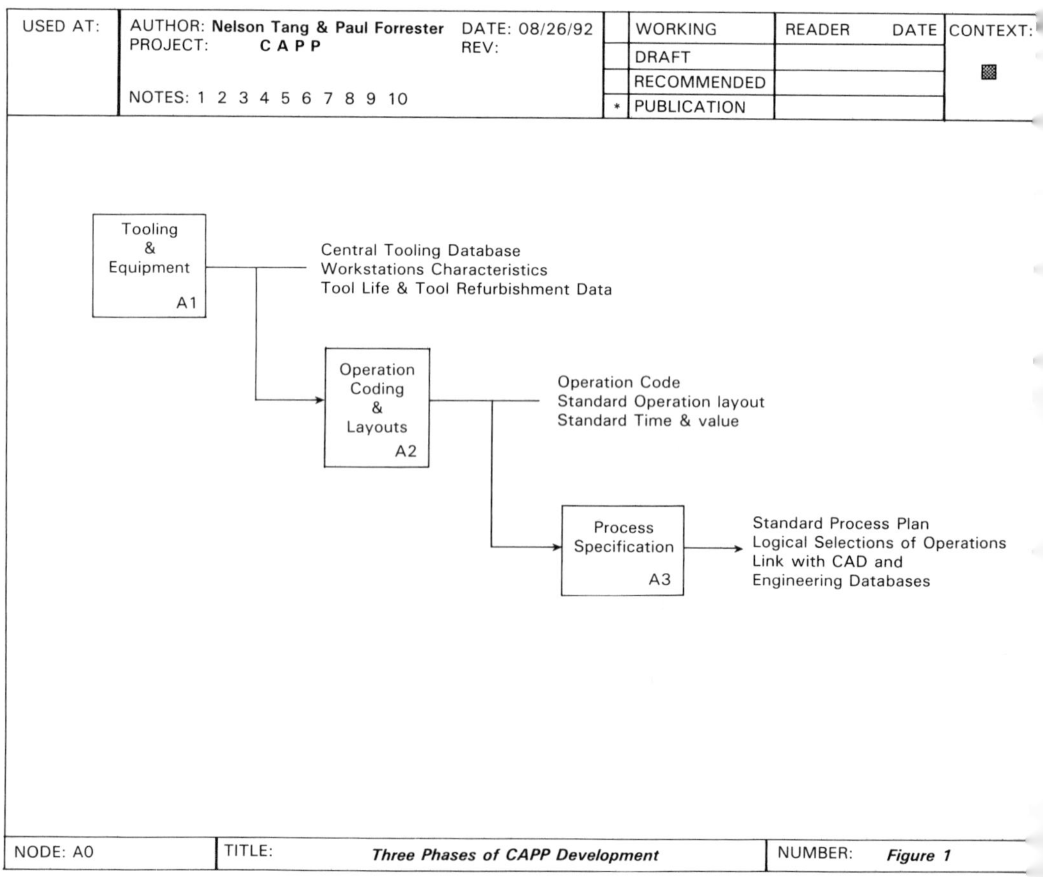

TITLE: *Three Phases of CAPP Development* NUMBER: *Figure 1*

aims of this work are not only to develop CAPP, but also to recommend a structured approach to improve integration of information flow between systems and to develop a manufacturing information system design methodology to analyze systematically user requirements, functional specifications and data structure.

Managing manufacturing information is more difficult than in the case of financial or commercial information; problems in the latter case are more often associated with volume and speed of transfer rather than with complexity and large variety, as in the former. In most organizational contexts, but particularly with advanced manufacturing technology implementation, system developments should not be seen in isolation and the interlinking of separate system modules and the identification of information interface points should be referred to at an earlier stage. For these reasons there is a need for a methodology to define and integrate the system planning with the analysis and design of the functional specifications and cross-functional communications,

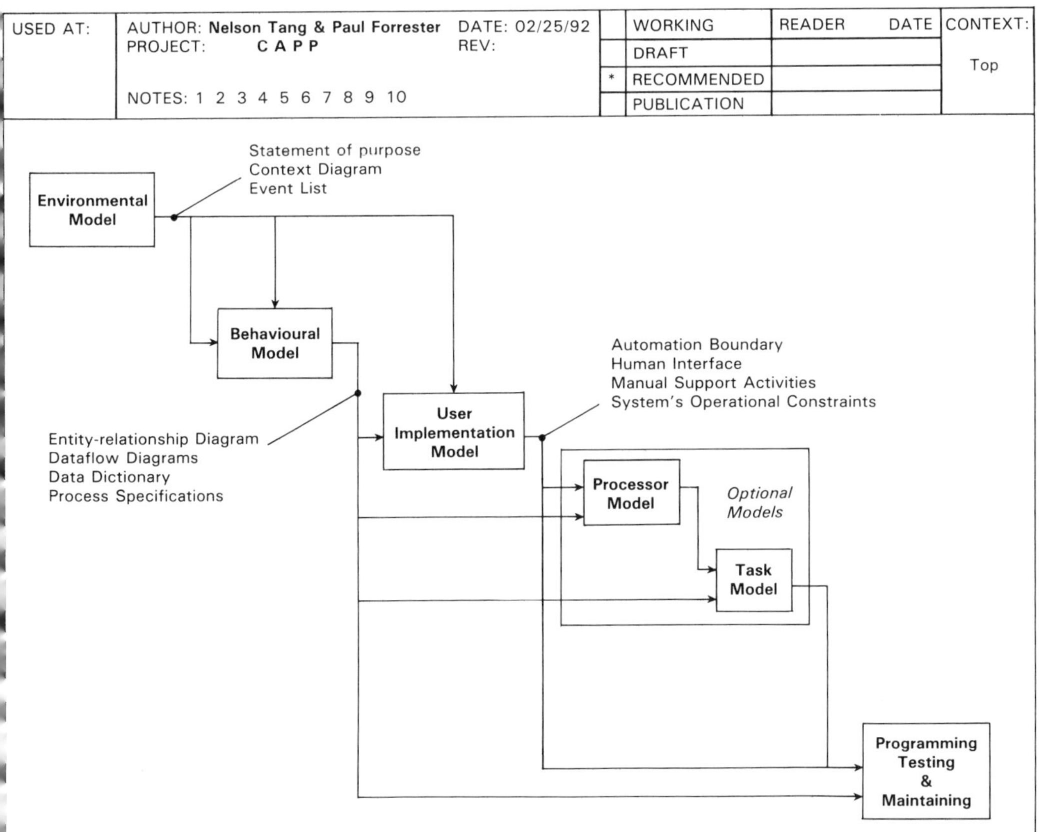

TITLE: *Three Models of MISADM* NUMBER: *Figure 2*

and it is necessary to specify software and hardware resources needed in the implementation stage.

As advanced manufacturing technology places increasing demands on information systems and also due to the complexity of manufacturing systems, the Keele Advanced Manufacturing Group (KAMG) decided to develop a new manufacturing information system analysis and design methodology. This can be used as a template for the development or integration of different manufacturing information systems and can be also used as a practical approach to guide analysts during the development process. This methodology is called 'Manufacturing Information System Analysis and Design Methodology' (MISADM) and it combines the structured analysis and design techniques and powerful modelling tools of both Yourdon and IDEF methodologies.

MISADM is divided into discrete stages incorporating well-defined models to describe and formalize the system being developed, each of these models being supported by related diagrams and tools. These models can be used as a reference and guide throughout the analysis and design processes and Figure 2 shows the recommended models to be developed in a sequential manner. Each model has documented output that serves as the input to the next model and the earlier models specify a behavioural abstraction defining what is the system being developed. This abstraction is progressively refined into a formal implementation of how the behaviour can be realized. These diagrams are produced using Design/IDEF Package (PC version) and have been found to be easily understood within the collaborating company, thus providing an excellent means of communication between the analysts and system users for the common understanding of problems and their analysis.

Source: This application was written by Nelson Tang, Keele University and is used by kind permission of Lucas Rists Wiring Systems.

■ 3.3 Design methodologies

What is a methodology?

In the context of this book a design methodology is defined as a 'structured collection of techniques and tools that assist the designer in the development of new processes'. It is useful at this point to draw a distinction between what is meant by 'models' and 'methodologies'. For the purpose of this book a methodology is seen as being a prescribed approach which offers the user a guide and the necessary means to implement a new system. A model, however, is seen as a framework which is descriptive and analytical rather than prescriptive. However, the two should not be viewed in isolation: indeed, this section will show that a methodology frequently makes use of conceptual models within its approach and often encourages the systems designer who may be looking for a 'quick-fix' or 'panacea' solution to perform more detailed analysis of the existing system and design process.

Methodologies from systems theory

A number of distinct methodologies have evolved from theorists in the systems area. For the purposes of this discussion two will be discussed. The first, proposed by Jaffe in the late 1960s, is given as an example of many similar methodologies that have been developed over recent years (Jaffe, 1967). The second, the soft systems methodology, is distinctive in its approach and has been evolved by Checkland from his original model of decision making and problem solving (Checkland and Scholes, 1990).

Jaffe's system design methodology

Jaffe's methodology concentrates predominantly on the design of internal systems, be they physical facilities, information systems or any other type. Therefore the methodology, illustrated in Figure 3.7, has the appearance of a closed systems analysis, although the details within each stage of the prescribed design suggest that some form of external analysis is required. The best description of the methodology would therefore be that it is of the 'partially-open' type.

The Jaffe methodology considers there to be two basic types of activity involved in systems design work: 'technical design activities' and 'design project management activities'. This categorization fits well with the intentions of this book: to provide guidance not only on the choice of systems hardware, software and organization, but also to consider the issues that arise when conducting a design project.

The Jaffe methodology first recognizes that there will be an initial conception of the system (existing and proposed) on the part of the user and that this should be identified and recorded prior to moving on to conduct a feasibility study. From this the technical design activities can commence with a definition of user requirements and continue on with the design and, eventually, proposals for the integration of the system. The technical design activities are punctuated by influences from ongoing design management activities, most notably the development of the design documentation as the system evolves and an evaluation of the new design. If the evaluation of the new system is positive, the design team should then recommend the design for capital approval. Jaffe adds a recycling routine to show that design activities can, and do, go through a number of reiterations both prior to the implementation of the system and following installation.

Jaffe's methodology has its merits. It distinguishes 'technical' from 'project management' activities and proves a means of identifying discrete steps in the design process to assist the project planner. Advocates of this type

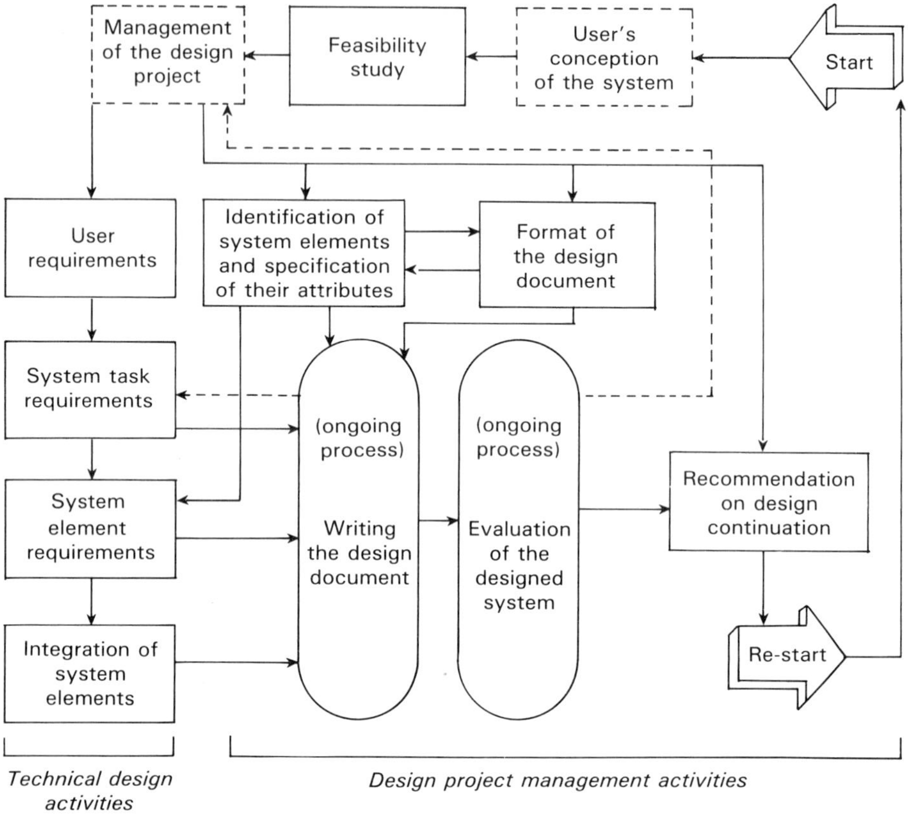

Figure 3.7 Jaffe's system design methodology.

of approach suggest that such methodologies are fruitful for directing attention to the existence of structures and relationships between elements in these structures during the design process. However, such approaches must be seen as having their limitations. First, they advocate and rigidly adhere to a step-wise progression of activities and events. This is an overly-clinical view of what occurs in practice. Although a neat progression might be planned, there will always be a high degree of feedback and recycling inherent within any system design project as a result of changed circumstances, modified requirements, etc. This leads to a second criticism, which is that Jaffe's methodology, albeit in common with the approach of many practising systems designers, focuses firmly upon the interaction of decision processes and events internal to the organization. It therefore makes little allowance for interference from the external environment. The important message here is that system designers must recognize the level of analysis at which they are working. If they are working at a relatively low level of conceptualization, a methodology such as Jaffe's may be appropriate, but it must be ensured that someone in the organization should, at the same time, monitor developments within the organization as a whole and the environment generally. In this way it is possible to monitor and detect any significant changes which might affect the emergent design.

Checkland's soft systems methodology

In some respects Checkland's soft systems methodology is difficult to separate from his basic model of decision making. However a distinction can be formulated as a result of the way that the soft systems approach has been used in industry. Far from its being a mere description of problem solving, decision-makers have plainly used the approach to structure and order their own design work. It appears that Checkland has achieved his objective of advocating a softer, more humanistic approach and understanding of the decision-making process (Checkland, 1981).

The soft systems methodology is suitable for the analysis of human and social aspects of decision processes, the higher levels of systems analysis within the Kast and Rosenzweig hierarchy of systems (Kast and Rosenzweig, 1970). Checkland's basic argument is that the soft systems approach is the general case of which the harder, more closed methodologies are special cases which underlie the overall decision process, and are invoked at lower levels of problem solving.

The soft systems methodology operates as follows. First the decision-maker should identify and then define and express the problem in the real world. Once the problem has been expressed, the decision-maker should seek to develop abstract models to represent the real world problem, either utilizing existing theories and models or by developing their own model of the system in question. This model should then be tested and verified within the real world by comparing it with the defined problem situation. If the model does not truly represent the real world, then the decision-maker must refine the existing model or develop new concepts, thus passing back into the abstract world of thought. If, however, the model is seen to be an accurate reflection of the problem situation, the decision-maker can then develop feasible and desirable changes to the real-world situation and then implement them. The implications of these changes should then be monitored and, if necessary, the whole process should begin again with a new expression of the problem situation and modified conceptual models developed.

Checkland argues that this methodology helps to clarify a process of decision making that is frequently viewed as a 'black art'. He argues that by understanding the soft systems model of decision making, and then adopting it in design activities, enables the decision-maker to distinguish between the different phases of the process, thereby leading to a more structured and analytical method of choice.

Structured analysis and design techniques

In the specialist area of manufacturing systems engineering, a number of structured design methodologies and techniques have recently been developed to assist in the design and analysis of complex advanced manufacturing systems. This chapter will now consider four methodologies suitable for use when developing new manufacturing facilities. Their principles can be extended to cover projects for conventional, rather than just 'advanced', systems. The methodologies under review will be SADT (Structured Analysis and Design Technique), IDEF (ICAM DEFinition), GRAI (Graphe a Resultats et Activites Interrelies) and CIM-OSA (Computer Integrated Manufacture Open Systems Architecture).

SADT methods

SADT, originally conceived by Ross of Softech USA (Ross, 1977), involves the identification of data flows and activities within a manufacturing or information system. The overall system is decomposed into a series of blocks at all levels in the system hierarchy. For each block the input, output, mechanism of conversion, and constraints and controls upon the activity under investigation are identified, as shown in Figure 3.8. At the top level of the system, the activity identified will be a collection of composite operations at a lower level of organization and so each block at the 'enterprise' level will be exploded into sub-blocks, which give a more detailed breakdown of the various activities. This process of decomposition can be continued to finer and finer levels of detail (Figure 3.9).

SADT provides a structured systems modelling approach to developing systems, and not only addresses the nature of all activities in an organization, but also sets these within a hierarchy so that they are investigated within the context of the overall system. The system designer adopting SADT can develop a model of the current situation and then use this to identify any problems or shortcomings within the existing design. The effects of any

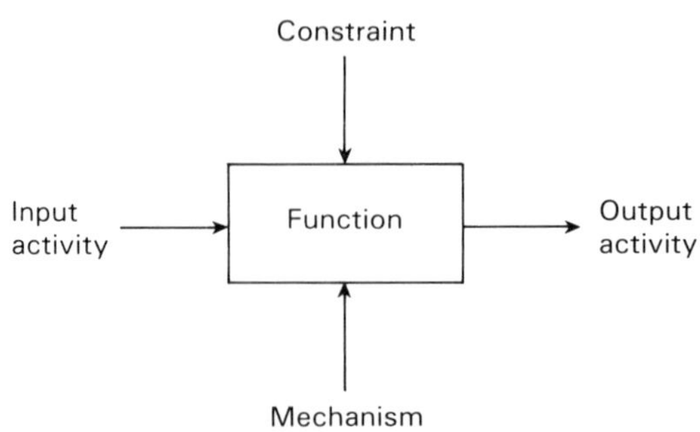

Figure 3.8 The form of an SADT block.

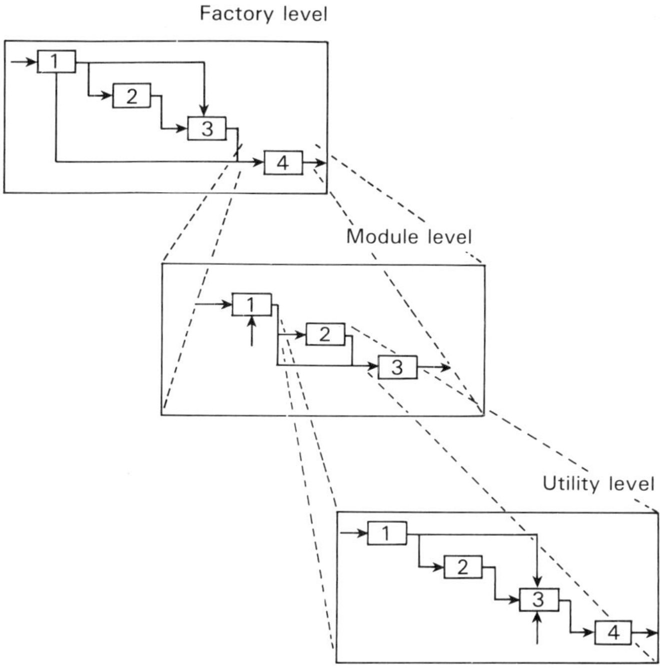

Figure 3.9 The decomposition of system activities within SADT.

proposed changes can then be mapped out and evaluated on the model prior to their introduction to the live system in order to evaluate their likely effects. Once changes are made to the system, the SADT model can then be revised and used in future system design work. (For more descriptions of SADT see Marca and McGowen, 1988; and Nicholson, 1991.)

The IDEF methodology

SADT was further developed and, in large part, provided the basic modelling techniques for the IDEF methodology. IDEF was initially devised within the US Air Force's Integrated Computer Aided Manufacture (ICAM) Programme to support its aircraft manufacturing activities. IDEF is a collection of SADT-type graphical tools and is now most commonly used to describe the architecture of a manufacturing system. It consists of IDEF-0, which models the function and relationships within the complete system, IDEF-1 for modelling information systems and data requirements and IDEF-2 for dynamic simulation of the behaviour of complex manufacturing systems. IDEF is most frequently used as a computer-based set of tools and the process begins with the development of an IDEF-0 model as a hierarchical representation of the manufacturing system.

Despite a widespread belief that IDEF is exclusively a computerized technique, it must be emphasized that only IDEF-0 was developed as a commercial computer-based tool. The use of IDEF is therefore mainly restricted to the manual construction of activity flow diagrams. Diagrams of the current system ('AS-IS' diagrams) are then analyzed to identify any inefficiencies and possible improvements to the design of systems. From this the proposed system diagrams ('TO-BE') can be drawn up and used as a blueprint for future systems development (see Ranky, 1990). Figure 3.10 shows a simplified IDEF-0 model for one level of decomposition within an organization.

The rationale behind IDEF is that using a graphical approach to the design of facilities enables the communication of current and proposed systems operations to all participants involved in the design or decision-making process. It is argued by advocates of the approach that IDEF, being a set of general purpose tools, is applicable to any area of manufacturing systems design. In practice the methodology appears useful, but somewhat limited. Flow diagrams are useful for analysis but not creative in themselves. The design of any system, or diagram representing the system, is only as good as the process designer's or systems analyst's skills permit (Yeomans, 1985; Ranky, 1990; Nicholson, 1991).

The GRAI methodology

The GRAI conceptual model has been described in detail in the modelling

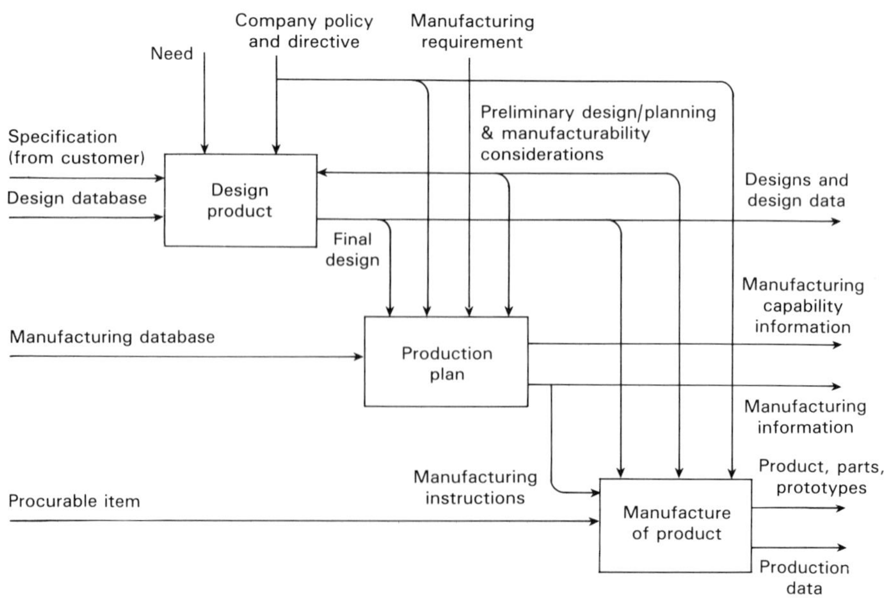

Figure 3.10 An IDEF-0 model for new product introduction.

section at the beginning of this chapter. Building upon this, the GRAI research group at Bordeaux have sought to develop a generic methodology for the development of the physical, decision and information system identified within the basic model. The GRAI method incorporates a series of graphical tools, but most widely used is the GRAI grid, an example of which is shown in Figure 3.11.

The GRAI grid retains the hierarchical character of the GRAI model, but within this hierarchical structure maps out activities through time. The methodology comprises two parts, a 'top-down' and then a 'bottom-up' analysis, both of which are mapped on the grid. The top-down analysis maps the activities between functions that occur from the corporate down to the operational level of manufacturing and maps out the transition from the start to finish of decision processes and information flows. In so doing the analyst seeks to identify any deficiencies that may occur within the organization processes and between decision centres. The bottom-up analysis uses 'GRAI nets' (Figure 3.12) to examine each decision centre in relation to its inputs and outputs, the dynamics of production and the constraints imposing upon the decision centre.

The GRAI methodology is used as an analytical means of investigating the production management and control system. The use of graphical tools enables detailed investigation of each decision centre within the overall context of the business and may highlight areas of inadequacy requiring improvement or redesign. The main criticism of the current applications of the GRAI production control design methodology is that, as with IDEF, it serves more

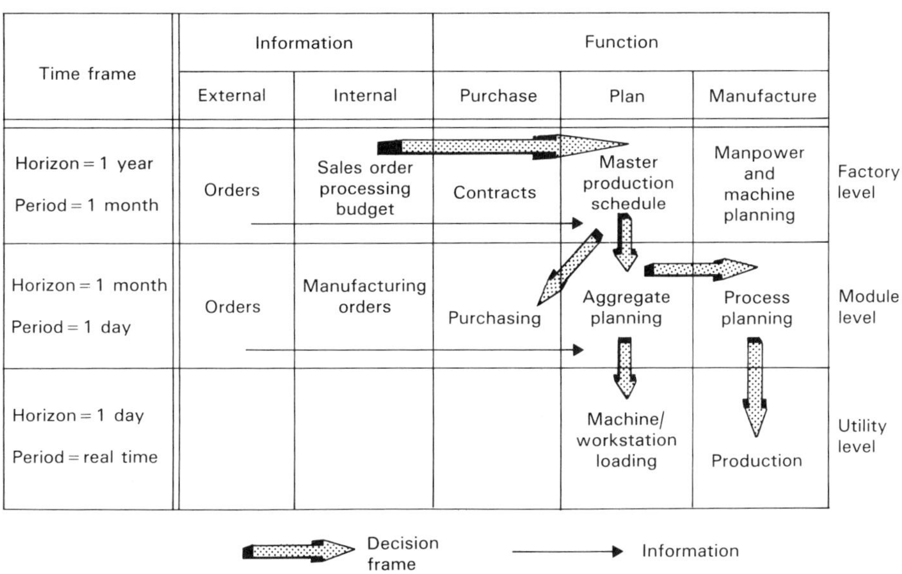

Figure 3.11 A GRAI grid for 'top-down' systems analysis.

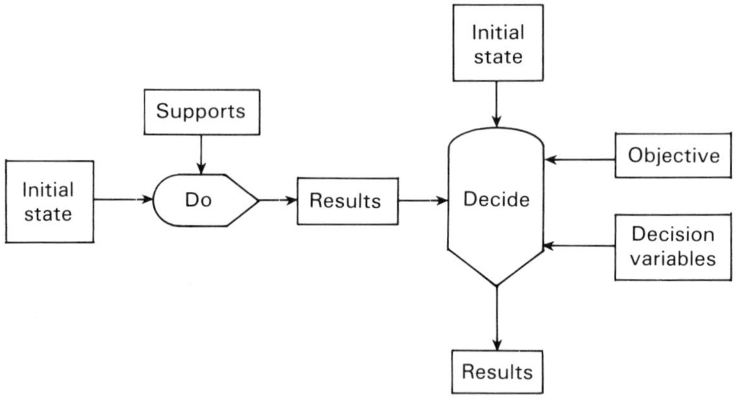

Figure 3.12 A GRAI net for 'bottom-up' systems analysis.

to represent current and proposed design configurations from which any deficiencies can be identified. It does not offer a good means of understanding the underlying design process for new systems. The GRAI model, however, can provide a structural framework for understanding the hierarchical composition of organizational decision-making. It should be employed therefore in conjunction with a methodology for creating an overview of the decision-making process for new systems. (For more detail see Doumeingts *et al.*, 1986; Roboam and Pun, 1989; Roboam *et al.*, 1989; Ridgway and Downey, 1991; Doumeingts and Chen, 1992.)

The CIM-OSA methodology

The aim of Computer Integrated Manufacturing (CIM) is to link all the activities of an enterprise into a consolidated management organization. It therefore encompasses every aspect of the business from corporate strategy to engineering, production, marketing and sales, purchasing, quality and financial systems. CIM Open Systems Architecture (CIM-OSA) is a methodology for the incremental progression towards CIM and was developed in the AMICE Project. AMICE, a reversed acronym for 'European Computer Integrated Manufacturing Architecture', is a consortium of nineteen major European companies and software houses which have collaborated under the auspices of the EC's ESPRIT (European Programme for Research in Information Technology) programme with the aim of developing guidelines and a generic reference architecture for the adoption and implementation of CIM. CIM-OSA is therefore widely seen as being a standard approach to the implementation of CIM. The objectives of the project have been summarized as follows (AMICE, 1989):

1. To enable fast, economic utilization of advanced technologies in industry.

2. To ensure long range, evolutionary CIM implementation and growth.
3. To enable and support independent development of CIM building blocks.
4. To guide the CIM users in designing, selecting, implementing and operating CIM systems, CIM subsystems and CIM system components.
5. To guide the CIM suppliers in the design and development of such systems that will aim for well-defined targets in developing new products and establish migration paths for existing applications towards more and more integrated systems.

CIM-OSA, represented diagrammatically in Figure 3.13, provides a framework comprising two distinct parts:

1. The 'reference architecture': the generic framework across the full range of operational and organizational contexts.
2. The 'particular architecture': the company specific portion of CIM-OSA requiring development by the methodology user.

CIM-OSA, it is proposed, provides a common language for describing the complex activities and information flows within manufacturing enterprises. It involves the use of specified 'building blocks', contained within the reference architecture, for what are termed 'enterprise modelling', 'information modelling' and the translation of these into an 'implementation model'. These blocks are in the form of constructs and views contained within the reference and particular architectures and cover such activities as product and process

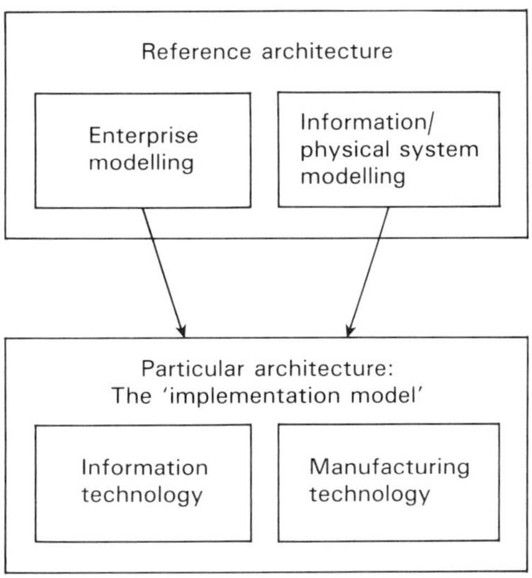

Figure 3.13 The CIM-OSA methodology.

development, production management, manufacturing control systems, quality management and maintenance.

When using CIM-OSA, each block is decomposed into its detailed functions with information flows graphically represented on SADT-type diagrams. The building blocks are contained within the reference architecture and comprise two main types:

1. The 'enterprise system' building blocks: the means by which an enterprise gathers and collates information on business requirements.
2. The 'physical system' building blocks: the specifications for the set of capabilities required to execute the enterprise functions.

The required enterprise system is called the 'enterprise model' and the physical implementation is termed the 'implementation model'. The implementation model is created from the enterprise model by a process CIM-OSA calls translation, hence the need for an 'intermediate model'. By using the building blocks contained in the reference architecture, methodology users will arrive at the particular architecture for their own organization.

Enterprise modelling involves the recognition of organizational structure and involves a disaggregation of the hierarchy (strategic view, tactical view and short-term operational view), identifying the requirements at each level. It is then proposed that the information flows must be mapped out within this enterprise model and so information modelling involves the representation of data flows between and within each of these levels. Finally, the implementation model uses the input/output type material when the information model is superimposed on to the enterprise model. Implementation takes the outputs of the enterprise model and arranges them into a logical configuration.

In essence, CIM-OSA seeks to establish:

1. Common understandings of CIM within and between organizations.
2. A generic architecture for reference by manufacturing industry as a whole enabling the development of particular architectures to structure individual enterprise activities.
3. Three modelling levels: 'enterprise', 'intermediate' and 'implementation'.
4. Four 'viewing' levels (function, information, resource and organization) to model, represent, structure and then optimize the content of the different activities of the enterprise.
5. Generic building blocks of a modular type for use by industry generally.

This framework is useful but has a number of limitations, not least of which is that it was developed on the initiative of the large companies within the consortium and by software suppliers. As a result it is geared mainly to the large-scale producer and suppliers of technical systems to these companies. Consequently the framework, and the systems approaches advocated, are limited in the extent to which they address the capital, human and technical resource constraints that impinge upon smaller-sized manufacturing companies because the modelling approaches and viewing techniques used in CIM-OSA are not easily understood by non-specialist analysts.

CIM-OSA then provides a useful framework for the development of standardized architectures for advanced manufacturing systems and does provide a standardized approach to the design of CIM infrastructures. However it cannot be seen as an all-inclusive methodology in its own right. In particular, the methods through the stages of design, development and implementation are not well defined and provide little by way of assistance to the user in this respect. So, whereas it provides a structure which can form the basis of a CIM system, it provides very little by way of guiding principles for the design of CIM systems. It provides no specific design tools and does little to direct the systems designer to those methods and techniques which may be of use. Finally, it has limitations in terms of its assumptions about the current situation in terms of manufacturing technology for all but the very large and leading edge companies. Its relevance and proven abilities for the advancement of CIM in, say, the medium-sized company must therefore be severely questioned. (For more information on CIM-OSA see Russell, 1987; AMICE, 1989; Marechal, 1989.)

Methodologies for developing manufacturing strategy

So far this chapter has considered methodologies that have focused on the development of internal information and manufacturing systems. Whilst linking these developments to the business needs of the organization, usually through structured decomposition, these approaches can be viewed as being relatively closed in relation to the markets and environment within which manufacturing organizations operate. This section reviews two structured approaches to the development of manufacturing strategies that can be used in conjunction with the systems analysis and design methodologies described above. A particular feature of these approaches is the way they link manufacturing facilities and production control choices with corporate policy and marketing strategy, and so with the markets and environments within which organizations operate. The two approaches reviewed here are the 'Cambridge Group's' methodology for developing manufacturing strategy (DTI, 1988) and Hill's framework for reflecting production/operations strategy issues in corporate decisions (Hill, 1985).

The DTI/Cambridge Group methodology

A methodology for developing manufacturing strategy has been devised at Cambridge University and has been widely publicized by the UK Department of Industry (DTI) under its 'Enterprise Initiative'. The methodology provides guidelines for practitioners in developing their objectives for manufacturing (DTI, 1988). The framework of the methodology is shown in Figure 3.14. It

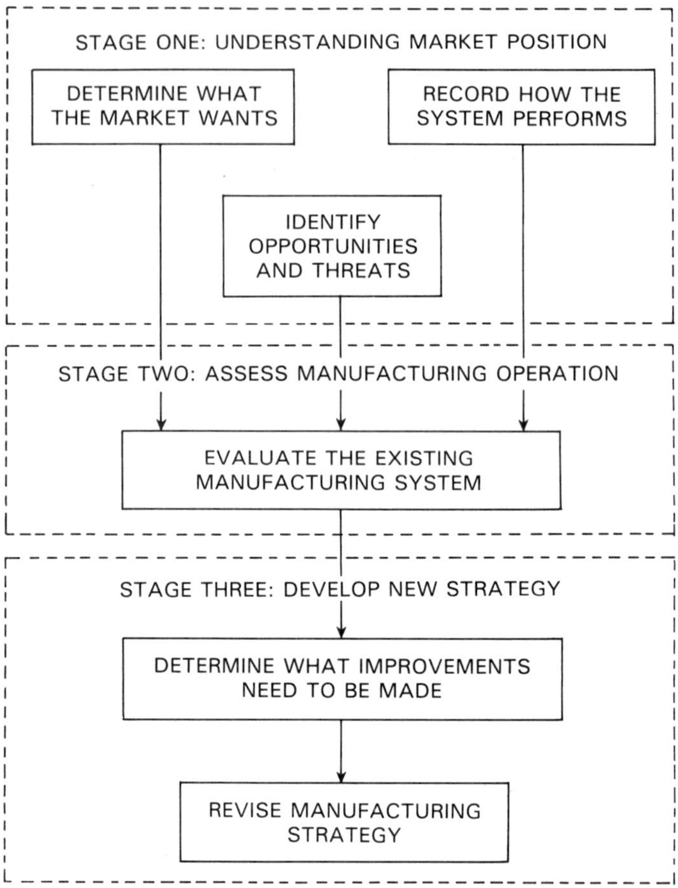

Figure 3.14 The Cambridge Group's methodology for the development of manufacturing strategy.

covers such issues as determining market requirements, competitor threats, performance of existing systems, and how to combine all these into a comprehensible and meaningful blueprint for production.

A key feature of the methodology is that, through the use of worksheets which the user completes, it not only performs an analysis of the market and competitive environment within which an organization operates, but also conducts an audit of existing manufacturing facilities and competencies. In this way it looks for areas in which system performance may be improved and to better align operations to the market needs.

The methodology involves three main phases, as shown in Figure 3.14. Stage One involves generating basic data on product families, which indicates the importance of the family to the business and the strength within current markets. The competitive criteria for each family is then determined (e.g.

price, delivery performance, etc.) and the organization's current performance against these criteria is assessed. Finally, potential areas of product profitability and vulnerability within the business are identified.

Stage Two involves an assessment of current manufacturing strategy and analyzes this in nine areas, namely:

1. Facilities: the manufacturing factories, including number, size, location and focus.
2. Capacity: the maximum output for each factory.
3. Span of process: the degree of vertical integration.
4. Processes: the transformation activities and the way they are organized.
5. Human resources: the people-related factors.
6. Quality: the means of assuring products, processes and people operate to specification.
7. Control policies: the operations planning and control guidelines and philosophies of manufacture.
8. Suppliers: relations and methods to ensure delivery of input materials.
9. New products: the mechanisms and processes for managing new product introduction.

Stage Three involves using the results of the analyses conducted in Stages One and Two and developing a new manufacturing strategy linking manufacturing activities with corporate strategy and market needs. The processes in this strategy formulation are identified as:

1. Select the most important product families, based upon assessments of the relative contributions to profits, growth potential of market share or overall market size, and current strength within markets.
2. Taking each family in turn, compare the market competitive criteria with the achieved performance.
3. Identify the manufacturing policies which contribute to any mismatch between the competitive criteria and the actual performance.
4. Identify the weaknesses of the policies.
5. Consider opportunities and threats.
6. Identify possible actions and strategic choices.
7. Repeat for other families.

In summary, the DTI/Cambridge Group methodology is a means of linking manufacturing systems choices with corporate and marketing strategy. The originators suggest that the methodology can be used in two main ways, both for auditing existing manufacturing activities in order to identify current strengths and weaknesses, and as a framework for reviewing manufacturing strategy and developing a new one if need be. As such, it not only provides a useful insight into the process of manufacturing strategy formulation, but also suggests a practical approach by which organizations might arrive at appropriate strategies for production and market-focused systems development.

The Hill framework

Hill's work is germane to any discussion on manufacturing strategy and its links to production systems design. Its main concern surrounds the notion of strategic management of the operations function (Hill, 1985). As his starting point Hill compares manufacturing management and performance in an international context through the extensive use of economic statistics. He then provides lessons within an overall framework for the translation of corporate policy and market strategy into 'process choice' and the development of the manufacturing 'infrastructure'. Process choice and the installation of the infrastructure can be viewed as the two essential components of production systems design, and is most relevant to the subject of this book.

Hill bases his framework upon the following premise (Hill, 1991, p. 21):

> The need for strategic difference in production/operations is, therefore, a prerequisite on which to build a sound and successful business. To accomplish this, it is necessary for business to recognize relevant issues at two levels:
>
> 1. How to develop relevant production and operations functional inputs to the corporate strategy debate.
> 2. To be aware of the different approaches at the operational level within manufacturing management so that appropriate consideration can be given to alternatives.

Hill summarizes his arguments and discussion within his 'framework for reflecting production/operations strategy issues in corporate decisions'. This framework is reproduced in Figure 3.15. Fundamental to the framework is an understanding of 'how products win orders in the marketplace', step three in his framework. As shown in the diagram, these include price, quality, delivery,

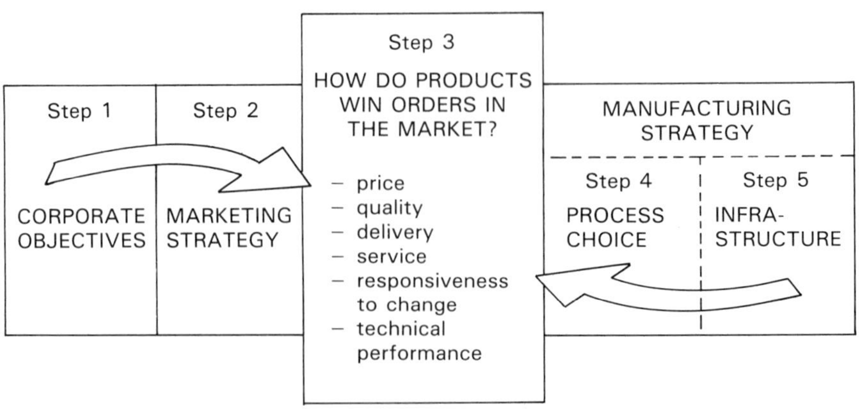

Figure 3.15 Framework for analyzing production/operations strategy.

service, responsiveness to change, and technical performance. In turn it is argued that the mix of these comes from two major sources, originating from the customer or market as reflected in corporate strategies on the one hand, or generated from the inherent features of the production system on the other (as indicated by the arrows on the diagram). The manufacturing organization must identify the order-winning criteria for its products, i.e. the correct balance between the factors detailed above, and then reflect these in corporate objectives, market strategy and production systems design.

The five steps in the framework are defined (Hill, 1991, p. 25) as:

1. Define corporate objectives.
2. Determine marketing strategies to meet these objectives.
3. Assess how different products/services win orders against competitors.
4. Establish the most appropriate mode to manufacture these sets of products or provide these sets of services – process choice.
5. Provide the infrastructure required to support the production/operations process.

Hill sees nothing unique about this process which, he says, is essentially the 'classical steps in corporate planning'. However, he regards the current situation as one where corporate decision-makers see only the first three as an iterative process with feedback loops at each stage of development. Process choice and infrastructure, the production systems design issues, are simply seen as linear and deterministic by most senior executives. It is argued that manufacturing strategy (i.e. steps 4 and 5) should also interact in an iterative manner with the first three steps: that manufacturing managers should enter the 'corporate debate'. He also sees manufacturing facilities as possibly providing areas of advantage which are frequently overlooked in strategic formulation, hence the arrow from steps 4 and 5 back to step 3 (Figure 3.15). This notion fits well with the view of manufacturing as a 'competitive weapon' (Skinner, 1978) and the concept of 'competing through manufacturing' (Hayes and Wheelwright, 1984).

The Hill framework demonstrates how decisions on production systems design are linked with decisions on market and competitive issues. Moreover it illustrates how excellence in manufacturing at the level of manufacturing systems can provide order-winning characteristics for products in ways which may not have been recognized by corporate decision-makers and marketing managers. The framework can be used in two ways. It can first be utilized for the assessment and evaluation of the effectiveness of manufacturing operations in relation to corporate objectives and product markets. Second, it serves as a guide for developing new, market-focused strategies and systems in manufacturing. As such it represents a significant contribution to the study of manufacturing strategy and process decisions.

3.4 Conclusions

To be useful and applicable to manufacturing systems design, any process model or methodology must adopt a number of approaches which make sense of the various influences on the system. This reflects the reality within organizations where numerous, often conflicting perspectives exist, all of which may have their own justification. However, it is difficult to envisage a single approach in the form of a generically applicable model or technique which can ever claim to cover all perspectives. The systems designer needs to be aware of these influences and conflicts, and must adopt a flexible approach in an attempt to encompass all likely occurrences throughout a systems design process.

Systems design must adopt to an open systems approach in order to recognize and evaluate the degree of influence exerted by environmental factors such as competitors' actions, market requirements and product and process technology. However, systems design is also a multilevel activity within any organization, so in some areas a partially-open or even a closed systems approach may be appropriate in order to give some appreciation of specific design events or process choices. Many of the activities of the process engineer, for example, may take recourse primarily to a closed systems perspective.

In developing systems there is a need to distinguish between the 'real-world' of the organization's current operations and the more abstract world of production systems analysis and design of proposed new systems. Having said this, however, the two should not be allowed to become divorced, otherwise a solution could evolve which may not be appropriate to the real-world situation. Application of the Checkland soft systems approach would seem suitable here. A soft systems methodology can be used to model at a conceptual level and relate these abstractions to the operating system. It has the additional benefit that it not only accounts for the hard specification of systems design, but also takes into consideration the effects of the human decision-making process, and therefore observes the influences of behavioural aspects upon systems design.

To gain a methodological balance in the appreciation of systems design activities both processual and structural aspects must be taken into account. The Mintzberg model of unstructured decision-making can be used to map the various stages of the decision process involved in systems design, and to recognize the extent to which feedbacks and multiple iterations affect this process. The GRAI model, on the other hand, can be used to structure and show the common decision interactions that occur between the various hierarchical levels within an organization's information and decision systems.

A number of structured analysis and design methodologies have recently evolved with an increased focus on the development of technologically advanced production systems, perhaps due to the need for creating order out of the complexity of systems design. These methodologies tend to be strong on the graphical aspects of representing of systems architectures, but weaker on issues such as managing the design process. SADT, IDEF and GRAI methods offer a means of representing manufacturing activities in flowchart form in order to analyze any deficiencies, but offer few guidelines in terms of guiding principles and prescriptions for the systems designer. CIM-OSA attempts to provide generic building blocks to enable the development of CIM systems that adhere to widely accepted standards and conventions. Once more, however, there are few insights into the process of system design work.

The chapter reviewed two recently developed frameworks assisting in the development of manufacturing strategies. The DTI/Cambridge Group's approach and the Hill framework provide a useful means of linking process and systems choices to business and market needs, thus they provide important concepts for the development of market-focused production systems.

The main conclusion from this chapter is that whereas generic models of the organization and its decision making (e.g. systems modelling, Checkland methodology, GRAI conceptual model) may be proposed, no single methodology could ever be seen to provide all the guidelines and framework for the development of manufacturing systems. Instead, a combination of these should be adopted at different levels of analysis and decision making within the organization. So, at the corporate level a strategic approach such as that advocated by the Cambridge Group or Hill could be seen as appropriate, whereas for the mapping of detailing data flows SADT or GRAI methods could be adopted. The majority of the models and methodologies described in this chapter have contributed in some significant way to the construction of the DRAMA methodology. The next chapter will therefore discuss the development and overall structure of the DRAMA design methodology which is then presented in detail in Chapters 5 to 14.

■ 3.5 Further reading

Checkland, P. B. and J. Scholes (1990) *Soft Systems Methodology in Practice*, Wiley, Chichester.

Hickson, D. J., R. J. Butler, D. Cray, G. R. Mallory and D. Wilson (1985) *Top Decisions: Strategic Decision Making in Organizations*, Blackwell, Oxford.

Nicholson, H. (1991) *Interconnected Manufacturing Systems: The Problems of Advanced Manufacturing*, Peter Peregrinus Ltd, London.

4 A Design and Implementation Methodology for Market-focused Production

Objectives of the chapter

On completion of this chapter you should:

1. Appreciate the principles underlying the DRAMA design methodology for the development of market-focused production systems.
2. Understand the structure and scope of the DRAMA methodology.

Chapter outline

The DRAMA methodology is the main output from a 3-year study into the design and implementation of flexible assembly systems within the electronics assembly industry. The methodology has been further developed to make it applicable for systems design and evaluation across a wider range of operational contexts than suggested by the original study. The result is a set of guiding principles for the design and implementation of market-oriented production systems, as presented in Chapter 5. This chapter tracks the development of the DRAMA methodology from its inception for electronics assembly systems through to its status as a generic methodology for systems design. The second half of the chapter then explains the structure of the methodology and its features showing the reader how DRAMA should be used in practice.

4.1 Background to DRAMA

The DRAMA process design methodology was originally devised by a team from the Innovation, Design and Operations Management (IDOM) Research Unit at the Aston Business School. Its development constituted the principal activity in a 3-year programme of research supported by the Science and Engineering Research Council under its ACME (Application of Computers to Manufacturing Engineering) Directorate. It is based upon the design and implementation of new manufacturing systems and the underlying decision processes. The work has been further developed since 1990 via joint research between Aston and Keele Universities.

International Computers Limited (ICL), a major manufacturer of mainframe computer products in the UK, has been the main industrial collaborator in the research. Consequently, DRAMA was synthesized from empirical work which investigated process design activity at ICL's main assembly plant at Ashton-under-Lyne, near Manchester, over the period 1985–90. During this time an integrated production system termed the 'Modular Assembly Cascade' was introduced into the plant (Bennett and Nagarker, 1988), hence the acronym used for the resultant methodology ('Design Routine for Adopting Modular Assembly').

DRAMA is not a set of rigid and mechanistic prescriptions so does not propose to its users a limited panacea of procedures. It can best be described as an empirical collection of 'guiding principles' that allows companies to analyze their competitive position and thereby design a production system appropriate to their needs, based upon a number of feasible design options. The design aspects of the methodology have been developed specifically from electronic assembly operations within a market-driven business environment, being the context within which ICL operates. However, the analytical components of the conceptual DRAMA model are generic in nature and can be used for analyzing strategic manufacturing decisions over a range of operational and organizational situations.

DRAMA has been modified and extended to facilitate its use as a generalized instrument for exploring hierarchies of organizational decision making and the links with production system design. Consequently, in addition to the original DRAMA process design methodology specifically for electronics assembly, its corollary, DRAMA II (Decision Rules for Analyzing Manufacturing Activities), has now been developed for the analysis and design of manufacturing systems in general.

Application 4.1 ICL's modular assembly cascade system

As its name implies, the cascade comprises a number of modules within which part of the total assembly operation is carried out. These modules are largely autonomous, their main constraint being the overall dimensions of the products or subassemblies they can handle. Both within and between the modules, 'just-in-time' principles are applied. Modules manufacturing larger dimensional assemblies 'pull' their requirements from those manufacturing smaller ones, hence the concept of materials 'cascading' down the various levels of assembly, as shown in Figure 4.1. The modules themselves derive their flexibility from making the production activities and material flow system as 'generalized' as possible. This has meant re-equipping the assembly areas with more general-purpose tools and investing in an ambitious programme of training to extend the range of operator skills and increase their ability to work in a less structured environment. Training in quality is also provided for all employees, both direct and indirect. This training, coupled with a 'zero defects' programme and the promotion of 'quality circles' has enabled a 'quality culture' to be fostered among all in the company.

The equipment used in the modular assembly cascade at Ashton includes automatic guided vehicles, computer-controlled cranes, an automatic

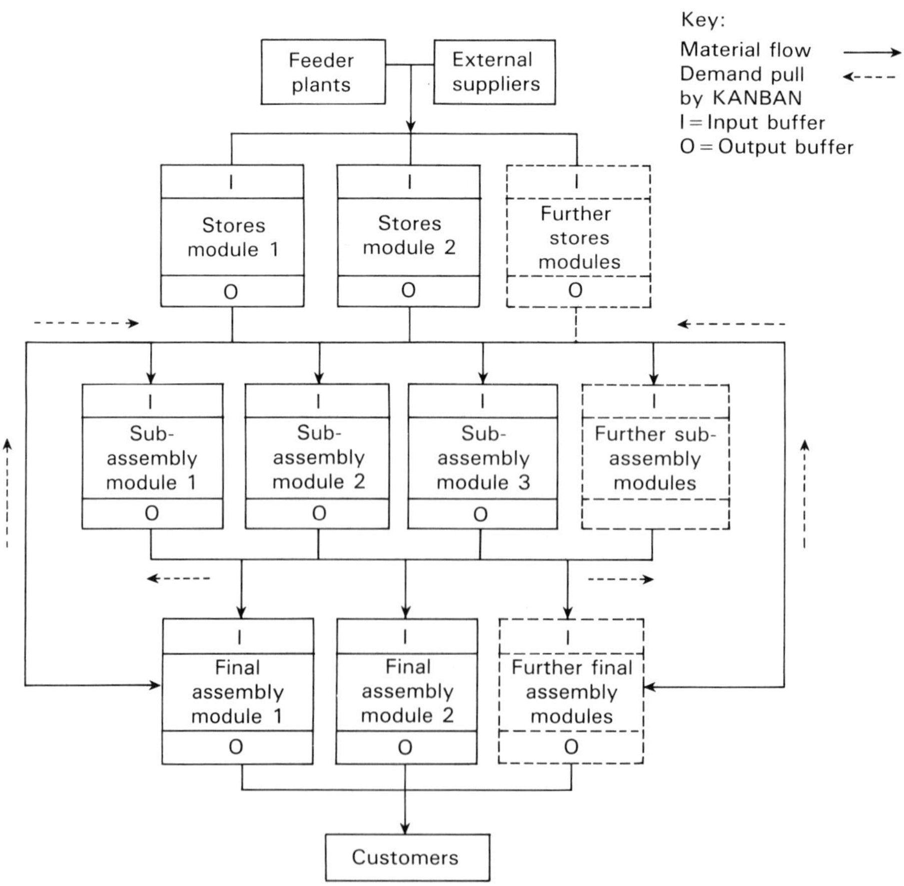

Figure 4.1 The modular assembly cascade concept.

paternoster and carousel stores, together with a hierarchy of computers for controlling the system. A unique feature of the material control system is the 'electronic kanban' technique, where the pulling method of a 'kanban' card (originally developed by Toyota) is replaced by the use of bar-coded containers together with automatic data capture and transmission.

An important point to note about the modular assembly cascade is that it should be viewed as a 'concept' rather than a fixed configuration of specific technologies. Therefore, other companies wishing to adopt the principle should appreciate that copying the cascade idea in a strictly physical sense will not necessarily provide the solution for their own particular needs.

Source: by kind permission of International Computers Limited. Taken from: D. Bennett (ed.) (1990), *Design Routine for Adopting Modular Assembly: A Process Design Methodology for Electronics Industries*. Aston Business School.

■ 4.2 The development of DRAMA

The research project

The DRAMA methodology was originally developed from the ICL (International Computers Limited) approach to designing and implementing production systems to meet its own needs resulting from a corporate change to being market-driven. The system studied was the previously mentioned 'Modular Assembly Cascade' designed and developed at the Ashton plant, for the manufacture and final assembly of mainframe and mini computer products. Figure 4.1 illustrates the Modular Assembly Cascade concept and Application 4.1 explores the basic principles of the concept in ICL. The original acronym for the methodology was, therefore, the 'Design Routine for Adopting Modular Assembly'.

The development of DRAMA has been in two distinct phases. The evolution of the original process design methodology for use in the electronics industries was the result of a 3-year programme of research conducted by the Aston Business School and supported by the UK Science and Engineering Research Council in collaboration with ICL. The first phase of the research is illustrated in Figure 4.2 and was conducted in the following way.

The design and implementation processes for new assembly facilities were reconstructed and documented by the research team. This reconstruction was retrospective for the module implemented prior to the commencement of the research and concurrent for the developments being observed in 'real-time'. The longitudinal study of developments at the Ashton factory was complemented by a detailed organizational study of the company as a whole. This explored the strategic choices made by the 'manufacturing and logistics'

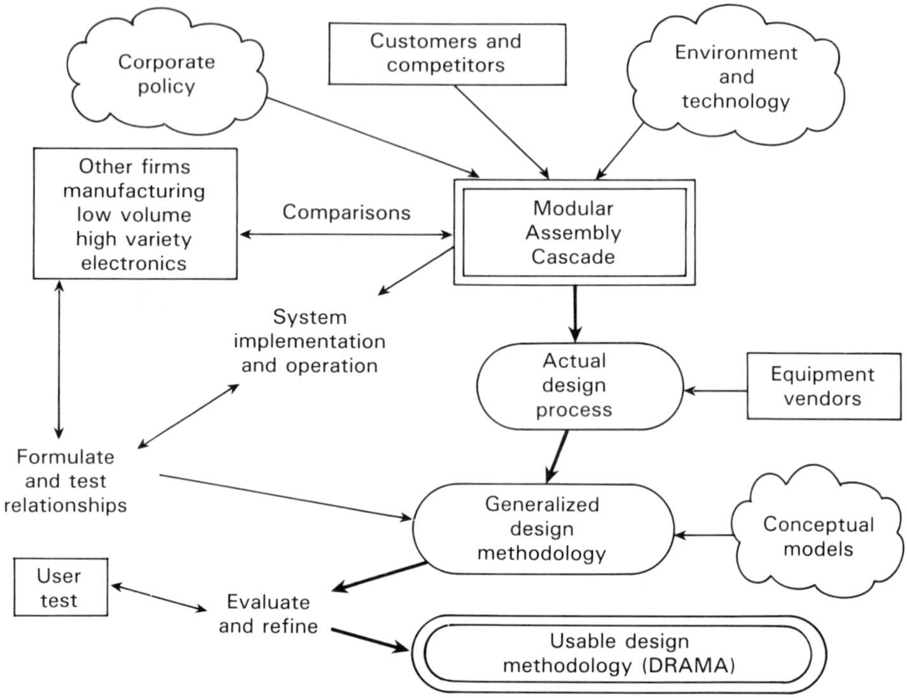

Figure 4.2 The development of DRAMA.

division within the recently formulated market-driven corporate policy. It also considered how Ashton's sister plants within ICL had translated manufacturing strategy to the level of production system design. Participative observation within Ashton was therefore supplemented by an extensive programme of semi-structured interviews throughout ICL, covering all levels of the organization from senior directors to shop floor employees. External analyses were also conducted to consider customers, competitors and the technological, structural and business environment within which the company was operating. Equipment vendor issues relevant to systems development were also explored.

To place into context the research conducted within ICL, supplementary studies were conducted in other companies manufacturing similar products for comparable markets. From the data, a number of relationships were identified, tested and incorporated into the methodology where appropriate. Existing conceptual models from the literature, as described in Chapter 3, were used at this stage to assist in the analysis of the observed decision processes. Later these also provided a framework for DRAMA, the design and implementation methodology, contained in Chapters 5 to 14.

Further development of DRAMA

DRAMA was originally developed specifically for use in the electronics industry. However, the basic model on which the methodology is based was considered to have utility beyond just electronics production. For this reason research on developing the generic aspects of DRAMA has continued beyond the ICL project. The objectives of this continued research have been to further validate DRAMA within the UK electronics industry and, second, to test DRAMA as a generic methodology for production systems design across a predetermined range of industrial and organizational contexts. The result has been the evolution of the DRAMA Design Routine into DRAMA II, 'Decision Rules for Analyzing Manufacturing Activity' and production systems design generally.

Following completion of the original project, work then continued on the methodology to make it more applicable to a wider range of operational and organizational contexts and to test the generic properties of the basic model

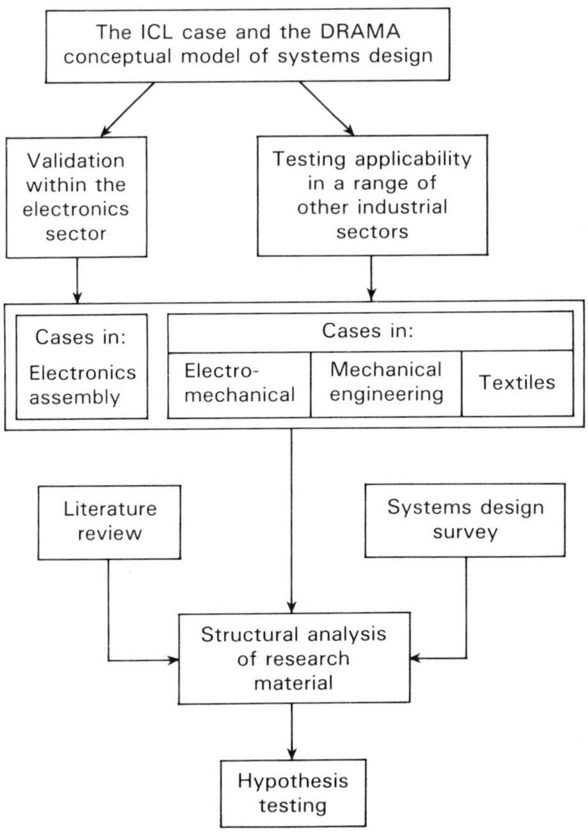

Figure 4.3 Developing DRAMA as a generic model.

and components of methodology. This book contains the results of this second phase of the research: a generic methodology which provides a guide to successful design and implementation of production systems which are responsive to the needs of the market, but efficient in manufacturing terms. Focusing upon the decision options and process choices facing the systems designer, the DRAMA acronym, because of its wider applicability, is now taken as standing for 'Decision Rules for Analyzing Manufacturing Activities' or 'DRAMA II'. The historical development of DRAMA and its further expansion into DRAMA II will be described in this chapter.

Figure 4.3 illustrates the research design adopted in developing the generic DRAMA model and methodology. The research has involved validating DRAMA within the electronics assembly industry and, second, testing DRAMA as a generic model for systems design. A total of thirteen case studies were developed and categorized into the four sectors identified in Figure 4.3, i.e. the 'electronics assembly', the 'electromechanical', the 'mechanical engineering' and the 'textiles' industries. These cases were analyzed within the framework offered by the DRAMA model, which gave the opportunity to test the validity of the model as a conceptual representation of systems design in this predetermined range of industrial sectors. The cases provided detailed descriptions of the nature of systems design within thirteen organizations. Given the depth of the work performed by the IDOM team within ICL, this represents a considerable amount of research material.

■ 4.3 Structure and content of DRAMA

Objectives and scope

The DRAMA methodology is essentially a set of guiding principles that assist the user to steer through the decision sequences that must be followed when designing a production system. It was based originally on electronics assembly. However, the model on which the methodology has been developed provides a structure which can be used as an analytical tool for examining the decision processes associated with production system design across a wide range of industrial contexts (Bennett and Forrester, 1991).

DRAMA does not take the form of a set of rigid and mechanistic prescriptions, so it does not impose upon its users a limited panacea of procedures. It can best be described as a pragmatic collection of 'guiding principles' that allows companies to analyze their competitive position and design a production system appropriate to their needs based upon a number of feasible design options. As a consequence of this, and in keeping with the process of system design in practice, the methodology deals with a complex situation. This complexity has been addressed, however, by structuring the methodology in a common format. The opportunity is taken here to describe the structure of the methodology and how it is used in practice.

DRAMA was initially developed for use in the high variety electronics assembly industry. However, the DRAMA model, upon which the methodology is based, provided a structure which, it is proposed, can be used as an analytical tool for generic production systems design.

The DRAMA components

The production system design process is seen as a dynamic combination of ten interdependent components that mesh together like a set of gears which drive the organization along, as shown in Figure 4.4. The DRAMA model separates the total production system design activity into these ten components which need to be addressed in the process of decision making. They are:

1. Market and environment.
2. Manufacturing strategy.
3. Organization.

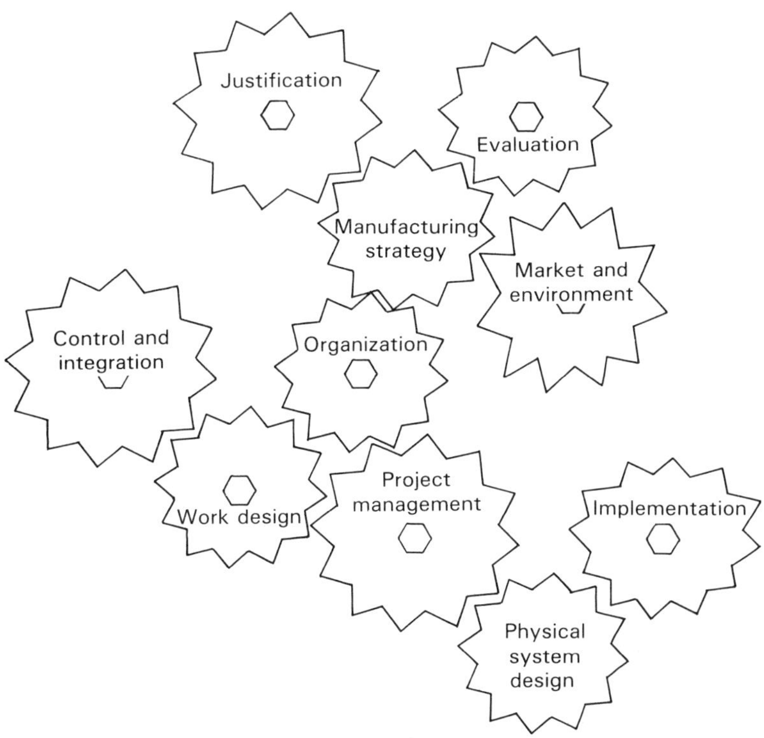

Figure 4.4 The meshing components of the DRAMA model.

4. Justification.
5. Project management.
6. Physical system design.
7. Control and integration.
8. Work design.
9. Implementation.
10. Evaluation.

Market and environment (Chapter 5)

This is the analysis of the economic, sociogovernmental, customer, competitive and technical factors to evaluate strengths and weaknesses, opportunities and threats. It involves codetermination of corporate objectives, policy and market strategy.

Manufacturing strategy (Chapter 6)

This is the manufacturing function's response to the market strategy, although it will also include an internal audit of performance. Manufacturing strategy formulation will involve decisions on make or buy, the degree of vertical integration and the setting of manufacturing targets.

Organization (Chapter 7)

This involves the design of the organization, both in terms of 'structure' (i.e. formal, recognized composition and lines of communication as shown on the organization chart) and 'state' (i.e. the culture, employment climate, flexibility, etc.).

Justification (Chapter 8)

Included here is the generation of the case for a new, or modified, production system. Justification is normally based upon a combination of financial arguments and strategic need in terms of qualitative advantages. Approval for new projects and capital expenditure is usually sought from senior management on the basis of this justification.

Project management (Chapter 9)

Once capital has been granted, system design projects need to be managed within the constraints of terms of reference, time-scales and budgetary constraints.

Physical system design (Chapter 10)

Involves the choice of production system 'hardware', plant layout and configuration at the levels of the factory, the operating unit (module) and equipment.

Control and integration (Chapter 11)

This is the design of the production system 'software', including operations planning and control, the computer systems and the integration of the system as a whole.

Work design (Chapter 12)

The choice of work organization within the production system addresses such issues as flexibility, responsibility for quality and operator tasks, etc. The 'humanware' of the production system is another part of work design.

Physical system design, control and integration and work design, now to be described, can collectively be termed the 'operating system' of the production facility.

Implementation (Chapter 13)

This draws together all the design work conducted in 1. to 8. and converts it into the installation of efficient and effective new production facilities and their acceptance by people in the organization.

Evaluation (Chapter 14)

This is, in effect, the 'control loop' for the systems design process, evaluation concerns the maintenance and improvement of production systems and the process of decision making within the organization. Production systems designs are continually evolving as a result of ongoing evaluations at all levels in the business.

In explaining the decision processes involved, DRAMA assumes a generally phase-wise progression through these components during a production systems design project. However, it is recognized that the components of design are interdependent and that decisions within one component will probably have an effect on a number of preceding and subsequent events in other components. Thus, a myriad of feedback and feed-forward loops exist between components, too numerous to represent on diagrams of the model, as a result of evaluations and reassessments throughout the design process.

To provide a more useful form for the systems designer, the DRAMA model has been represented in a structured and diagrammatical way, as illustrated in Figure 4.5. It is recognized throughout that the components of DRAMA are interdependent and that design decisions have an effect on preceding and subsequent components, thus initiating a myriad of feedback and feedforward loops. This notion corresponds closely to Mintzberg's decision process model, which sought to assign structure to the longitudinal process of strategic decision making (Mintzberg *et al.*, 1976).

Figure 4.6 shows the conceptual form of DRAMA as it progresses through the case, assessment and design methodology tracks. This method of investigating the current context and detailed activities of what might be seen as a 'leading edge' manufacturing organization and then relating this to theory and practice elsewhere in the industrial sector represents a generally applicable approach to formulating principles and guidelines for future design activity in that industry. This approach has been used in a range of industries and has led to the development of the generic DRAMA methodology contained in the next chapter of this book.

Decision-making domains

DRAMA provides a model of organizational decision making which explores the effects of participation in the production system design process at different levels of the manufacturing business. Three such 'domains' of decision making are identified, 'strategic', 'tactical' and 'operational', and these are shown in Figure 4.7.

The 'strategic domain' embraces the activities of senior management down to the level of individual plant units. Market and environment, manufacturing

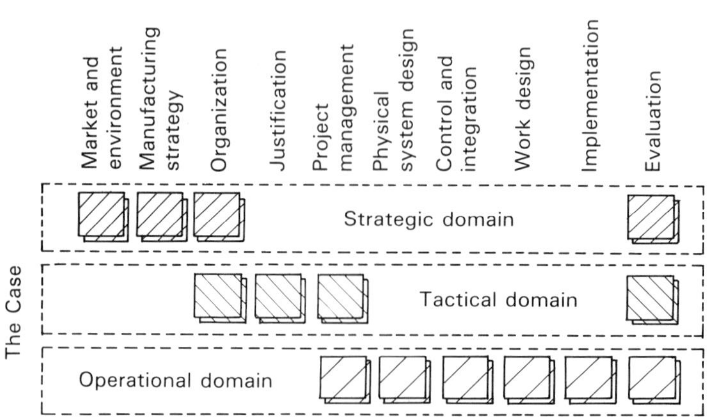

Figure 4.5 The DRAMA conceptual model of production systems design.

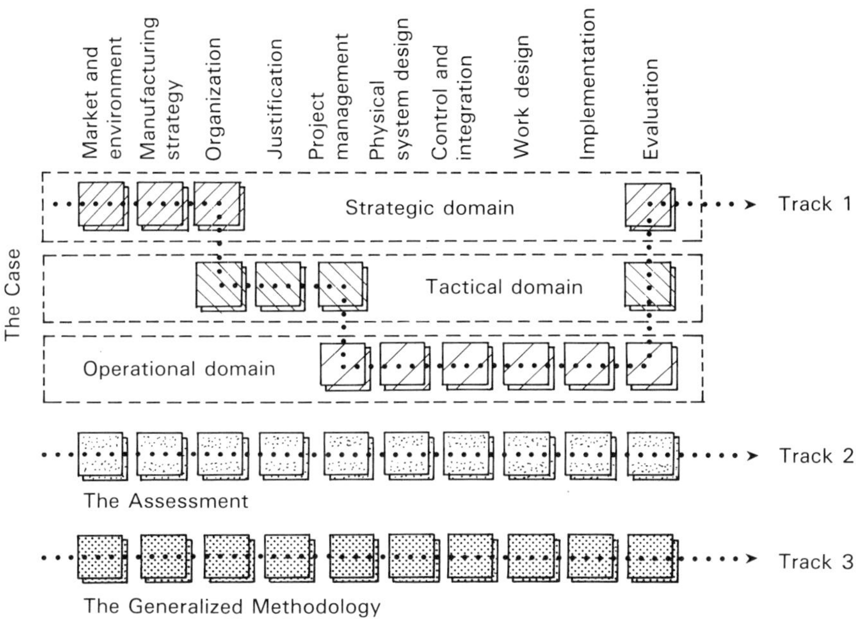

Figure 4.6 Conceptual form of the DRAMA components.

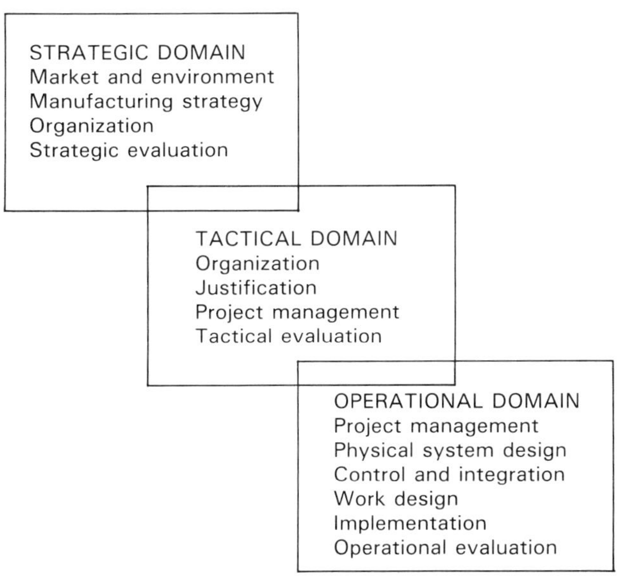

Figure 4.7 Contents of the decision domains of DRAMA.

strategy, organization and evaluation are identified as components within the strategic domain and can be considered together when focusing upon the strategic issues of systems design. The 'tactical domain' spans what might be termed the 'middle management' levels, where senior management decisions are translated down into plant level procedures. Components identified within this domain are organization, justification, project management and evaluation. The 'operational domain' relates to the lower organizational levels where detailed production decisions and systems operation take place. The operational domain includes project management, physical system design, control and integration, work design, implementation and evaluation. Thus, as illustrated by Figure 4.6, organization and evaluation span the interface between the strategic and tactical domain, while project management and evaluation span the tactical and operational domains.

This disaggregation of organizational decision making into domains provides a hierarchical dimension to the DRAMA model to supplement the longitudinal analysis provided by the sequential components. Design activity and decision-making routines can thereby be tracked at different levels of the organizational as well as laterally through time.

Presentation of the DRAMA methodology

The DRAMA methodology detailed in Chapters 5 to 14 is presented in the form of a narrative supplemented by flowcharts, decision trees and checklists. Each of the chapters contains a review of theory and best practice in relation to each of the design components of DRAMA. The methodology for each component is then presented. Figure 4.8 illustrates the procedures followed

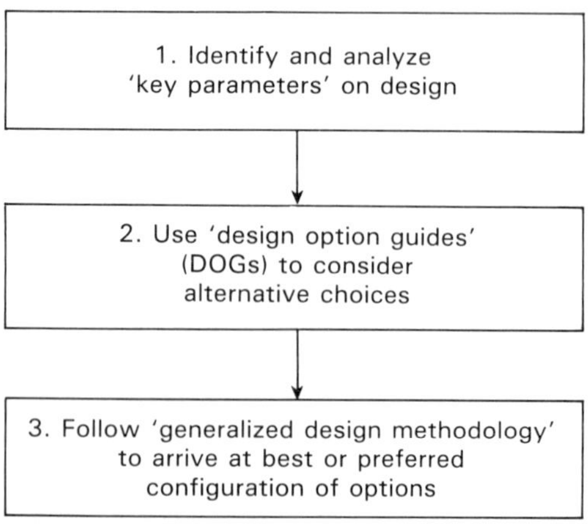

Figure 4.8 Development of a process design methodology for a DRAMA component.

when using the methodology to design production systems. As each component presented in this standard format Figure 4.8 should be considered in conjunction with the ensuing descriptive text.

Key parameters

'Key parameters' are the inputs to each DRAMA component. They are items which serve to shape design or configuration of each component, as identified by an assessment of current theory and practice for each methodology component. Key parameters are influential factors shaping design and are prerequisite areas of analysis or audit prior to embarking upon the detailed design of the component under review. Within the DRAMA methodology, key parameters may comprise a combination of outputs from previous DRAMA component designs or they may come from the 'environment' (e.g. market, process technology, etc.) which impinges upon design decisions.

Key parameters are identified for each component within the methodology. There must be an agreement by all in the design team on the nature of each key parameter prior to starting to design the components. This is essential as they serve to provide terms of reference and the available scope for decision-makers and system designers alike.

Design option guides

The user is now introduced to the available design options for the components. This is done through the use of 'design option guides' (DOGs), the form of which is shown in Figure 4.9. The DOG offers a list of 'result criteria' from which the user selects a 'principal result objective', a preferred and weighted compilation of the criteria suggested. The design options are then presented together with their 'related features', grouped according to whether they are advantageous (shown as positive) or disadvantageous (negative) in relation to the result criteria. The expected result from each option can, therefore, be compared with the principal result objective developed by the user. Where there is an apparent gap between the two, the user will recycle to consider other options. The best single, or preferred, combination of options can eventually be identified.

The choice between alternatives is, therefore, an iterative procedure that permits the selection and modification of options until the preferred configuration, as governed by the user's own result objectives, is achieved. The number of DOGs varies between DRAMA components, but tend to proliferate as the user moves down towards the operational domain of decision making. This is what one might logically expect from operational design options, although they are not so far-reaching in their effect as strategic choices, are certainly more numerous. The DRAMA manual comprises a total of 27 DOGs with each representing a particular element of design.

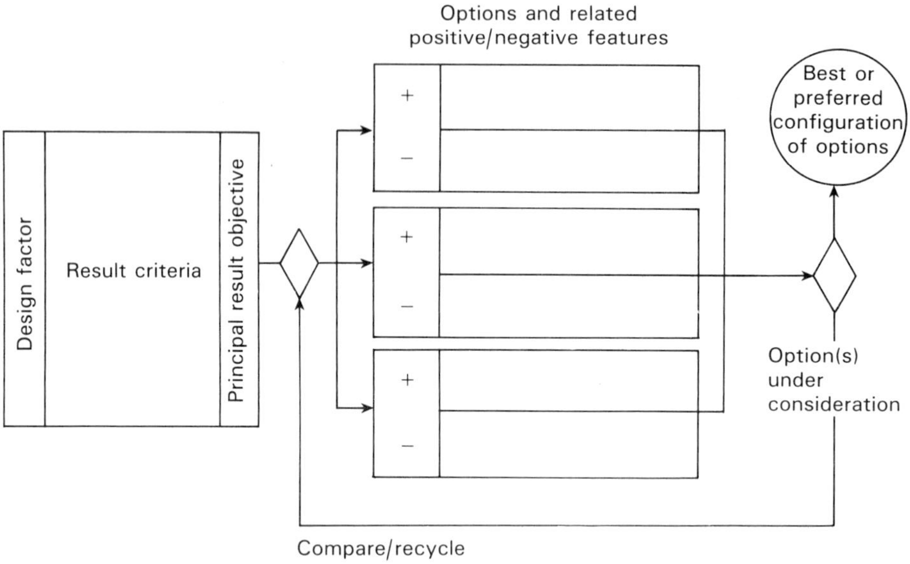

Figure 4.9 The basic construction of design option guides.

DOGs are not a highly qualified 'expert system', but they do provide a decision aid for designers of assembly systems that allows for options to be identified and evaluated in a comprehensive and structured way. Compare this with the rather arbitrary nature of system design commonly prevalent within organizations, where decision-makers and designers frequently have a myopic view of this process.

Generalized methodology flowcharts

The methodology for each component is summarized in flowchart form to provide a succinct representation of the recommended design process and decision routines. The 'generalized methodology flowchart' places the elements of design analysis and choice into context. Key parameters are indicated as inputs and the main areas of design choice are highlighted. Outcomes of the design process are also indicated where they occur and a strong emphasis is placed upon the value of evaluation, feedback and the recycling of information regarding the performance of design components for future design modifications and improvements.

The flowcharts show the design of production systems to be a continual and iterative process, being influenced by persistent internal and external change and thus never coming to a permanent conclusion. The analogy of design components as meshing gears in Figure 4.4 is still appropriate, therefore, whereby all are continually turning, sometimes at different rates, but always subject to change from other gears.

4.4 Features of DRAMA

A particular feature of the DRAMA methodology is that it was originally developed on the basis of a major case study analysis and the experiences and lessons learned by other organizations when designing assembly systems. Its structure also ensures that the user considers the importance of co-ordinating all functions of the business and the need to understand the relationship between different levels of organizational decision making.

DRAMA comprises a modular structure by virtue of its division into components. This enables the methodology to be adapted and used for specific needs, the user only selecting those components that are of immediate interest. Coupled to this, the 'domain' concept of decision making and the incorporation of this into DRAMA provides different levels of application within the manufacturing organization. This disaggregation of the design process into components and the recognition of decision domains allows both a longitudinal and hierarchical analysis to be conducted. It also permits separate consideration of each domain and/or component, while the input and output parameters of each component provide the linkages between strategy, tactics and operation. DRAMA assists in understanding the relationship between corporate strategy and the design of operations systems by offering an empirical model for design which incorporates all levels of manufacturing decision making.

Existing models and methodologies have been incorporated or referenced within the methodology where relevant. The incorporation of these and systems design approaches in the development of DRAMA means that the basic model is multiperspective. The DRAMA methodology advocates an open systems approach in the analysis and design of components in the higher domains of managerial decision making, while a closed systems approach is adopted when progressing down to the lower domains. This fits well with the Kast and Rosenzweig notion of a 'hierarchy of systems' approaches whereby higher levels of sociotechnical analysis demand a more open approach (Kast and Rosenzweig, 1970). The research methodology employed when developing and using the DRAMA model was adapted from Checkland's soft systems approach with its movement and feedback between the real and abstract world. A soft systems approach would also be relevant whenever the realities of actual human decision making are being considered, as is the case when using DRAMA.

DRAMA advocates a top-down and market-driven approach to strategy formulation and its translation to the level of operational design. However, it also recognizes more readily than other theories that bottom-up and lateral forces are extremely important in shaping and evolving strategy and eventual system design. DRAMA also recognizes a correlation between the production system design process and a phase-wise and cyclical movement through the model's components, from 'market and environment' to 'evaluation' and back. It recognizes that there is a host of interconnects between components both within the same domain, for example in the simultaneous consideration of the 'physical system', 'control and integration' and 'work design' and hierarchically, such as 'organization', affecting 'manufacturing strategy' and the functional design of the operational system.

DRAMA's principal asset, therefore, is that it offers a comprehensive framework of enquiry when considering the design of production systems. It recognizes direct and indirect linkages between the market and strategy in the higher domains of organizational decision making and the decision-making processes that occur in the tactical and operational levels. It offers a flexible model by virtue of its modular design, which enables the design process to be divided into a number of separate components. DRAMA, then, relates to production systems design in its wider sense and includes consideration of market, organizational, technical and behavioural aspects.

■ 4.5 Conclusions

DRAMA (Decision Rules for Analyzing Manufacturing Activities), is a practical methodological approach which offers a framework of enquiry linking a strategic analysis to the design of production systems. Its flexibility, which is derived from a modular structure, allows users to select and focus on those components of the design process that are of immediate concern to them. The methodology has evolved as part of a detailed research project where a case analysis was conducted using participative observation spanning three years. The result is an underlying model to the methodology, which displays realism and recognizes both the hierarchical and longitudinal nature of decision making when designing production systems.

This chapter has described the development of DRAMA, an industry specific design methodology, into DRAMA II, a generic analytical approach. The generic methodology serves as a tool for designing, auditing and refining production systems. The ensuing chapters will detail the methodology for each of the ten DRAMA design components.

■ 4.6 Further reading

Bennett, D. J. (ed.) (1990) *Design Routine for Adopting Modular Assembly: A Process Design Methodology for Electronics Industries*, Innovation, Design and Operations Management Research Unit, Aston Business School.

PART III
The DRAMA Methodology in Operation

5 Market and Environment

Objectives of the chapter

On completion of this chapter you should:

1. Understand the external and internal processes of analysis on the basis of which an organization may formulate corporate objectives and market strategies.
2. Be able to develop appropriate statements of strategic intent for the organization which, in turn, will constitute a major input to the design of following DRAMA components in the development of market-focused production systems.

■ Chapter outline

This chapter considers the strategic and market analyses which should be conducted by an organization in the development of corporate policy and in stating their market and product strategies. The first section contains a review of established techniques and analytical frameworks which are available for use in developing strategy within the business. This is followed by a discussion of more recent views on the need for market-focus and customer orientation within manufacturing operations. The DRAMA methodology for market and environment is then presented. The key parameters within the formulation of strategy are identified as customers, competitors, technological developments in products and processes and wider sociogovernmental aspects which impinge upon the manufacturing organization. The purpose of this component of DRAMA is to facilitate the development of appropriate corporate objectives and organizational direction. The methodology for Market and Environment comprises two design option guides (DOGs) relating to 'corporate policy' and 'market/product strategy'. The two resultant statements of intent represent the end-point of this particular methodology section and provide the major context in the design of other DRAMA design components. The chapter ends with a review of the process of decision making and design in developing corporate and market strategies and is illustrated in a generalized methodology flowchart analysis for the market and environment.

5.1 Theory and current practice

The 'market and environment' component of DRAMA involves not only the analysis of external factors, but takes this a stage further by developing the corporate policy and marketing strategy for a manufacturing company. Decisions taken at this stage shape those later on in the design process for new production systems. Such decisions are therefore largely within the domain of senior executives within the organization. A large body of literature on approaches to the development of corporate policy and business strategy has evolved during the period since 1970. So, before discussing recent trends in the development of market-focused corporate and market strategies, three established theories of strategic analysis will be discussed. These are the 'Boston Consulting Group matrix' (see Shanklin and Ryans, 1981; Daft, 1991), Porter's model for marketing strategy (Porter, 1985; Miller, 1988) and the method of strategic analysis and policy formulation developed by Johnson and Scholes (Johnson and Scholes, 1989).

The Boston Consulting Group matrix

The Boston Consulting Group (BCG) matrix is the most widely adopted form of portfolio strategy analysis. Portfolio strategy is that part of corporate strategy which plans an organization's mix of products and businesses. It attempts to fit together parts of the company, its 'strategic business units' (SBUs), in such a way as to produce synergy and competitive edge. It identifies that companies usually require a balanced mix of SBU investments, with some being high risk and with growth potential, and some having low risk but guaranteed returns, and so on.

The BCG matrix is illustrated in Figure 5.1. It considers SBUs across two dimensions, their market growth and their share of the overall market for the products they produce. Market growth is the rate at which the industry is growing, whereas market share defines the size of the SBU in the industry within which it operates. The combinations of high and low market growth and market share generate a two-by-two matrix with four categories for a company's portfolio. Each of these types will now be considered.

Stars

A 'star' is an SBU within the organization which has a large market share in a rapidly expanding market. The organization should consider investing funds in this business because of its potential for future growth and the consequent opportunity for future income and profit generation. If market share is maintained or increased, the star will generate positive cash flows even when the industry matures and growth slows down.

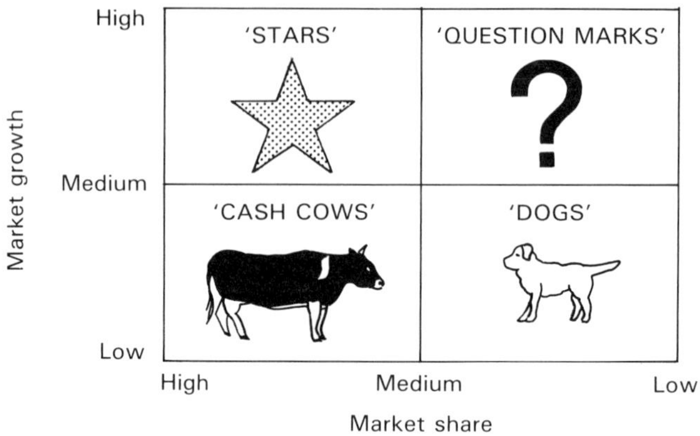

Figure 5.1 The Boston Consulting Group matrix.

Cash cows

The 'cash cow' is where a high market share is held in a mature industry. Heavy investments in production facilities, research and development and marketing are no longer required, so the organization should 'milk' the cash cow SBU for the money it can invest in its star and question mark businesses.

Question marks

The 'question mark' (or 'problem child') is an SBU in a rapid growth industry, but with a relatively small share of the overall market. The question mark is a high risk business: it could become a star if the proceeds from cash cows are invested in them, but could also fail. Alternatively the business could be divested to another company if thought too risky or not key to the organization's overall future plans.

Dogs

The 'dog' is a poor performing SBU: it has a small market share in a slow growth market. The dog is unlikely to yield large profits or any potential for future cash generation for the organization. As a result there should be no investment in the business and should only be kept by the company if in profit. Otherwise the business should be sold off.

The BCG matrix provides a means for companies to map their range of SBU and so provide a visual representation of the state and portfolio of the overall business. It is usual for SBUs to be represented by circles within the matrix, their position representing the growth and market share and the size of the circle indicating the relative size of each SBU. Using the BCG matrix, then, companies are able to evaluate which businesses should be milked for funds, which have future potential and should be invested in, and which should be divested as high risk or loss making businesses.

The Porter categorization of marketing approaches

Porter has identified four types of business strategies adopted by companies, namely those of 'differentiation', 'cost leadership', 'focus' and 'stuck-in-the-middle'. The first three of these he identifies as effective strategies, whilst stuck-in-the-middle represents a poor strategy where a business is 'master-of-none' of the three effective strategies. In fact 'stuck-in-the-middle' represents no real strategy at all, with the organization generating no distinctive competitive advantage. The effective categories of strategy are now discussed.

Differentiation

This type of strategy represents an approach that attempts to distinguish the products offered by an organization from that of its competitors. The product is advertised and presented to the customer as being unique and different in terms of quality, the features offered or in some other way. Organizations adopting a differentiation strategy are commonly categorized or should seek to develop strong marketing abilities, good co-ordination between functions, creative flair and innovation, strong skills in research and development, and a reputation for quality and technical leadership.

Cost leadership

This type of strategy occurs where the organization seeks to provide the product to the customer at a lower price than its competitors. Thus, in its operations, it pursues a strategy of developing highly efficient facilities, aggressive cost controls and reduction programmes, and economies of scale in production. Such an organization is characterized by tight financial management, close supervision of labour, process engineering skills and product design for ease of manufacture.

Focus

The third type of effective marketing approach identified by Porter is really a subset of the differentiation and cost leadership strategies. It occurs where the organization pursues one of the above strategies in a distinct geographical market or for a limited target group of customers within the overall market. Thus the focus strategy can invoke any of the organizational and operational circumstances discussed above for differentiation and cost-leadership depending on the type of focus strategy being adopted.

The Porter categorization provides a potentially useful and easily understandable framework for companies formulating their competitive strategies. They are able to analyze their current position and determine the type of strategy they are currently pursuing. If this is of the 'stuck-in-the-middle' category, the strategy can then be adapted to one of the effective types. There are areas of similarity between Porter's work and the classification of strategic approaches put forward by Miles and Snow (1978). The differentiation strategy is similar to the 'prospector' trait of Miles and Snow, cost-leadership can be akin to the 'defender' and the focus approach is like the 'analyzer' type. Miles and Snow's fourth type, the 'reactor', is not a proactive strategy at all and so can be compared with the 'stuck-in-the-middle' approach.

The Johnson and Scholes framework

The Johnson and Scholes framework is intended to assist in the analysis and development of corporate strategies. Whereas the BCG matrix and the Porter classification provided typologies of strategic approaches and their associated characteristics, the Johnson and Scholes framework is more concerned with the process of strategy formulation. The framework, illustrated in Figure 5.2, comprises three main stages: strategic 'analysis', 'choice' and 'implementation'. Each of these will now be considered in turn.

Strategic analysis

Analysis involves the collection of data and information from the environment plus an internal evaluation of the organization. Environmental analysis includes evaluation of the market and customer requirements, the collection of competitor information for benchmarking and other analyses, an appreciation of the prevailing social conditions including the labour market, and familiarization with legal and government aspects. The internal analysis involves consideration and understanding of the multiple values, expectations and objectives that reside within the organization and the capital, labour and other resources that are available.

Strategic choice

Following from the analysis stage, the choice of strategy for the organization must be made. There are three steps identified for this. First, strategic options need to be generated and so a high degree of lateral and novel thinking is required at this stage. Second, these alternatives must be evaluated and screened so that the most feasible and appropriate strategic options can emerge. Finally, the strategy must be selected from these preferred options.

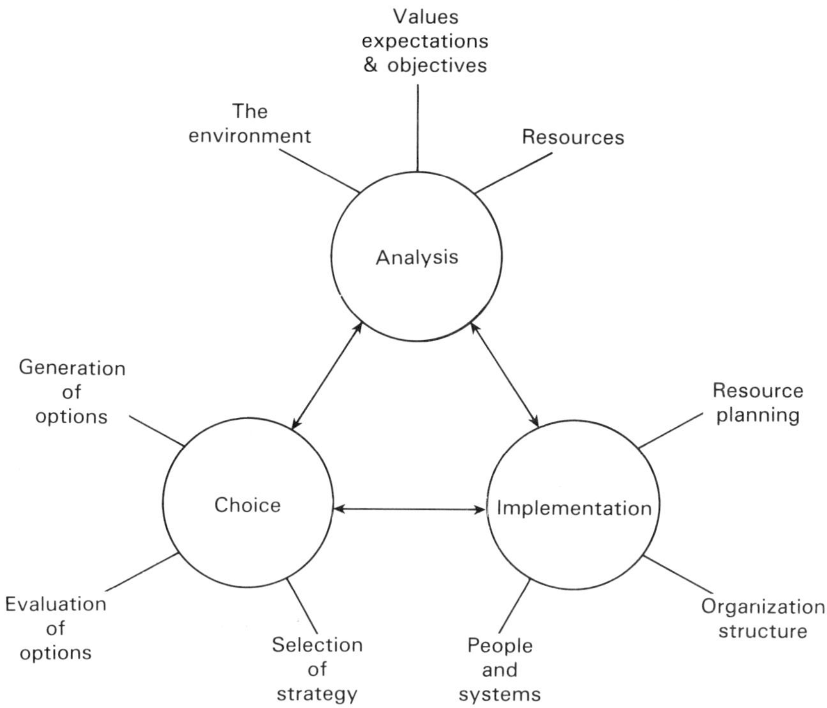

Figure 5.2 Johnson and Scholes framework for strategy formulation.

Strategy implementation

Once a decision on a particular strategy has been made, this must then be implemented and put into practice within the organization. Three main considerations are identified within the framework. First, implementation must take into account the existing human resources, corporate culture and organizational processes and these should be adapted and modified where necessary to align with the new strategy. Likewise, the organizational structure should match and be appropriate to the strategic direction of the organization. Third, implementation must include a high degree of resource planning on the part of the strategist to ensure that plans and policies are put into operation with the minimum of disruption and resistance.

The Johnson and Scholes model provides a pragmatic framework to assist in the formulation and implementation of strategies and the overall management of this process. It also offers a means of evaluating existing strategies for the analyst and practising manager alike.

Market-focused enterprises

In the strategic management literature there has been a growing tendency to advocate the need for manufacturing companies to be market driven and use quality, service and time-based competition to achieve competitive edge. The need for market focus is now widely acknowledged and has been identified by writers from the 'management excellence school', most notably by Peters and Waterman, 1982 (see Peters and Waterman, 1982; Peters, 1987). In the area of strategy and marketing Porter and Kotler can be seen as supporting this general prescription (Porter, 1986; Kotler, 1986), as can a number of manufacturing specialists (for example, Hayes and Wheelwright, 1984). The consensus in this literature would appear to be that manufacturing operations should respond to the identified needs of the market rather than developing products which are 'pushed' on to customers using price and technical specification as the sole competitive factors (Bessant, 1988). In this context the term 'competing through manufacturing' has now come to the fore where quality, service and response are included as competitive levers in addition to more traditionally acknowledged order winning criteria such as price and technology.

Until recently the dominant approach to the design and operation of manufacturing systems has been the quest for economies of scale by producing standardized products in large volume. Increasingly this is being replaced by a policy of offering a wide variety of products or volumes or, alternatively, to configure product designs directly to customer order. As a result, volumes per design variant have been reduced in many industries. In addition, the life-cycles of product designs have been substantially reduced for almost all products as manufacturers seek to provide customers with the features they desire from the product (see Bennett *et al.*, 1992, for a fuller discussion on these points).

The interface of an organization with its environment is a key factor in strategic, organizational and production systems development. The organization in fact interfaces with the environment at all levels of decision making. The GRAI conceptual model, described in detail in Chapter 3 and illustrated in Figures 3.4 and 3.5, shows this phenomenon clearly; for example, middle and junior levels of engineering management that interact with the technological environment, surveying those technologies and methods which seem most advantageous to pursue in the design and continual development of production facilities. However, most discussions regarding organizations and their environment tend to concentrate upon customer and competitor analyses and the appropriate response through the development of corporate policy objectives at the strategic level of decision making. Despite the drive for market focus, it remains essential that companies should continue to monitor its technological environment in order to capitalize on any new or evolving manufacturing technologies or equipment that would result in competitive edge.

It is now common for manufacturing companies to stress the theme of increasing emphasis on the use of market-driven objectives. Beyond the overall statement of corporate objectives, however, there exist subtle differences at the next, more detailed, stage of strategy formulation: the development and implementation of market strategy. A basic distinction can be drawn here between those market strategies of demand 'range', such as the production of large numbers of product variants and the customizing of products, and those which emphasize the need for 'response' including the reduction of design and manufacturing lead-times and improving delivery reliability (Slack, 1983; Cooke and Slack, 1991). The optimization of levels of quality and service is central to companies adopting such strategies, particularly those which place equally high demands for both range and response in their strategy.

The models of strategic analysis presented earlier in this chapter and the review of current notions of market and environment provide means for structuring and analyzing external factors. The analysis of strategies and markets is a complex and bewildering exercise for manufacturing companies. Reasons for this are numerous and include the dynamic nature of competition, the variety and conflicting demands of customers and the vast array of options in terms of product and process technology. These factors mean that companies have to consider and respond to a large number of ever-changing factors when determining corporate objectives and market strategies. The term 'strategic response', with an emphasis on reaction, would seem an appropriate notion to sum up the responding activities of organizations to their environments. There is an argument, therefore, for saying that the strategies pursued, whether proactive or reactive, are best viewed as the company's 'response' to its perceived market and competitive pressures.

The DRAMA methodology for Market and Environment provides a set of guiding principles for those manufacturing companies wishing to design and develop market-focused production systems. This implies the need for a fundamental choice prior to using the methodology: that is whether the company should be predominantly market-driven or, alternatively, technology-led. All companies are driven by their market to some degree and the term technology-led does not imply a total disregard for market and customer needs. The difference in the two approaches is more a difference in emphasis. A large number of Far Eastern and US companies invest heavily in product technology research and development and can be viewed as technology-led. Indeed, this is a viable option where technical performance and marketing the product as a 'leading edge' item is seen as important. The majority of smaller companies, including nationals in the UK and Europe, suffer cost disadvantages and low economies of scale. They find it difficult to invest resources in fundamental product research, and also consider the risks of such an approach too high to bear. In these cases the technology-led strategy is inappropriate or just infeasible, so they stress the need to be market-driven and so reflect this in their stated corporate objectives.

Deciding upon a market-driven or technology-led strategy has important implications in the use of this book which, as the title suggests, concentrates upon market-focused production system design. DRAMA is not intended to be a methodology for technology-led systems design. Rather, it is aimed at those companies who wish their production systems to be predominantly market-focused. Despite this, however, many of the principles contained within the methodology make it applicable to all organizations, whatever orientation they may be. It thereby provides pragmatic approaches enabling an organization to improve their analyses and customer satisfaction in the markets they serve. So, whether market-driven or technology-led in their strategic approach, the argument remains that all organizations should seek to embrace and cultivate as much market-focus in their operations as they can.

Corporate policy provides a framework of intent regarding the aims, objectives, growth aspirations and profit levels. The next stage from formulating corporate policy in DRAMA is to develop the market strategy. The market strategy should state in specific terms the approaches to be taken towards the market and dealing with customers. The market strategies of most companies will seek to minimize response times to the market, maximize the range of feasible and effective activities in the marketplace, and to optimize quality standards and service levels. Inherent in these objectives, however, are the perceived trade-offs between these different goals in the allocation of resources (see, for example, Hill, 1985). Whereas each should be optimized in the perfect world, the reality is that manufacturing companies place varying emphases on each to arrive at the best mix of objectives for the market and organizational conditions that pervade in their business. For example, one manufacturer may be preoccupied with reducing the lead-time from the initiation of product design to the commencement of manufacturing, whereas another may seek to provide the potential to cost-effectively configure its basic product designs to whatever specific customer demands require. The evidence suggests however that quality and service in truly market-driven organizations remain as priority factors in all cases.

The DRAMA methodology for market and environmental analysis and design is now presented and culminates with the development of stated corporate and market objectives and strategies.

■ 5.2 Key parameters

Identification of key parameters

DRAMA identifies the key factors influencing decisions on market and environment to be those of 'customers', 'competitors', 'technology' (product and process) and 'sociogovernmental factors' (Figure 5.3). To be successful, an organization should take account of its customers within its selected

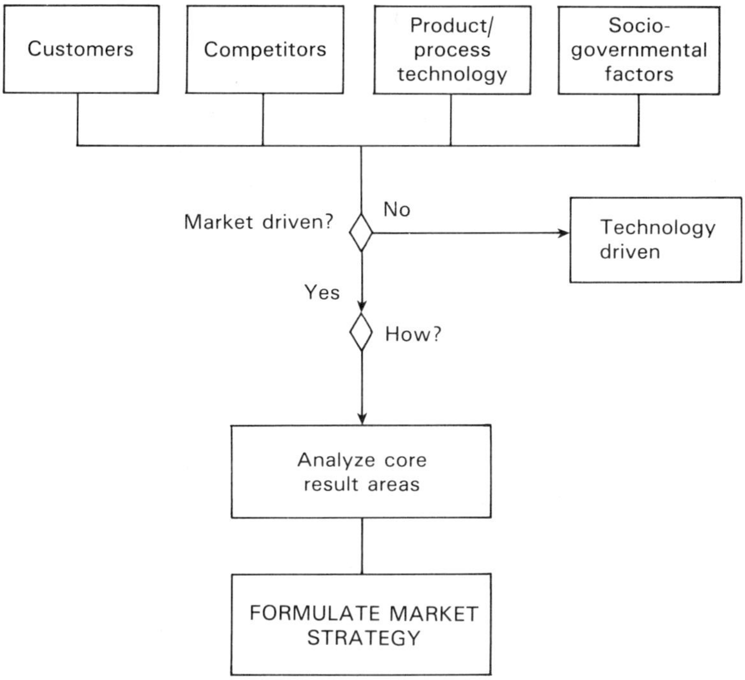

Figure 5.3 Key parameters for market and environment.

markets, whilst at the same time being aware of the competitive threat of rival companies. The design of products and manufacturing processes should involve consideration of the latest technologies available and will require scanning of developments outside the organization. The wider issues of society and government policy and legislation also need to be considered and adhered to wherever necessary for the continued success of the company into the future. Factors include 'green' environmental issues, product liability, health and safety at work and the demographics of the available workforce. In designing a new production system a detailed analysis of each of these parameters is crucial if the facilities are to operate in true market-focused mode. Each of the key parameters for market and environment are now considered in turn.

Customers

It is widely accepted that without serving the needs and aspirations of customers, organizations are eventually destined to failure. Without a pool of satisfied customers, companies will not achieve the profit and growth objectives they have set as long-term goals. Thus, formulation of corporate objectives and market strategies must take into account some form of customer analysis, whether these customers are current or potential future purchasers of goods and services. Here market research may be a particular means of obtaining the views of the customer base, although the need to listen and respond effectively to customers directly and in 'real-time' upon demand is now widely suggested as a necessary prerequisite to obtaining more information on market needs and behaviour patterns.

Competitors

Organizations should not only look to the market, but also be aware of the activities of those companies with whom they have to compete for business. Evaluating the competition, however, is often a difficult task for the obvious reasons that companies do not wish to divulge to rivals their own strategic and market plans and ideas for the risk of losing market share or leadership. In some industries, competitor analysis is easier than in others. For example, in the car industry or consumer product markets, analysis of product design may merely involve the purchasing of competitors' products and analyzing them. In other markets, for example defence equipment or large-scale project-type operations, this form of analysis is not so easy as the company will not normally be in a position to purchase competitors' 'products'. Frequently customers in the marketplace who exercise choice between different companies' products are in the strongest position to evaluate competitive information. In some cases customers are keen to supply such information in return for quality improvements or price reductions in the products they are purchasing and so should be seen as a key resource available for the collection of competitor information.

The above discussion assumes that it is possible to identify competitors who, generally speaking, are operating within the same manufacturing industry. However this might not always be the case. Indeed, the biggest threat often comes from competition outside the established or identified industry within which an organization operates. An often-cited example of this is the Swiss precision watch industry, which for a long period of time did not identify Far Eastern electronic quartz watch manufacturers as their main competitors for quality and accuracy until it began to lose markets to these 'high-tech' producers. (Some interesting discussions on the analysis of competition both from within and outside established industries is provided by Oakley, 1984, which is summarized in Bennett *et al.*, 1988.)

Application 5.1 Customer and competitor analysis at Lucas Rists

Lucas Rists Wiring Systems is the largest indigenous UK manufacturer and supplier of wiring harnesses to the automotive industry. However, in global terms, the company is a relatively small operator, a fact which is becoming increasingly critical as the industry moves towards world-wide competition and product designs for cars become standardized (although in many variants) across the various markets (witness Ford's 'World Car', the 'CDW27', to replace the Sierra). In such circumstances it is now apparent that 'big is beautiful' in the eyes of the major car manufacturers when making the choice amongst component suppliers: after all, it requires huge resources and a world-wide network of manufacturing facilities and distribution channels to supply components and parts to all plants and markets. This has had major ramifications within the harness industry, Lucas Rists being no exception.

The industry is characterized by a small number of large suppliers plus a large number of medium-sized national companies. The largest independent manufacturer is Yazaki of Japan; other large players include Sumitomo and United Technologies, in addition to Reinhagen, part of General Motor's wiring business. These companies are generally acknowledged as being capable of supplying the car assembly industry in more than one of the three major markets (Japan and Pacific Rim, Europe and the USA). The majority of the national companies are European and most notably include a number of German companies, plus Lucas Rists and Volex in the UK.

To survive it is apparent that these smaller suppliers must grow on an independent basis, through merger, by acquisition or through collaboration with the larger manufacturers. Given that independent growth would involve too long a time-scale in the cut and thrust of modern business, companies have tended to adopt one or more of the other three strategies. For example, Labinal has grown from a national to an international business through acquisition, most notably Cavis of Italy, formerly the in-house harness manufacturing business of Fiat. However, this strategy is fraught with difficulties of maintaining control within the business. Despite the attractions of the mergers options and the discussions that appear to have taken place between companies in the industry, very few have actually taken place, perhaps as a result of the syndrome of 'who is taking over whom', a frequent deliberation and obstacle for prospective mergers. Therefore relations between companies have tended to be confined to the level of collaboration, the German national companies being particularly attractive to others in the industry because of their perceived technological knowledge as a result of supplying the likes of BMW and Mercedes.

However, collaboration between medium-sized companies has its limitations in terms of becoming a world player: after all most of these companies are confined to supplying the European market only. Collaboration and partnership with an industry giant would seem desirable, but to date the nationals have had little to offer and attract the larger companies into such a relationship. Lucas Rists is the exception to this rule, however, and has cultivated a close relationship with Sumitomo of Japan, culminating in the establishment of a joint venture in the UK, 'Lucas SEI Wiring Systems Ltd.' The main manufacturing facilities for Lucas SEI are at Ystradgynlais in South

Wales. For the reasons of growth, world presence and adoption of Japanese manufacturing practices, the advantages of the partnership for Lucas Rists are readily observable. For Sumitomo, however, the joint venture company had attractions. First, although having a significant presence in the Far East and US markets, Sumitomo had no manufacturing presence in Europe. The Lucas Rists partnership has given them this toe-hold, important not only for the fact of operating in all three major world markets, but also with the coming of the single European market in 1992 and the need to demonstrate a European pedigree. The second factor is that the majority of Japanese assembly plants to be set up in Europe have been established in the UK, and Lucas Rists, being the largest UK harness manufacturer, was an obvious attraction. Finally, Lucas Rists is the major supplier to the Rover Group, who have tightened their ties with Honda of Japan, to whom Sumitomo is the major supplier in Japan and the US.

This collaboration has given Lucas Rists a substantial advantage over its European rivals, none of whom have yet developed a similar relationship with a large Japanese company. Lucas Rists is able to maintain its independence in the UK market and take advantage of the Sumitomo links in terms of design and manufacturing expertise.

This pressure for growth and the insistence on global supply has come from the major world automotive companies. These manufacturers, the customers of the harness manufacturers, are demanding closer links and long-term relationships with a small number of suppliers, those that are able to demonstrate their abilities to deliver on time every time with consistent product quality and with the flexibility to respond at short notice to changes in product design or delivery schedules. In return for these demands, the car assembly company offers loyalty to its suppliers, help and advice in the resolution of problems and an end to multiple sourcing and the playing-off of suppliers against one another. Under such circumstances, growth alone is not sufficient; it must be married with effective and efficient operations, so companies have adopted a strategy of just-in-time management and continual improvement to effect such positive change.

Source: by kind permission of Lucas Rists Wiring Systems Limited.

Product and process technology

The caveats of concentrating upon technological drive in process and product design at the expense of other factors has been discussed in some detail already in this chapter. However, a market-focused organization ignores technological change at its peril. Developments in terms of new products or advanced manufacturing processes can create a competitive edge in markets for those companies that are quick to adopt and adapt innovations. Product advances can provide the customer with additional features, improved performance and economy of use, whereas new process technology can improve quality and reduce manufacturing costs and, therefore, lower prices and/or increase profit margins. This does not mean that all new technologies are good technologies. The organization should counteract any pro-innovation bias with an ability to track, adapt and develop only those technologies that are both appropriate to their business needs and to the markets within which they operate. This balanced approach to technological change should be incorporated within the strategies and objectives that are set by manufacturing factories.

Sociogovernmental factors

A major parameter influencing the development of corporate and market strategies is that of social and government influenced change. Companies should ensure that they dispatch the requisite sensing 'devices' into their environments to detect such changes, and have the internal mechanisms to best react to them. Social and government factors take many forms but Oakley's categorization of these into 'changes in attitude, fashion and taste', 'legal changes', 'political trends' and the effect of 'diminishing natural resources' is useful for those organizations seeking to identify the important determinants upon their business (Oakley, 1984).

■ 5.3 Selection and configuration of design options

Having analyzed the main influences upon the organization from the market and environment, the next stage is to consider the design options open to the organization in the configuration of corporate and market strategy to match business requirements. The specific design choices for strategic corporate response to the demands of the market and environment are contained within two design option guides (DOGs). DRAMA recognizes the process of strategic choice as comprising two basic steps, namely:

1. The development of corporate policy (see the DOG in Figure 5.4).
2. The development of market/product strategy (see DOG in Figure 5.5).

Each of these stages will now be discussed in turn.

Corporate policy

The DOG shown in Figure 5.4 guides the decision-maker through the formulation of corporate policy and suggests the 'result criteria' that should be

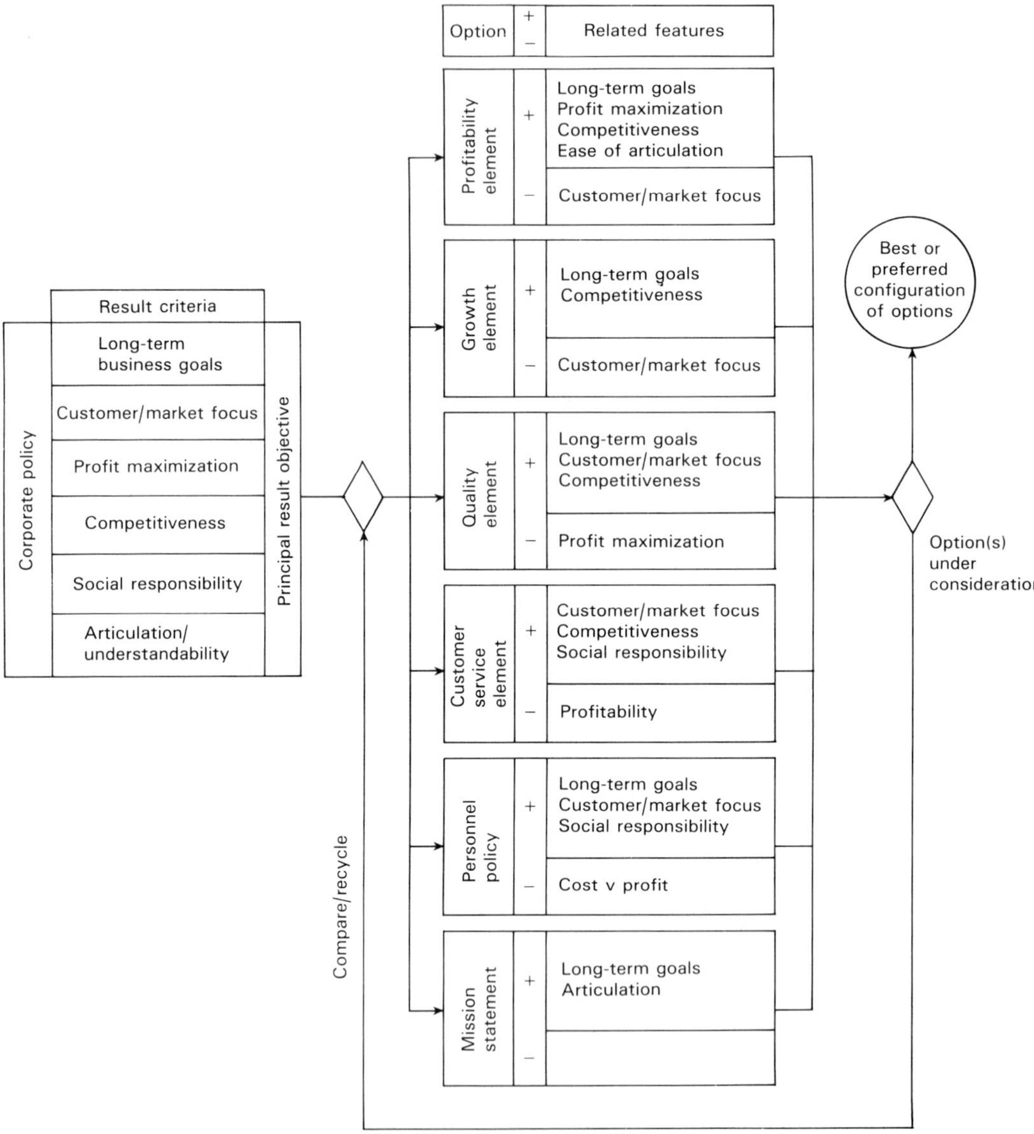

Figure 5.4 Design option guide for corporate policy.

considered and weighted accordingly at the commencement of this process. These criteria include:

1. The need for a statement of long-term business goals.
2. The degree of emphasis on customer and market-focus.
3. The need for profit maximization.
4. The continued competitiveness of the organization.

5. Social responsibility to employees and the wider community and environment.
6. The need for the corporate policy to be well articulated and capable of interpretation by all relevant people and parties.

There will normally be conflict between the parallel and concurrent achievement of these items, so trading-off and compromise is required between these criteria on the part of the decision-maker in order to arrive at an appropriate balance. This balance, the 'principal result objective', is therefore a considered configuration of the criteria where an appropriate weighting is given to each. So, for example, if the company is under pressure to deliver short-term returns and contribution to cover costs, then profit maximization might be put top of the list of criteria. If, however, the organization is in an industry where environmental issues and the need to avoid adverse publicity are at stake, then social responsibility may have a relatively high rating as a strategic issue.

Six options are then offered by DRAMA for corporate policy and the related positive or negative features associated with a dependence on each of these options are stated in Figure 5.4. Note that these features relate to the result criteria identified previously. Each of these options is now discussed in turn.

Profitability element

The majority of manufacturing companies aim to make a profit. Even where the organization is a 'not-for-profit' one, the corporate objectives must still recognize the need for returns and 'loss avoidance' wherever possible. The question is what weighting is given at the policy level for profitability, as decisions made at this stage can influence the degree to which financial cost-control and budgeting systems have an influence over activities within the organization. The arguments for maintaining a high profile in policy for profit-making are clear and well supported, but other considerations need to be added to any corporate policy. Of particular note here is that statements on profit and an over-concentration on this aspect does not provide an indication of any customer or market-focus.

Growth element

Like the profit element, growth is undeniably a valid aspiration in most organizations and focuses the company towards a sense of purpose in achieving long-term goals and the need to maintain competitiveness. Again, however, it provides no mechanism in itself regarding ways and means of ensuring market orientation.

Quality element

Corporate policy statements should have something to say on quality. In contrast to the profit and growth elements of policy, quality can be used as a means of emphasizing market and customer considerations. Companies are increasingly stating quality as a top priority within the organization, giving it a very high profile within the objectives of the business. For statements of intent in quality matters to be effectively translated into action will require substantial costs in the training and development of awareness of personnel within the company.

Customer service element

Like quality, the customer service element can be used to complement the growth and profit aspirations whilst providing customer focus. The organization should be clear about its policy towards customers, so that a unified approach to the satisfaction of the market is attained by all functions within the company. Customer service and quality are also alike in terms of whether investment and concentration of effort in these areas should be seen as a cost and therefore detrimental to profits, or whether, in the long term, costs are outweighed by the benefits that accrue. The general answer is that a correct balance should be attained between costs and perceived value, although more specifically this point of balance will vary considerably from one organization to another.

Personnel policy

Personnel policy acts as a term of reference for the employment and treatment of people. Such a policy is not really an option: rather it should be recognized that the way in which the personnel policy is stated and operated can have profound effects upon the nature of the organization as a whole. Some companies stress the need for flexibility and adaptability of employees to the needs of customers and the market, whereas others may closely associate this with an emphasis on customer service and quality assurance. Thus, the policy of employment and people management emanating from the top can exert a positive influence upon the development of market-oriented approaches and attitudes throughout the organization.

Mission statement

Now in fairly common use within manufacturing companies, the mission statement is a concise and articulated statement on the main purpose of the organization. Normally containing distilled objectives from the five design options identified above, the mission statement can provide a means of communication of corporate policy to customers and employees alike. Some organizations see the mission statement as being so important that they expect everybody within the organization to be able to quote the mission verbatim. However, it must be recognized that the effectiveness of mission statements depend upon the extent to which all in the organization accept this as a collective objective and purpose.

The methodology user should now select one, or more probably a configuration, of these options and relate the associated features to the previously determined principal result criteria. The process of design is, therefore, iterative and involves recycling of design option configurations and adjustment where necessary. Eventually a best or preferred configuration of design options should be developed and formally documented as the corporate policy.

Market and product strategy

Following the development of a stated corporate policy, the user develops a market/product strategy using the DOG shown in Figure 5.5. Relative priorities should be established for the identified result criteria, which range from profitability, levels of turnover and costs to the requisite levels of quality, customer orientation and flexibility that the organization displays in the marketplace. The design options are now detailed.

High volume versus high variety manufacture

A fundamental consideration in developing market strategy is whether to manufacture in high ('standardized') or low ('wide variety') volumes. The former has advantages in terms of economies of scale but offers little by way of choice for the customer. High variety manufacture, however, can provide for the specific needs of the customer, but traditionally suffers in terms of costs and, potentially, in the degree of standardization in the product or process.

Customization

Linked to the volume versus variety decision is the question of whether products should be customized to meet a variety of market needs, or even produced as 'specials' to satisfy the needs of individual customers. Customization takes the variety concept a stage further being customer-driven in that the customer decides upon the product configuration instead of a predetermined range being offered by the manufacturer.

Product and geographical 'nichemanship'

One option as part of a market strategy is to focus the organization's efforts by aiming to serve distinct niches or segments of the market either on a

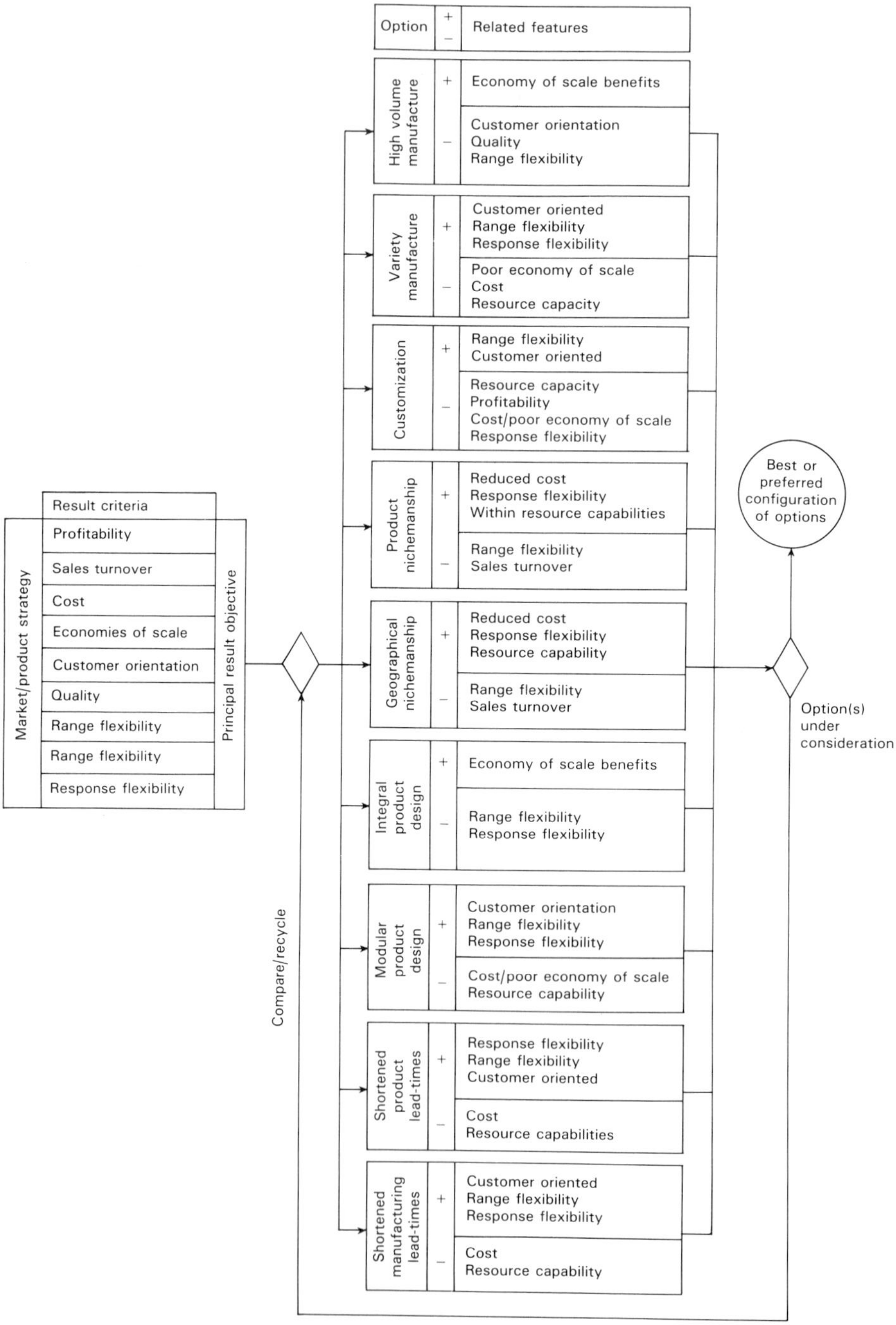

Figure 5.5 Design option guide for market/product strategy.

product or a geographical basis. Such an approach enables the company to address specific customer needs and, within its niche, to achieve economies of scale and customer awareness normally only available to larger manufacturers. The disadvantage most frequently cited is that of a lack of scope and range for production, which may leave the organization vulnerable should its chosen markets or customer bases come under increased competitive threat.

Integral versus modular product designs

Fundamental decisions on detailed product designs must also be addressed in combination with the above design options because of the close link between products and production facilities. A choice needs to be exercised between designs whereby components are 'integral' or where these are fitted together as 'modules', so enabling modifications and high variety manufacture through add-ons to the basic design.

Product and manufacturing lead-times

Finally, a decision needs to be made regarding the extent to which the organization wants to concentrate response to market as a competitive weapon by striving to reduce 'time-to-market' for new products ('design lead-times') or customer 'order-to-receipt' times ('manufacturing lead-times'). In the market-focused organization, to be able to reduce design and manufacturing lead-times is often seen as paramount and a key strategic edge, although the constraints in terms of cost and resource utilization must be recognized in this approach.

■ 5.4 Generalized methodology

The overview of the design process for corporate policy and market strategy using the DRAMA methodology is shown in Figure 5.6. The development of high level corporate strategies and objectives is primarily based upon the analysis of market and environmental factors. The approach suggested by DRAMA in the development of corporate and marketing strategy is now presented as a step-wise process:

1. **Consider customer needs and evaluate competitors.** Market research and direct interaction with customers can be used in connection with the former, and the technique known as 'benchmarking' can be used for the latter.
2. **Conduct an internal audit of market competencies.** This should involve a review of existing policy and market strategy, whether explicitly stated or

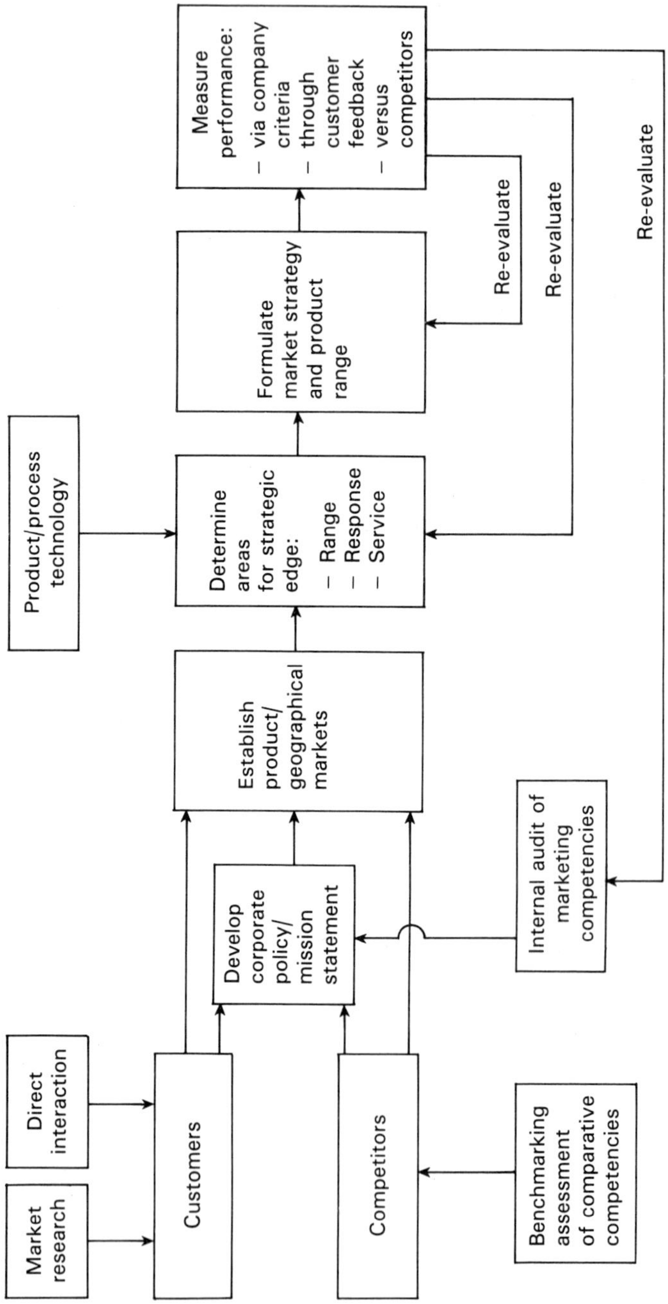

Figure 5.6 Generalized design methodology for market and environment.

conducted as custom and practice, and an evaluation of the extent to which the strategy and actual performance matches the needs of the market.

3. **Compare and combine the findings of the internal audit with customer and competitor analysis.** This will facilitate the drafting of the mission statement to accurately reflect the company's fundamental strategic response and approach to satisfying profits, growth and customer aspirations.

4. **Establish markets.** Which products should the company sell and within which geographical markets should it operate?

5. **Analyze the organization's position, given the above, and evaluate the company's areas of strategic edge.** A number of methods may be adopted here. The information could be plotted on a BCG matrix or analyzed within the Porter or Johnson and Scholes frameworks, which were reviewed earlier in this chapter. Alternatively the Hill framework, described in Chapter 3, could be used to establish the order-winning criteria of products and the extent to which the company supplies these. In the latter case, this analysis could then be continued on to the analysis and formulation of manufacturing strategy detailed within the next DRAMA component as shown in Chapter 6.

6. **Formulate your market strategy on the basis of the results from steps (1) to (5).** Functional strategies, including that for manufacturing (see Chapter 6) will hinge around the market strategy. Indeed, it forms an important key parameter to the design of all DRAMA components in the development of market-focused production systems. The precise product range will also be determined at this stage.

7. **Articulate your market strategy.** The strategy should be written in a form that facilitates a comparison between it and any subsequent performance. Quantification should be used wherever possible, although non-quantifiable targets and goals can also be of value in providing clarification on future direction.

8. **Measure performance and compare with strategic targets.** Performance should be measured not only using the company's own criteria and data, but also through customer feedback information and in relation to industrial competitors. The process leads naturally to a periodic re-evaluation of the company's market competencies, areas of strategic edge and the suitability of the existing strategy. In this way the market strategy can evolve iteratively and in line with developments in the market and environment.

The key factor in the design of the market and environment component of DRAMA is the need to make a 'structured response' at corporate level to current circumstances and to any new opportunities and threats that may arise.

Analysis of the environment and the development of a precise and universally comprehensible market strategy are two necessary pre- (not 'post-') requisites for any company seeking to design a market-focused production facility. Formulating corporate objectives and market strategy involves the analysis of external factors and key internal competencies to determine the strategic direction of the company. However, thorough analysis at this level and an appropriate and intelligent strategy for the organization do not, in themselves, guarantee corporate success. The key is in putting these strategies into operation. The rest of the DRAMA methodology deals with the development of production systems within the context and confines of a market-driven strategy.

Application 5.2 ICL: Market and environment

Operating in a market where product innovation and technological change are key factors leads to a temptation to concentrate on the technical aspects of product and process design. This was, in effect, the ICL of 'old' prior to 1980 when advanced new products were 'pushed' on to customers without any substantive regard to their individual needs. The initiatives of ICL senior managers in the early 1980s led to a reorientation of corporate policy with the aim of creating adaptability and responsiveness within the company to external market and industry factors. The technology-driven base of the company was reviewed and the decision was made to induce the development of market-led operations. In this new approach product and process technologies remained important, but only in so much as they satisfied customer needs or capitalized upon strategic opportunities in the market. The objective of ICL's corporate policy is highlighted in the corporate mission statement:

> ICL is an International Company dedicated to applying Information Technology to provide profitable, high value customer solutions for improved operational and management effectiveness.

The change within the manufacturing division in response to the new corporate direction has been summarized by a senior ICL executive as follows:

> . . . the process of adopting a 'can-do' approach towards meeting market and customer requirements which absolutely challenges conventional British manufacturing management thinking. . . . What we have done, by working on a cascade of changing mindsets in management, is thrown out old ideas, turned them over and said that it is possible to be flexible and it is possible to be more responsive. Thus cycle times, process times and lead-times can and will be reduced.

The key points of ICL's response to market and environment factors are now summarized:

1. ICL has, in the past, been small in relation to its major international competitors. However, the acquisition of ICL by Fujitsu in 1990 now gives ICL the back-up of being part of the five largest information technology companies in the world.
2. ICL has changed from being predominantly a technology-led company to one which now advocates a market strategy based on the provision of solutions to customers. The main stimulus for this was a change of company senior management in 1980.
3. ICL's adoption of a market-driven policy is a common approach in the dynamic computer and 'downstream' electronics industries. It is now rare for companies to compete merely on the basis of price and technical product specification.
4. A niche approach to market strategy, in which the company focuses upon a discrete number of product and geographical markets, has allowed ICL to compete with a degree of success in thinly sliced segments.
5. ICL has been influenced by the international movement towards increased democracy and codetermination which has taken place within society generally and, notably, in the computer industry. It has sought to increase levels of employee participation in decision making and, since 1988, has been seeking to develop a single status organization through 'harmonization'.
6. 'Bottom-up' technological innovation with the initiation of the Ashton plant's 'cascade' system and the formulation of a business case for investment in the new process would seem to have preempted the top-down translation of the market-driven policy into the manufacturing operations. It serves as an example of how manufacturing can, in its own right, contribute to the order winning criteria of products.

Source: by kind permission of International Computers Limited.

5.5 Conclusions

This chapter has considered the decisions made and policy set at the top level of the business. Such policies set the context for the development of production systems, so their process of formulation and the reasoning behind them should be understood by the system design team. In addition to a general discussion on theory and practice, three approaches to analyzing market and environmental factors were presented to provide a framework for corporate decision-makers.

The DRAMA design methodology for this component involves making decisions about corporate objectives and then developing an appropriate market strategy. These are matters for strategic debate within the company and, if intelligently developed, should serve to dictate and direct decisions made in lower levels of the organization for some considerable time in the future. It is the market strategy in the DRAMA methodology which lays the foundation upon which the manufacturing strategy and the organization design to support this will be decided.

■ 5.6 Further reading

Kotler, P. (1986) *Principles of Marketing*, Prentice Hall, Englewood Cliffs, New Jersey.

Pearson, G. J. (1990) *Strategic Thinking*, Prentice Hall, Hemel Hempstead.

Sharplin, A. (1985) *Strategic Management*, McGraw-Hill, New York.

6 Manufacturing Strategy

Objectives of the chapter

On completion of this chapter you should:

1. Understand the means by which corporate objectives and market strategies are translated into objectives and strategies for manufacturing.
2. Be able to develop a manufacturing strategy to facilitate the initiation of market-focused operations.
3. Make informed decisions on the most appropriate concepts and philosophies to adopt within manufacturing operations.

■ Chapter outline

This chapter considers the translation of corporate and market objectives into manufacturing strategies. It begins by reviewing existing theory and practice in what has been a developing area of study over the last twenty years. In particular the frameworks provided by Hill and the Cambridge Group are presented in detail to provide a theoretical basis for reviewing current practice in the development of manufacturing strategies. The key parameters influencing and constraining the formulation of manufacturing strategy are identified and then analyzed in detail. Design options guides are presented to cover decisions concerning the organization of manufacturing, quantification and target setting for strategic objectives, and the development of an effective quality management system. These three elements are seen by DRAMA as being the key constituents of manufacturing strategy. Finally the methodology for manufacturing strategy is summarized in the generalized methodology flowchart in section 6.4.

■ 6.1. Theory and current practice

The manufacturing strategy paradigm

Until recently the terms 'manufacturing' and 'strategy' used in conjunction with one another would have seemed contradictory. Manufacturing decisions were perceived as being low-level, day-to-day choices and there was little debate on manufacturing operations at the strategic level of the business within most organizations. There was a predominant attitude that strategy was something that accountants and marketing specialists were concerned with, to the exclusion of manufacturing managers. Production personnel were simply not involved with strategy formulation and analysis, rather they concentrated on the operational planning and control of activities with a limited time horizon at lower levels of business analysis and operation.

An interesting keynote address was given at the Sixth UK Operations Management Conference at Warwick, in 1990, by Wickham Skinner, regarded by many as the pioneer of the manufacturing strategy paradigm. He reported that the term 'manufacturing strategy' was first used at the Harvard Business School in the late 1940s where, at that time, he was a young research student. For twenty years his use of the term was considered to be a contradiction in terms and he had frequent disputes and arguments with those who saw manufacturing as having little or no consequence in competitive terms. The words 'manufacturing' and 'strategy' were seen as incongruous. Surely it was the intelligent strategic use of marketing and finance that gave an organization its competitive advantage and manufacturing operations simply got on and produced goods as cheaply and fast as they could. Indeed, using Skinner's words, manufacturing was typically seen as a 'millstone' around the corporation's neck: a function that used up substantial quantities of the company's resources, always seemed to be on the defensive when requests were made or changes required, and never providing marketing with what they needed.

The idea of formulating manufacturing strategy, however, has gained momentum since the publication of Skinner's article in the *Harvard Business Review* entitled 'Manufacturing – Missing Link in Corporate Strategy' (Skinner, 1969). Skinner argued that manufacturing could and should be linked to corporate objectives and marketing strategies. He suggested that the links between manufacturing effectiveness and corporate success go far beyond merely ensuring high efficiency and low cost and should include other considerations such as quality, product and process focus, service and delivery, etc. He was strongly critical of conventional methods of operations management which were based upon the principles of scientific management and Taylorism and advocated a top-down rather than a bottom-up approach.

Skinner concluded that manufacturing operations can be either a 'competitive weapon' or a 'millstone' for a company, and seldom neutral (Skinner, 1978, 1985). He argued that by adopting a strategic approach to managing operations the organization can achieve competitive edge over its rivals. By not paying sufficient attention to manufacturing and its links with corporate objectives, however, the production operations can impede corporate success even when corporate objectives have been well stated and an effective marketing strategy has been formulated.

Following the lead by Skinner there have been numerous calls for the design of manufacturing operations to be considered as an important element in the corporate strategy of an organization. The late 1970s and early 1980s saw the realization in the USA and Western Europe of the emergence of global competition, particularly from Japanese companies. Many reasons have been cited for the success of Japanese and, more recently, of other manufacturers around the Pacific Rim, but one of the most publicized is that of their effective manufacturing operations. The Harvard Business School took up this cause and published many papers which attempted to identify why the Japanese had attained manufacturing superiority (Abernathy *et al.*, 1981; Hayes, 1981; Wheelwright, 1981).

The work of Porter (described in Chapter 5) was also germane to this new view of manufacturing and also demonstrates how corporate management texts can provide lessons for manufacturing managers wishing to adopt a top-down and market-oriented approach to operations (Porter, 1985). Others have also advocated the virtues of deriving competitive advantage from the approach to manufacturing taken by exemplars most notably in Japan and the US (Schonberger, 1986; Hayes *et al.*, 1988).

A key issue in these works was the attempt to extract the lessons from the successful manufacturing economies of the world to apply these and rejuvenate the competitiveness of American companies. Robert Hayes and Steven Wheelwright carried their ideas forward and, in 1984, published their book *Restoring Our Competitive Edge: Competing Through Manufacturing* (Hayes and Wheelwright, 1984). Rather than merely stating the malaise of Western manufacturers, this book attempted to provide guidelines for senior executives wishing to compete through manufacture and so achieve levels of corporate effectiveness comparable with the best in the world. These are contained in their 'framework for analyzing manufacturing effectiveness'. The framework presents four stages in the process of improving manufacturing effectiveness in global markets. As well as being descriptive and explanatory, the framework is intended to provide guidance to managers when attempting to evolve their manufacturing strategies and operations (Hayes and Wheelwright, 1984; Wheelwright and Hayes, 1985). The stages identified in the framework are as follows:

Stage 1. *Minimize manufacturing's negative potential – 'internally neutral'*: This is where senior managers view the manufacturing function as neutral, that is to say it is incapable of influencing competitiveness success. The emphasis here is to minimize any negative influence that manufacturing may have upon corporate and market developments.

Stage 2. *Achieve parity with competitors – 'externally neutral'*: Senior managers see manufacturing as important only in so much as it matches the effectiveness and efficiency of competitors in the same industry. Here the emphasis is upon following industry practices in managing the workforce, avoiding large-step innovations in product or process technologies and viewing economies of scale and efficiency as the most important factors in production.

Stage 3. *Provide credible support to the business strategy – 'internally supportive'*: Senior managers expect manufacturing to support and strengthen the company's competitive position. They see the contribution of manufacturing as deriving from and dictated by corporate objectives. They ensure decisions made in manufacturing are consistent with corporate and marketing strategies, that strategy is translated in terms meaningful to manufacturing personnel, and are proactive in developing longer-term manufacturing strategies at the functional and factory levels.

Stage 4. *Pursue a manufacturing-based competitive advantage – 'externally supportive'*: This is where senior executives see manufacturing capabilities as a significant influence upon overall competitiveness. Manufacturing strategy is not merely determined by internal corporate and marketing strategies, but allows manufacturing executives to have a meaningful role in contributing to the development of the company and its strategies as a whole. In some cases this will result in the realization that manufacturing is the key competitive weapon within the organization.

Managers can identify at which stage their companies are in this framework and therefore identify changes that need to be made within the organization to progress to the next stage.

An important contribution to understanding Western manufacturing decline was provided by Terry Hill in his thesis on the strategic management of the manufacturing function (Hill, 1985). Writing at the same time as Hayes and Wheelwright, Hill also drew international comparisons in the management of manufacturing. His conclusion was that the pressure on the manufacturing function in many businesses was such that the emphasis has been upon how 'to manage reactively, and to be operationally efficient rather than strategically effective' (Hill, 1985, p. 27). He argued that manufacturing executives must begin to think and act in a strategic, not reactive manner. 'The purpose of thinking and managing strategically is not just to improve operational performance or to defend market share. It is to gain competitive advantage . . .' (Hill, 1985, p. 27).

In the mid-1980s term 'world class manufacturing' (WCM) was beginning to come to the fore to describe the operations of the most successful companies. Foremost amongst those authors spreading this term was Richard

Schonberger. First, Schonberger had attempted to reveal that Japanese excellence in manufacturing was not culturally bound and gave some insight into the methods adopted by Japanese companies (Schonberger, 1982). Picking up on his theme of blending total quality control and just-in-time management approaches, Schonberger then attempted to show how companies could achieve world-class status (Schonberger, 1986). Schonberger's work has been criticized as being highly prescriptive and not providing adequate framework for the analysis of production operations to enable the development of manufacturing strategies in given situations. However 'world class manufacturing' as a concept has found favour and caught on widely amongst practitioners as a goal to be attained. In this respect, the following (adapted from Hayes *et al.*, 1988) provides a useful definition of what WCM status means:

1. **Becoming the best competitor.** Being better than any other company in your industry in at least one aspect of manufacturing.
2. **Growing faster and being more profitable than competitors.**
3. **Hiring and retaining the best people.** Having workers and managers who are so skilled and effective that other companies are continually seeking to attract them away from your company.
4. **Developing an excellent engineering staff.**
5. **Responding quickly and decisively to market changes.** Being more flexible than one's competitors in responding to market shifts or price changes, and getting new products into the market faster than they can.
6. **Practising 'simultaneous' or 'concurrent' engineering.** Being able to develop products and processes in parallel and so cutting time to market down below that of competitors.
7. **Continually improving facilities, systems and people skills.** The emphasis on continual improvement is probably the ultimate test of a world-class organization.

As far as the link with production system design is concerned, there have been few attempts to translate manufacturing strategy into decision processes for technology and work organization. Exceptions to this are Hill (Hill, 1985), who addresses the matter within his work on 'process choice' and Kantrow, who has developed the idea of the 'strategy-technology connection' (Kantrow, 1980). In addition, Schroeder has considered the impact of production system technology on the development of corporate strategy (Schroeder, 1990). The DRAMA methodology can be used to close the gap that is seen to exist between the manufacturing strategy literature on the one hand, and the production system design tools and techniques on the other. This current mismatch, and the contribution of DRAMA to closing the gap, is illustrated in Figure 6.1.

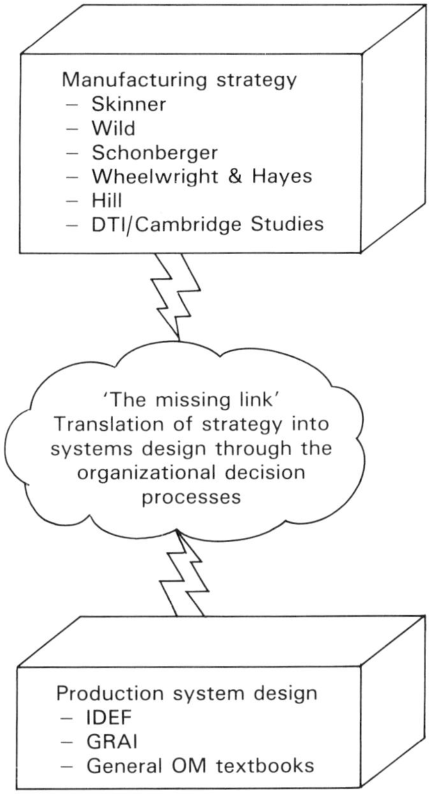

Figure 6.1 Gaps in the theory of production system design.

It is acknowledged that the various contributions to the literature reviewed above have raised the profile of manufacturing within organizations and have highlighted the importance of manufacturing in the achievement of corporate objectives. However, most of this work tends to be largely promotional and highly prescriptive without providing any meaningful structure for the detailed analysis of manufacturing decision making. Two approaches to the analysis and formulation of manufacturing strategy have recently been produced, however, which seek to provide practical guidelines in the development of strategies for production. Already introduced in Chapter 3, they are Hill's framework (Hill, 1985) and the Cambridge Group's methodology (DTI, 1988).

Hill framework for analyzing manufacturing strategy

The Hill framework relates well to the previous attempts of Wild who had considered the role of manufacturing in business policy and set up a policy framework, a feature of which was the interaction of the operations management function with the external environment at all levels (Wild, 1980). The framework for reflecting production and operations strategy within corporate objectives was described in detail in Chapter 3 (see Figure 3.14). It identifies a number of issues in the formulation of operations strategies. First, there is a need for 'strategic difference' in operations as well as in marketing strategy. This recognizes that different parts of the manufacturing system and those parts of the system serving different customers or markets should be operated separately and involve distinctive process and infrastructure choices to initiate a high degree of market-focus. Second, is the need to link marketing and operation so that their strategies interrelate and complement one another. Third, there is the need to identify and distinguish between 'order-qualifiers' and 'order-winners'. Qualifying criteria are those that get a company's product into the marketplace or on to customers' shortlists and keep them there, while order-winning criteria give the product superiority and thereby competitive edge over the offerings of competing organizations. To be successful, therefore, companies should seek to convert order-qualifying criteria that merely get them recognized as a competent supplier of manufactured goods into order-winners so that their products are more attractive to customers than those of their competitors. Typical order-qualifiers and winners identified in the framework are:

1. Price (cost).
2. Design.
3. Quality.
4. Delivery speed.
5. Delivery reliability.
6. Demand flexibility.
7. Product/service range.
8. Colours/patterns.
9. Economy of use.
10. Design leadership.
11. Technical support.
12. After sales service.

Not only are these criteria seen as being affected by the top-down imposition of corporate objectives and marketing strategy, but the features of the production process and its control infrastructure can also contribute and add to the order-winning qualities of the company. In recognition of the link between corporate success and manufacturing decisions, many manufacturing companies have established corporate policies which are market-driven, and have realized the role of manufacturing performance in the achievement of objectives.

The Cambridge Group's methodology

The work of the Cambridge Manufacturing Group (DTI, 1988) was presented in detail in Chapter 3. The resultant methodology for manufacturing strategy proposes that an organization should start with a strategic view of the business as a whole and then develop a business strategy, product strategy and manufacturing strategy prior to planning the manufacturing facilities. The methodology is in the form of a work-book, which enables the strategist to analyze and develop manufacturing strategy by completing a number of 'work-sheets'. The outcome is a refined or new strategy which accurately reflects market position, growth potential and customer needs and that the organization should now have a better understanding on what it has to do to serve its chosen markets more effectively.

Quality, service and human resource management

Hill and the Cambridge Group provide an interesting insight into the activity of strategy formulation and translation. However, they adhere to a top-down rather than an overly market-deterministic view of manufacturing system development. A further criticism is that they say very little about the subject of human resource management and organization, which invariably contribute to the success or otherwise of any manufacturing strategy. Modern concepts of human resource management, which are seen as directly influencing quality and service performance have achieved prominence (see for example, Storey, 1991). Quality programmes are frequently used as the vehicle to change the attitudes and beliefs of employees, and so enable operational activities to be more market-focused and customer-oriented.

Within manufacturing strategies the market-driven approach has led to a partial retreat from traditional scientific management and 'Fordist' techniques which seek to exploit economies of scale, minimization of cost and maximize labour productivity. This approach, it is now widely acknowledged, neglects other factors which may contribute to the wider business needs of the organization. The traditional approach did not place high emphasis upon the notion of continual improvement. Typically, optimization of operations was sought, usually to minimize total costs of inventory, but little attempt was made to improve parameters upon which such optimization was based, such as reducing costs of ordering and so lowering the economic order quantity. More recently, particularly as a result of the much publicized experiences of Japanese manufacturing companies, effort has been placed on reducing or removing constraints to operational performance rather than operating at optimal levels within these constraints (Schonberger, 1986).

The areas in which continual improvement is sought are numerous, but some examples will be given here. Recently, much attention has been given to the reduction of lead-times within manufacturing companies. In terms of product design, the faster a company can move from the original design specification, or from a request to modify an existing design, to manufacture of that product, the more attractive the products are likely to be in the market.

Similarly, short manufacturing lead-times, from the placing of an order to receipt of the goods, will attract customers who, given the choice, would rather usually prefer delivery sooner rather than later. Much attention has also been given to increasing the level of operational flexibility and widening the range of tasks which can be accomplished within the production system in order to satisfy the specific and specialized needs of individual customers. Costs remain an important factor because of their direct relation to profit and selling prices. In addition to improved labour productivity, however, much greater importance is now given to the reduction of costs through lowering inventory levels and increasing stock turnovers.

Quality also has an important part to play in the manufacturing strategy of the market-driven company. Quality can be viewed in two different ways: in a product sense and, more holistically, in an organizational context. Quality and reliability are important attributes for any product and levels of sales are often determined by a company's past reputation in this respect. However, quality in its wider organizational sense is often seen as fundamental in establishing and maintaining the appropriate culture and attitudes within the organization. Total quality management and its associated methods and processes are, as a result, now seen as important components of manufacturing strategy.

Application 6.1 Development of manufacturing strategy at Black and Decker, Spennymoor

Black and Decker opened their plant at Spennymoor in the North-East of England in 1965. The factory was extended in 1970 and today is the company's largest plant worldwide, being a global manufacturing and design site. The factory produces high volume domestic electrical tools and appliances and exports 70 per cent of its output.

Manufacturing Engineering Manager, George Eldrington, explains that there are two foundation-stones that support the overall policy, the so-called plant philosophy.

Foundation-stone one: 'People are the key asset'

The major elements of this foundation-stone are:

1. Respect for the individual.
2. Progressive single status employment conditions including no clocking in or out and the same canteen facilities and holiday entitlements for all employees.
3. Continual people development where a large budget allows for excellent training facilities and an apprentice scheme for young entrants.
4. Management credibility.

Eldrington says 'We like to see ourselves as a people organization' and reports that the plant has excellent employee relations – not one hour has been lost due to dispute since its opening.

Foundation-stone two: 'Distinct competencies'

This enables the plant to manufacture only those products and perform only those operations at which it is proficient. These have been identified as:

1. The production of high precision components.
2. Manufacture of series wound motors.
3. Product assembly.

Plant philosophy

Given the above, the plant philosophy comprises three elements:

1. To continually improve the quality of everything we do and everything we make through total quality management (TQM).
2. Treating everyone as a customer or a supplier.
3. Providing and demanding total customer satisfaction (TCS).

Manufacturing strategy

On the basis of the plant philosophy, a five-year manufacturing strategy for Spennymoor has been developed. This has four main drivers:

1. **Quality.** Improved quality has been effected by a new 'build methodology' where the emphasis has been to maximize operators' work content and increase individual operator responsibilities for quality.

2. **Flexibility.** The aim is to improve flexibility in terms of:
   (a) fast changeovers of processes and lines from one product to another;
   (b) improved responsiveness to customer demand;
   (c) the introduction of assembly lines with multiple product capability;
   (d) the ability to vary production rates.

3. **Cost.** Whilst quality and flexibility have become increasingly emphasized over recent years, cost is still recognized as a key competitive factor. Efforts towards cost minimization are based around increased asset utilization of floorspace and capital plant and equipment.

4. **Environment.** People are viewed as the key asset, so efforts are made to get the best out of operators, administrative staff and managers by providing a high quality workplace conducive to effective working.

In summary, the overall objective is 'to achieve and sustain Black and Decker, Spennymoor and its products as the best in the world'. The efforts taken over recent years, particularly the focus upon people, customers and quality, are now placing the plant in a position where this is becoming a reality.

Source: by kind permission of Black and Decker Limited.

6.2. Key parameters

Identification of key parameters

The initiation of a market-driven corporate strategy will typically call for a customer-focused response from the manufacturing function and the operation of a demand pull system at the operating level. The key parameters to the formulation of manufacturing strategy therefore include the corporate policy and market strategy developed in the previous DRAMA component, market and environment. However, as shown in Figure 6.2, the process of manufacturing strategy formulation is not merely a top-down and deterministic process. There must also be some form of internal audit of competencies within the company to ascertain the current state of manufacturing and to what degree existing processes and infrastructures are capable of supporting corporate objectives. Given this analysis, the organization should be able to determine:

1. Areas where there exists a strategic edge over competitors.
2. Areas where this is lacking and needs to be developed.
3. Recognize that the best policy may be to contract out some parts of the business.

An important decision, therefore, is to decide what 'levels of manufacturing' the company will pursue, i.e. all processing in-house, subcontraction of component processing, or buying in of assembled products. Each of the key parameters shaping the development of manufacturing strategy will now be discussed.

Corporate policy

Corporate policy provides the overall context for the development of manufacturing strategy. In the market and environment section of DRAMA, corporate objectives were translated into meaningful strategies for marketing, but the development of manufacturing strategy should refer back to source objectives in addition to being driven by the market strategy. The reason for this is that corporate policy will contain objectives for cost minimization,

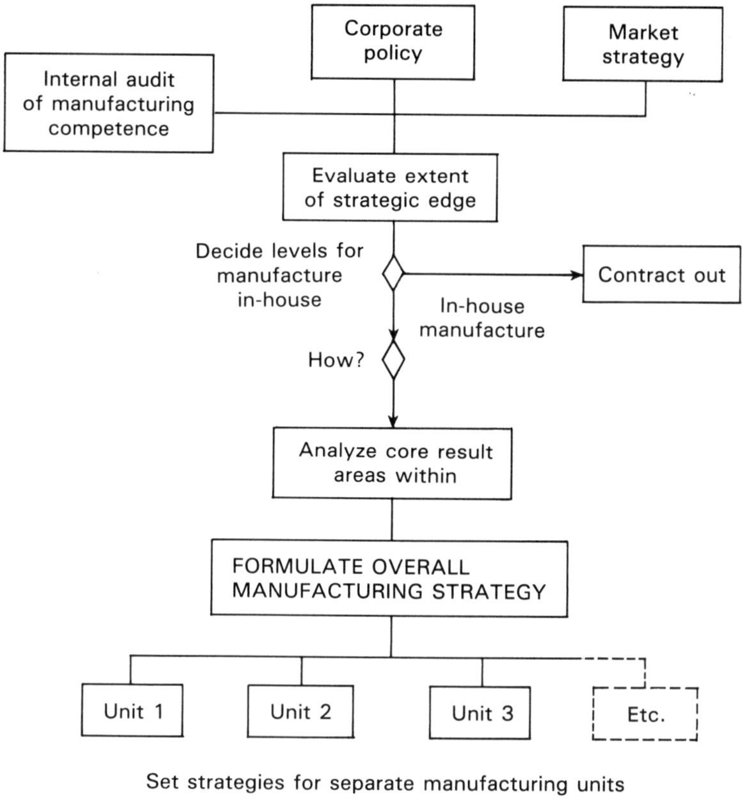

Figure 6.2 Key parameters for manufacturing strategy.

efficiency and production potential which need to be satisfied by production operations whilst not really being applicable for inclusion in statements of intent in dealing with customers.

Market strategy

It is important that manufacturing activities should support and complement the market strategy described in Chapter 5. Manufacturing strategy formulation should therefore take into account the criteria upon which products are competing within the market and so set terms of reference for the subsequent development of market-focused production systems.

Manufacturing competencies

Manufacturing strategy should not be developed in total isolation from an understanding of the current state of production operations within the business. There is no point in developing grandiose strategies and objectives for operations which, if achieved, would result in competitive edge when the capacity to implement such plans does not exist within the business. Therefore the existing competencies must be analyzed and taken account of in evolving manufacturing strategy to ensure that the organization does not attempt to take too great a step at a time in seeking to achieve competitive edge.

Manufacturing strategy is the first step in translating corporate objectives and policy into detailed decisions for operations. Once an overall strategy for manufacturing has been formulated, it is then important to disaggregate this down into separate strategies for operating units in the business, then subdividing these down further wherever appropriate. If conducted correctly this should provide clearly defined targets and mechanisms for their accomplishment. In so doing, these unit strategies facilitate the measurement of performance for all parts of the business against overall manufacturing objectives.

Application 6.2 ICL: Manufacturing strategy

In the 1980s, in response to the new market-driven corporate policy, senior ICL manufacturing managers realized that a more flexible, responsive and cost-effective functioning of manufacturing was required. Furthermore, with the increasing trend in the electronics industries to subcontract manufacturing to Pacific Rim countries, it became apparent that ICL's manufacturing operations had, in the words of one Company Director 'no divine right to an automatic existence'. The strategic choices for ICL Manufacturing Operations were seen as follows:

1. To do nothing, i.e. to maintain traditional practices and eventually close UK manufacturing plants.
2. To move manufacturing 'off-shore' immediately and obtain labour cost reduction and, potentially, economies of scale.
3. To go through a process of developing and implementing a new manufacturing strategy which would give a competitive edge to the company.

ICL chose the last of these and now develop five-year manufacturing plans based upon a number of key factors:

1. Cost reduction through better utilization of facilities, manpower and space.
2. Reduced 'time-to-market' through the integration of design and manufacturing functions, simultaneous engineering concepts and the elimination of functional barriers and communication problems.
3. Improved flexibility of response with an ability to respond to marketing changes and customer needs within weeks rather than months.

The steps suggested by ICL executives in their post-rationalization of manufacturing strategy formulation are as follows:

1. *Agree mission statement:* i.e. manufacturing should manufacture to supply products to customers' requirements and not to put into stock.
2. *Recognize the 'business drivers':* in the case of ICL the drivers identified were cost, productivity, quality, inventory holding and service.
3. *Develop business plans for manufacturing as a whole and for individual production units:* five-year plans were agreed.
4. *Implement strategy using two key concepts:* integrated manufacturing and total quality management.

The above illustrates a very logical development of manufacturing plans within the company. The key observations of the actual process of manufacturing strategy evolution within ICL over a number of years are now presented:

1. In the early 1980s the market-driven corporate policy was adhered to quite blindly and manufacturing tended to follow marketing's every whim. In 1985, however, manufacturing operations became more proactive in analyzing its activities and developing internal strategies and plans for manufacturing units.
2. ICL's manufacturing and logistics functions have remained viable by providing the company with a strategic edge in terms of response to market needs, flexibility and quality. Any uncompetitive areas of production, most notably the assembly of high-volume low-variety products where ICL finds difficulty in achieving economies of scale, have been subcontracted to external suppliers. More recently, high-volume products such as PCs have been produced in the Nokia Data factory in Finland, which ICL acquired in 1991.
3. The modular assembly cascade system at Ashton acted as a catalyst in the translation of the corporate mission to manufacturing operations and the development of manufacturing plans within ICL generally.
4. Manufacturing activities are subjected to measurement against a strict and quantified overall manufacturing strategy, which comprises overall targets for cost, productivity, inventory, quality and delivery (service) performances. These are then refined into individual factory and production unit strategies.
5. 'People'-oriented strategies are now being pursued to complement the quantified business targets. These have the objective of optimizing the quality awareness and business understanding of employees so as to develop change and flexibility as cultural norms. Extensive programmes of training, especially in the area of TQM, have been the main vehicle in this process.
6. It is now widely accepted by ICL executives that 'flexibility of service', 'time-to-market', competitive delivery lead-times and product/service

reliability are the key areas where the company maintains a distinctive edge over its competitors.

Source: by kind permission of International Computers Limited.

6.3 Selection and configuration of design options

Within the DRAMA methodology, manufacturing strategy is seen as comprising three main choices:

1. A decision concerning the number of levels and consequent organization of manufacturing (see DOG for 'Overall Organization' in Figure 6.3).
2. The setting of targets and quantifiable measures for measurement and evaluation purposes (see Figure 6.4).
3. Choices regarding the establishment of a quality management system (see Figure 6.5).

Each will now be considered and explained for the methodology user.

Manufacturing organization

At the top level of manufacturing systems design, a decision needs to be made concerning the degree of vertical integration of the facilities (whether they should deal with downstream activities only, or incorporate upstream feeder processes as well). Once this has been decided the fundamental organization of manufacturing must then be determined. Figure 6.3 shows that the decision upon which this choice is made is determined by the desire for management control, for customer range and response flexibility, maintaining quality standards and the cost of manufacture. The correct balance of these various result criteria needs to be decided to arrive at a principal result objective. The relative strengths and weaknesses of each design option should be compared to arrive at the best of preferred configuration of options.

The levels of manufacturing

In terms of the levels of manufacturing, the choices are categorized into three: contract-out all manufacturing, contract out upstream raw material and component processing activities (i.e. downstream manufacturing in-house only), or vertically integrate within one organization (downstream and upstream manufacturing).

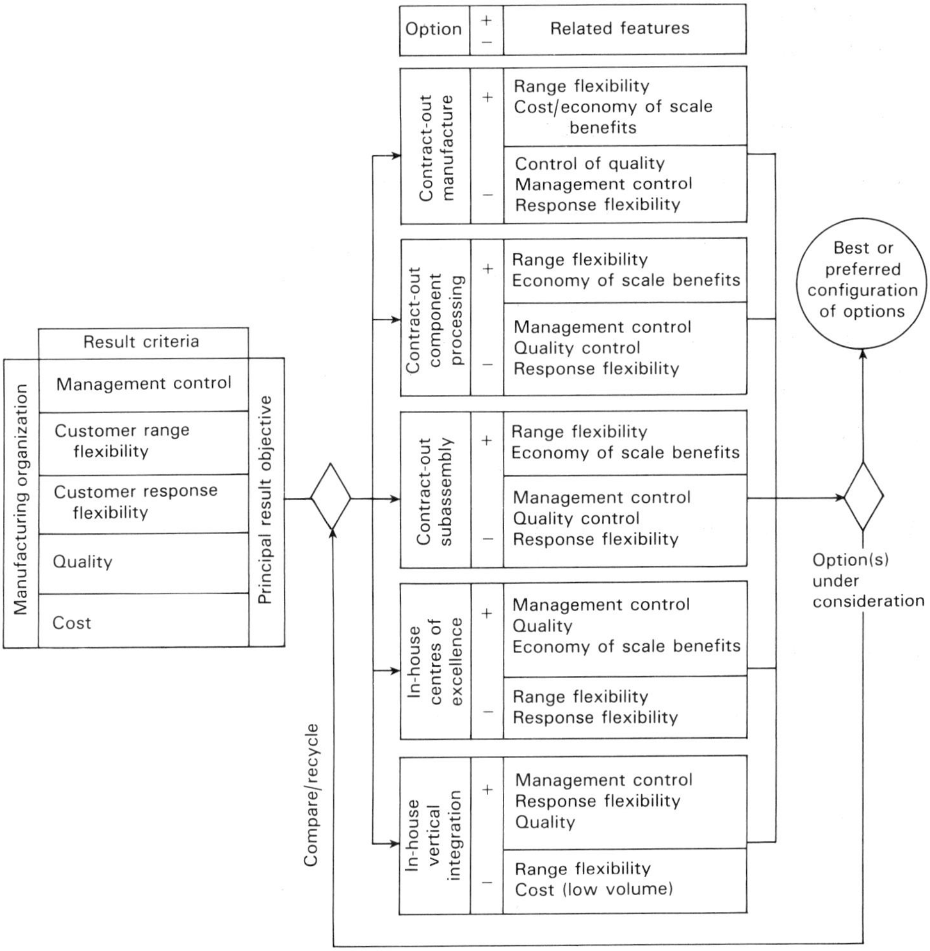

Figure 6.3 Design option guide for overall manufacturing organization.

Centres of excellence versus plant sufficiency

In terms of the basic organizational choice, a decision needs to be made between 'centres of excellence' and 'plant sufficiency'. The centre of excellence choice, normally confined as an option to the multiplant operator, involves factories:

1. Concentrating on discrete operations or products.
2. Being a centre of knowledge and expertise in this type of processing.
3. Seeking to achieve economies of scale compared to the situation where such operations are replicated across many sites.

Plant sufficiency in contrast involves multilevel and in some cases multiproduct manufacture in one factory.

In-house vertical integration

Linked to the options given above, the company must decide upon the degree of vertical integration it deems necessary within its own operations. A high level of integration involves the running of upstream raw materials and component processing activities in addition to downstream assembly and final product operations, whereas a low degree confines manufacturing to downstream only. Advantages of vertical integration include management control over the supply chain, adaptability and response to changing needs and quality assurance through the increased control. However the range of products possible may be constrained by not having the flexibility to buy in from suppliers in the short term and there may also be cost disadvantages in terms of a lack of economies of scale compared to producers who supply materials and parts to a number of final producers. (See Kumpe and Bolwijn, 1988, for a fuller discussion on vertical integration in manufacturing.)

Setting strategic objectives

Once the basic 'make or buy' and production organization decisions have been made, the company should decide on the targets and measures it will use to evaluate operations in relation to the business requirements. Here the strategic manager is looking for various indicators as shown by the result criteria. Having selected a balance and ranking between the result criteria shown in Figure 6.4, the decision maker must then develop relative targets and measure indicators. Five broad categories of targets are identified in the DOG:

1. Cost-based.
2. Efficiency-centred.
3. Output-based.
4. Quality-oriented.
5. Service-centred.

Normally, dependence on one of these measures does not give a full and true picture when evaluating performance. Therefore the options presented in this DOG will all be adopted, although the degree of dependence and ranking of each would vary substantially from company to company.

Cost-based objectives

Cost-based objectives are those which are stated and performance is measured in financial terms and include areas of management accounting such as materials, labour and overhead costs and their variances. Their advantage is

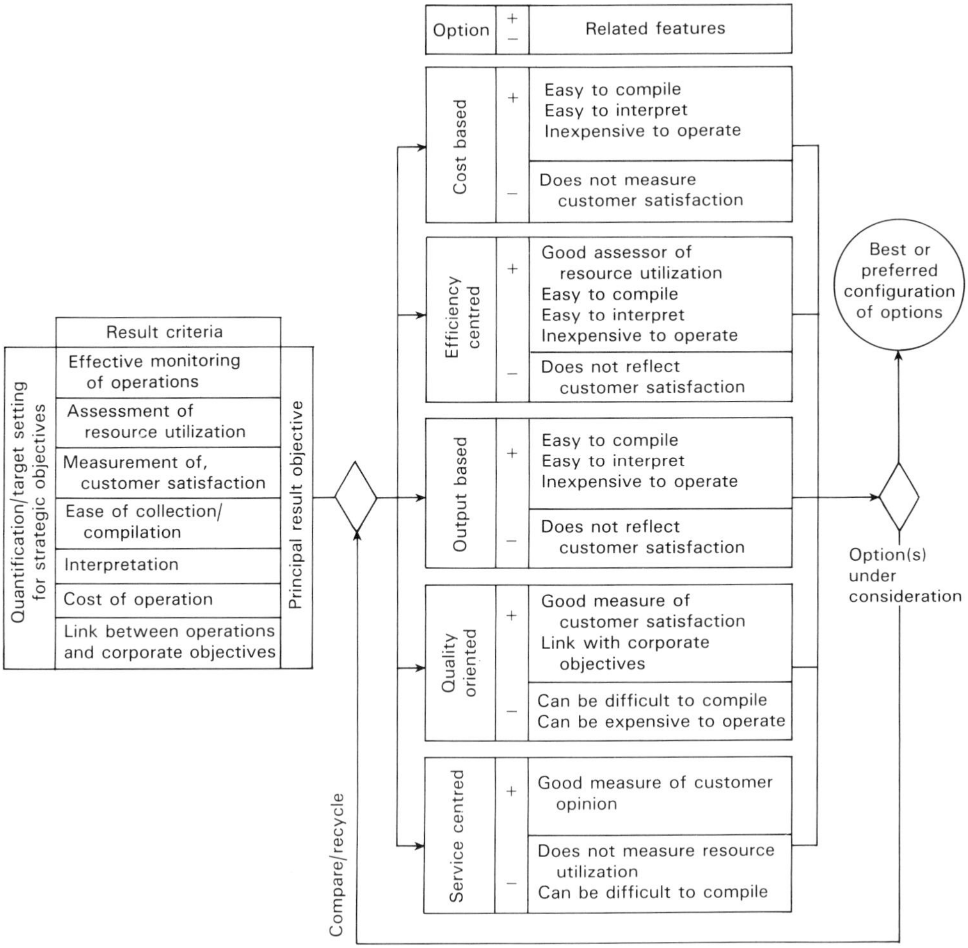

Figure 6.4 Design option guide for setting strategic objectives.

that they are relatively easy to compile and interpret and are normally inexpensive to operate, feeding off data and information used for such essential activities as purchasing and payroll processing. The main drawback is that financial measures are not direct measures of customer satisfaction and so can be seen as concentrating upon inputs to the business.

Efficiency-centred objectives

Efficiency-centred objectives are concerned with achieving satisfactory levels of productivity within the business and are usually stated in terms of percentage running time out of total time, material yields, etc. In so doing they provide a means of assessing the efficiency of resource utilization but, like cost-based objectives, they relate mainly to the satisfaction of internal requirements rather than those of the customer.

Output-based objectives

Measures of volume produced remain widely used in industry and are important in so much as determining whether actual capacity, net of routine capacity losses such as planned maintenance, is in line with the level of customer demand. Where discrepancies do occur, actions should be taken to discover the reasons for this either by removing capacity losses or increasing capacity available through new facilities or increased utilization. However, other than satisfying the level of demand, output-based objectives cannot be seen as a measure of customer satisfaction for quality, response and the like; they are also a cruder measure of resource utilization than efficiency- and cost-based measures.

Quality-oriented objectives

The setting of targets for quality achievement provides the basis for a more direct measure of customer satisfaction than those options presented above. They should also be linked with the stated corporate and market objectives for a market-driven organization which should set the guidelines for the level of quality expected of products in the markets in which the organization operates. However, quality is sometimes difficult to measure by variables, and so for quality assessment by attributes it is sometimes difficult and expensive to compile meaningful and appropriate figures on quality performance.

Service-centred objectives

The measurement of customer service, like quality, is a more direct measure of output from the production system and its adherence to customer needs. Service-centred objectives can include targets for the proportion of customer orders delivered on time, the level of orders met ex-stock and the response of the organization in days or weeks to requests for design modifications. Service-centred objectives are becoming increasingly popular in market-focused businesses, but must be used in conjunction with other measures which are better at assessing costs, efficiency and resource utilization.

Establishing a quality management system

Finally, developing the quality issue in manufacturing strategy, the company must agree at the highest level the type of quality management system it should adopt. The criteria for choice and the various options available to the manager are shown in the DOG in Figure 6.5. A market-focused manufacturing

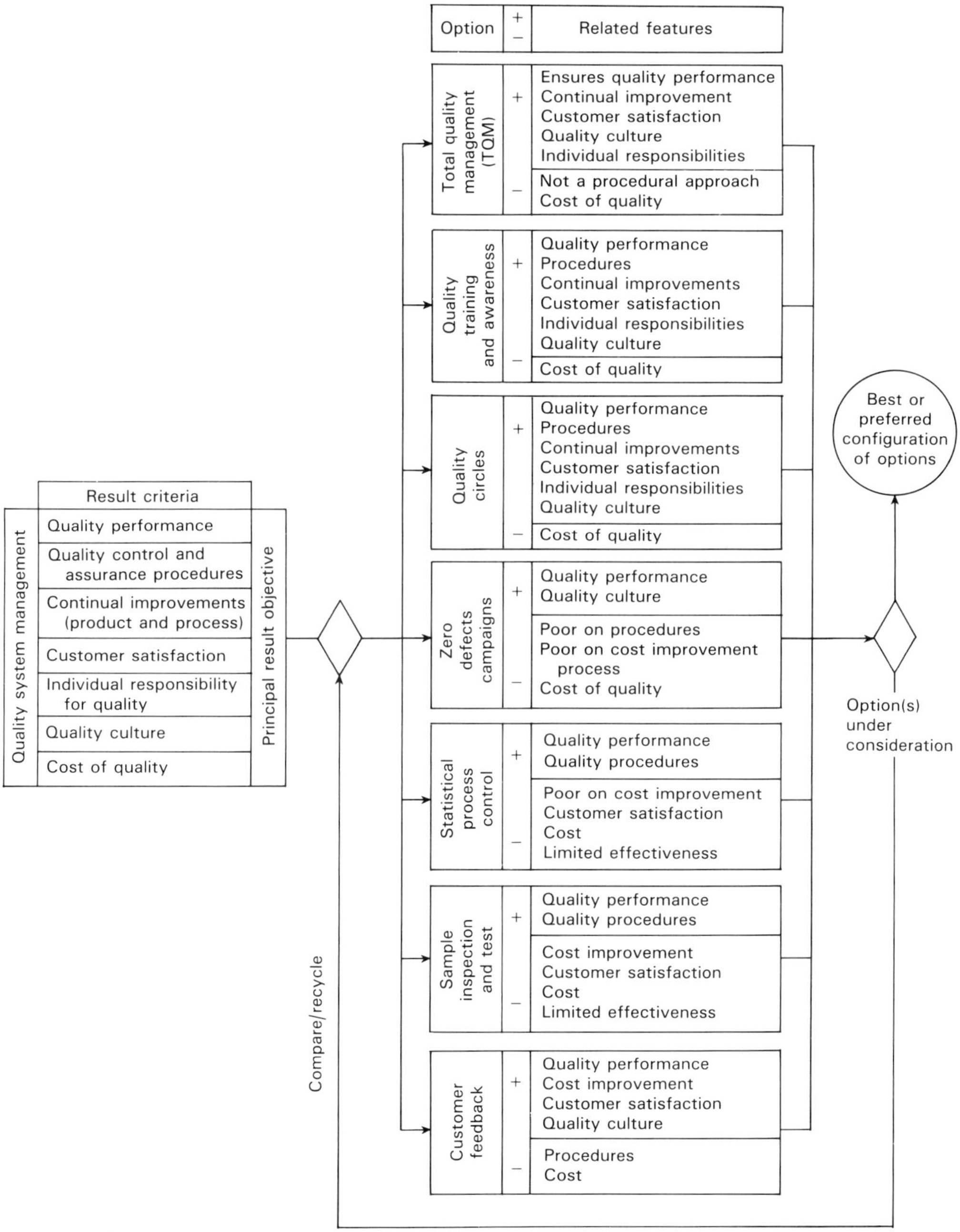

Figure 6.5 Design option guide for the establishment of a quality management system.

company would be expected to adopt some form of total quality management (TQM) system and so the first option is expected to be selected in response to this need to be customer-oriented. The other six options could be seen as subsets to TQM and therefore restricted to lower level decisions concerning mechanisms and techniques. However, making choices on these at the strategic level indicates a level of top management commitment to each and so result in a higher chance of their being successful.

Total Quality Management (TQM)

Total quality management can be seen as an organization-wide commitment to quality achievement and customer service. The focus is on ensuring high levels of quality performance through continual improvement, inducing customer satisfaction, developing an ethos within the organization of dedication to quality, and encouraging people within the business to assume individual responsibility for the quality of their own work.

Quality training and awareness

Quality training is a mechanism to improve performance within the organization but, in common with all forms of training, a decision must be taken regarding the optimum amount of investment that should be set aside for training, beyond which the costs begin to outweigh the advantages.

Quality circles

Quality circles are voluntary groups of employees (6–12 people) who meet on a regular basis. Each group has a leader and meetings tend to follow a formal structure. The overall programme within a company is usually co-ordinated by a 'facilitator' and there is close liaison with management. The quality circle approach assumes quality-conscious employees and offers no direct financial incentive to circle members. The benefits of quality circles can be identified as improved employee relations, improved quality, increased motivation and responsibility for quality, employees more receptive to new methods and ideas, and that it provides a forum for discussion and communication within the organization. Quality circles also have a number of related caveats in their use. Senior management commitment is critical to their success and so attention should be paid to engendering and communicating this. Second, there can be resentment from employees, a feeling that the workforce is doing management's job, and also resentment from managers, who see circles as eroding their prerogatives. Finally, circles can be seen as a 'secret society' by those not participating in the programme.

Zero defects campaigns

A zero defects programme aims to reverse the attitude of 'that's good enough' to one that strives for perfection in quality and service: a 'right first time' attitude. The approach recognizes two ways to change attitudes: through product/customer awareness and through quality awareness. In zero defects the individuals are shown that their own tasks are important. Zero defects is a campaign-based approach and therein lies its drawbacks. There are arguments that this approach is simplistic, takes no account of individual motivation factors, requires a high degree of co-operation from all and does not tend to infiltrate suppliers. Nevertheless, many manufacturing organizations claim to have used a zero defects approach to raise initial awareness in quality matters and their importance in the business prior to embarking upon a wider programme of quality management. The zero defects approach can therefore be seen as a valid component of a larger package of quality management initiatives.

Statistical process control (SPC)

SPC is used for the quality control of processes by variables and makes use of 'process control charts', as devised by Dr Shewhart of Bell Telephone Laboratories in the USA. SPC provides a structured and statistical means of controlling quality, but does have a number of limitations to its use. Most notably, the use of control charts does not in itself engender an ethos of continual improvement and, strictly speaking, should only be used where the distribution of variables to be measured conforms to a normal distribution.

Sample inspection and test

The inspection and testing of samples from outside suppliers, and of those parts and products manufactured internally, can be used as a means to control quality both of variables and, where measurement on a scale is difficult or impossible, by attributes. One of the most widely used forms of maintaining quality standards using sampling is that of acceptance sampling where a decision to accept or reject a large batch of items is based upon the information derived from inspection or test of a (relatively) small number of samples. Like SPC, sample inspection and test is invaluable in many manufacturing operations in maintaining and controlling quality at the operational level and therefore has its part to play in the context of the wider quality management system. However, care must be taken that the rationale for decisions concerning acceptable quality levels (AQLs) is based upon an accurate interpretation of what is acceptable to the customer. Furthermore, the organization must attempt, wherever possible, to increase the tightness of their tolerances in order to continually improve the quality of its products and processes.

Customer feedback

In addition to the internal programmes and quality control measures indicated above, customer information from the market also provides an important component in most companies' quality management systems. It provides direct feedback on customer satisfaction and any quality problems that might be occurring in the market, many of which may be impossible to detect or identify using just internal mechanisms of quality control and assurance.

6.4 Generalized methodology

Figure 6.6 gives the generalized methodology flowchart for the development of manufacturing strategy. The diagram shows that a top-down formulation of manufacturing strategy is, by definition, influenced by the corporate policy and market strategy choices made at the previous stage of design in DRAMA, market and environment. However, the influence of the performance and strategic edge provided by the existing manufacturing system should not be overlooked so it is important to perform an evaluation of current manufacturing performance. The process of strategy formulation is as follows:

1. **Perform an internal audit of manufacturing competencies.** The results from this can be compared to other organizations and competitors using benchmarking techniques and thereby the performance gap between current performance and business needs can be identified.

2. **Decide upon the levels of manufacturing.** Consider the areas of strategic edge provided by the current operations and the requirements laid down by the corporate policy and market strategy. Use the DOG in Figure 6.3 to assist in this choice.

3. **Determine the manufacturing strategy.** Set appropriate targets and objectives for critical design attributes using suitable measures and indices from those shown in the flowchart and the DOG in Figure 6.4.

4. **Develop an appropriate and relevant quality management system.** Use the DOG in Figure 6.5 to implement a suitable quality management system.

5. **Measure performance versus objectives.** Use the results from this in 6. and 7.

6. **Re-evaluate manufacturing objectives and the quality management systems.** Do this in the light of performance recordings.

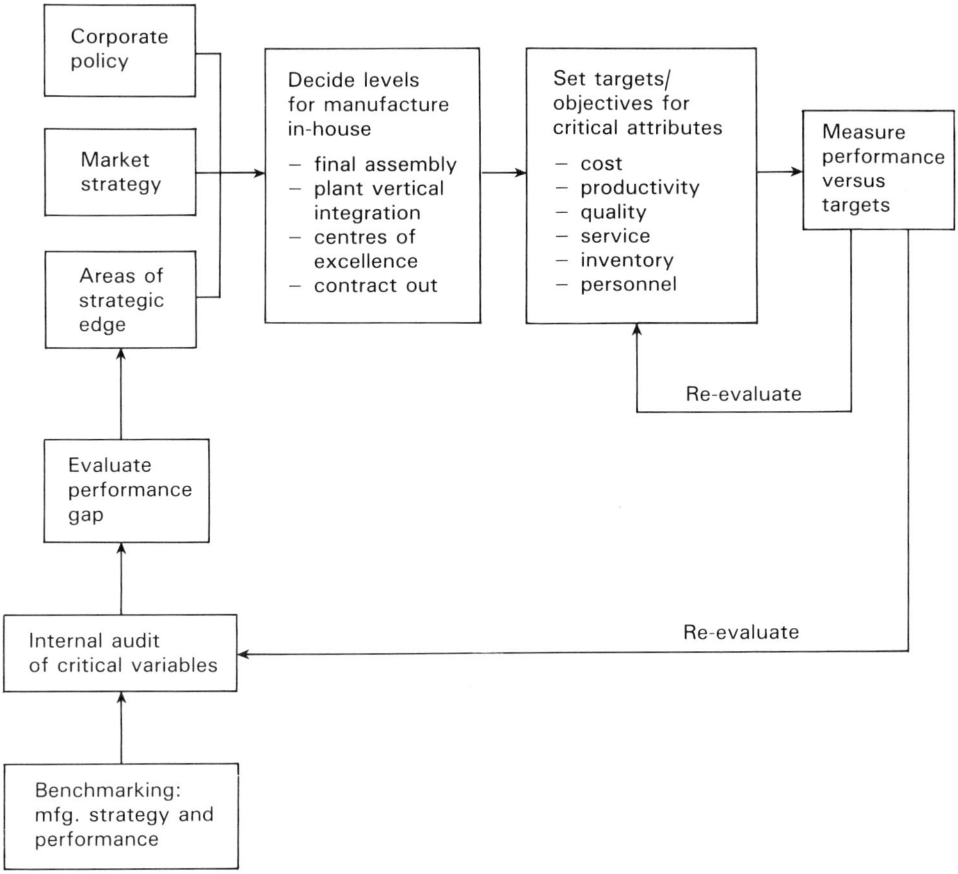

Figure 6.6 Generalized design methodology for manufacturing strategy.

7. **Periodically reappraise manufacturing systems.** This should be done against the competition and in the light of corporate objectives. Identify any perceived strengths and weaknesses and modify manufacturing strategy and design options in the DOGs where necessary.

In conclusion, the development of a coherent and explicit manufacturing strategy which reflects adequately the market requirements and internal proficiencies of the company is a prerequisite for any organization wishing to evolve and maintain market-focused production operations. Without comparison to competitors together with a structured process of design to ascertain fundamental production organization, the targets and measures to use and the components of the quality system, there will be little or no basis for judging manufacturing performance at a future point in time. This lack of performance judgement and indication of strategic direction would result in a lack of control for senior managers and cause difficulties when modifying and fine tuning the organization's approach to production.

6.5 Conclusions

This chapter has considered the process of manufacturing strategy formulation and has provided guidance upon how this may be conducted. The literature on manufacturing strategy was reviewed and seems to suggest that a strategic, that is to say top-down, and market-oriented, approach to manufacturing should be taken by managers so that quality and service become integral parts of the strategy alongside factors such as cost and efficiency. The range of products and services required by the market and the response demands in terms of time to market and order response are critical areas in which the manufacturing company should seek to satisfy its customers. This calls for flexibility within manufacturing to cope with these demands.

In developing manufacturing strategy it is important that decision-makers understand and have analyzed the main inputs or key parameters to the process of strategy formulation, most notably the corporate policy and objectives of the organization, its market strategy and an internal audit of the competencies within the manufacturing and supporting functions of the business. In so doing, an appropriate and realistic strategy can be developed and from this strategies for separate manufacturing units can be set.

The DRAMA methodology suggests that manufacturing strategy should be thought of as comprising options in three areas, namely those concerning overall manufacturing organization, the setting of objectives for operations and the development of a quality management system. By doing this, the organization will set the context for the subsequent development of new production systems and also provide the measures and targets by which the success of system operations can be evaluated once designs have been implemented.

6.6 Further reading

Ettlie, J. (1988) *Taking Charge of Manufacturing*, Jossey-Bass, New York.

Goldratt, E. M. and J. Cox (1989) *The Goal: Excellence in Manufacturing*, North River Press, New York.

Hill, T. J. (1985) *Manufacturing Strategy: The Strategic Management of the Manufacturing Function*, Macmillan, London.

7 Organization

Objectives of the chapter

On completion of this chapter you should:

1. Appreciate the nature of decisions and alternative approaches that comprise organization design.
2. Be able to identify and select appropriate organization structures from a list of options.
3. Be able to develop organizational conditions that are conducive to good operating performance across a range of different situations.

■ Chapter outline

Organization design in the DRAMA methodology represents the path between strategic decision making and the middle domain of tactical decisions. Corporate level organization design can be viewed as an over-arching activity, to then be translated down to lower levels in the organization concerned with plant organization, departmental organization and so on. An extensive body of literature in the field of organizational behaviour has been developed over a number of decades and some of the key issues in the developed theory and practice of organization design will therefore be presented in this chapter. A key observation is that structure and the organization's underlying processes and culture are key determinants of corporate performance. The various influences and implications of organizational choice are therefore considered in detail. The DRAMA methodology for organization is then presented. The intention of this component is that the designer should develop the two strands of organization identified above – structure and state. The methodology therefore comprises one design option guide for 'organization structure' and another for 'organization state', which relates to the prevailing conditions and cultural aspects associated with design choices. The end objective of this section is to assist in the design and implementation of organization structures and states that enable a high degree of responsiveness and flexibility to market demands.

■ 7.1 Theory and current practice

Organizational decisions and systems performance

It is accepted that market-focused organizations should be able to sense and interpret intelligently market and environmental factors, and for their structures to be flexible enough to adapt both to customer requirements and competitive opportunities and threats. It has also recently become clear that a strongly uniform customer-oriented corporate culture is a key determinant of market success for manufacturing companies. Associated with this, quality assurance and the various means of providing high levels of customer service should pervade decisions on organization in the same way as it influences decisions in the other DRAMA design components.

Theory and knowledge is now relatively well advanced on the subject of technical aspects of production systems design. However, the organizational issues associated with new technology are more confused and, in many cases, conflicting. One notable piece of work in this area is that of Ettlie, who calls for an adoption of 'appropriate' as opposed to merely 'advanced' product and process technologies to correspond with the development of organization design (Ettlie, 1988).

Despite the attractions of Ettlie's concept of 'simultaneous design', he provides little by way of guidelines for the manufacturing decision-maker wishing to take advantage of new technology but also aware of the importance of maintaining and developing the 'right' organizational form.

This section now considers the wider body of management literature and writings on organizational behaviour which seem applicable to the debate on manufacturing organization design. Three areas of the literature are identified: organization structure, innovation and organizational culture.

Organization structure

The major debate in organization theory has been in response to the vexed question of what exactly determines structure. Clearly there are many influences upon organizations and these differ between companies, but if the most important determinant or determinants could be identified and proved then this would provide a means of reorienting structures and production organizations towards the market. The five most widely proposed determinants are corporate strategy, size, technology, the environment and power/control. Each of these will now be explored (the five determinants of structure are considered in detail in Robbins, 1990).

The strategy–structure relationship

Strategy represents the most traditional view on what determines structure in that decisions on structure are merely seen as rational responses to the stated objectives of the organization. However, more recently, strategy has been seen more as just one of a number of influences. Some important pieces of work have studied the relationship between strategy and structure. Chandler, in a major study of the US's largest companies concluded that structure does indeed follow strategy (Chandler, 1962). Considerable support remains for this position, although more recently the generalization of his findings has been questioned as a result of his definition of 'strategy' and the sample of very large companies that he studied (Grinyer *et al.*, 1980). More latterly Miles and Snow have analyzed the strategy–structure relationship (Miles and Snow, 1978). They developed a typology which distinguishes between four types of corporate strategy: 'defenders', 'prospectors', 'analyzers' and 'reactors'. They argue that classifying organizations as one of these strategy types does have predictive properties in the range of industries they have studied. Despite these claims, a number of criticisms have been voiced concerning the strategy imperative. Most notably it has been argued that, contrary to Chandler's argument, strategy might be determined and constrained by structure and not the other way (Keats and Hitt, 1988).

The size–structure relationship

Despite difficulties in defining what exactly is meant by 'size', it has been argued that size can be a major determinant of structure (Hickson *et al.*, 1969; Blau and Schoenherr, 1971; Meyer, 1972). The general theme of the argument is that the larger an organization becomes, the more specialized and formalized the constituent parts of the business become. However, this assertion has been criticized, and it has been argued that size may be related to structure, but does not cause it (Argyris, 1972). Also, the size–structure relationship could be merely a tendency and does not necessarily represent best practice in the way organizations should design structures (Peters, 1987).

The technology–structure relationship

A major contribution to the technology imperative was the contribution of Woodward who first identified three types of production technology (unit, mass and process) and then identified relationships between these three types and the organization structure of the company (Woodward, 1965). Other advocates of the link between technology and structure are Perrow and Thompson (Perrow 1967; Thompson, 1967). Recent evidence seems to support some form of technological imperative within smaller organizations, but the effects are less pronounced in larger enterprises where there appears to be very little relationship between technology and structure (Hickson *et al.*, 1969).

The environment–structure relationship

Different organizations face varying degrees of environmental uncertainty. It has been said that structural design and change can be used as a means to reduce this (Robbins, 1990). Burns and Stalker entered the debate on environmental influence by proposing two types of organization: 'mechanistic' and 'organic' (Burns and Stalker, 1961; see also Oakley, 1984). They argued that those organizations facing stable environments should adopt mechanistic structures, whereas those in uncertain and dynamic situations should be organic in nature. Other notable authors have identified the existence of a complex relationship between the organization structure and the environment (Emery and Trist, 1965; Lawrence and Lorsch, 1967).

The power–structure relationship

There is an argument that strategy, size, technology and environment represent only a small part in determining structure. Child, in his work on strategic choice, recognized that managers have considerable scope in the decisions they make concerning organization and argued that power and control are the main determinants of organizational structure (Child, 1972, 1984; see also Clegg, 1979). It has been said that rational influences, namely the four factors mentioned above, account for only 50 to 60 per cent of the variability in structures (Robbins, 1990). The remainder are as a result of the power network and political factors within an organization.

Views are mixed, therefore, regarding the major influence upon organizational structure. However, Peters and Waterman in large part turn much of these arguments upon their heads by saying that we should be investigating the actions and major determinants in those companies seen as successful rather than passively observing structure determination in companies generally (Peters and Waterman, 1982). The argument runs that those companies with superior performance possess structures that are adaptive to change and which closely track developments in the market and environment. This view coincides with the theme of this book on the development of market-focused production systems. However, it is maintained here that the five determinants of organization structure should be borne in mind as potential and probable influences upon the organization of production systems and management structures.

The general theme in the structure literature is that any company wishing to develop market-focused operations systems should seek to base their organization around some form of 'product-focused' or, if resources are limited, 'matrix' organization, as opposed to the more conventional organizational forms based upon functional or process differentiation. Additionally the use of 'flatter' organization structures is widely advocated to replace the hierarchical and multiple-level designs historically used by conventional organizations. However, there appears not to be complete agreement on all the issues and, if one considers the organizational forms prevalent in manufacturing companies today, there seems to remain a reticence to change to more radical structures. The question remains whether this is as a consequence of inertia, or whether it represents a conviction held within companies that control is better achieved through conventional designs. Until this question is resolved, process-oriented, functional and other traditional forms of structure will remain as feasible organizational options.

As already mentioned, adaptability and flexibility are key concepts in the design of organizations and are particularly important considerations for companies wishing to reflect market trends in tastes and requirements. It has been argued and widely accepted that organizations should be willing and able to adapt to any new circumstances. Ackoff suggests that matrices of responsibilities should be used for this purpose and offers a method of 'multi-dimensional' organizational design (Ackoff, 1977). Flexibility is the other key concept, but this is open to many different interpretations. Some commentators have considered flexibility as meaning few rules and little structure (Hemphill and Westie, 1950) whereas others understand it as the ability to change existing rules and structures (Pugh *et al.*, 1968). On a practical point, Trist *et al.* (1963) provide a number of methods for the creation and evaluation of effective flexible structures. Furthermore, Atkinson sheds considerable light on organizational and labour flexibility in the modern company (Atkinson, 1984) in his model of the 'flexible firm' (see Chapter 12 for further discussion).

Innovation theory

The second area of concern in the organizational literature is innovation theory. Much of the work in the field of innovation arises from the work of Rogers, and concentrates on the diffusion of innovations within society and throughout organizations (Rogers, 1983; Clark, 1987). Recently, some more prescriptive literature has appeared, which offers ways in which companies can mobilize their resources in order to be responsive to innovative new ideas and to be able to adopt them quickly. The work of Kanter is a case in point. She argues that for companies to be successful they must be innovative in a number of ways. First, they should be able to pull resources together quickly on the basis of short-term recognition. Second, they should have a host of 'exterior sensing mechanisms' and, third, they should have more surface exposed to the environment (Kanter, 1984, 1989). Kanter goes on to say that for managers to be able to initiate new ideas, organizations must have a participative management style, loosely defined jobs that require people to stretch for power and resources, and collaboration (not suspicion) across functions. She advocates more widespread use of what she terms 'integrative' organization structures to achieve this. The integrative approach is where multi-disciplinary groups, matrix organizations and team working are the norm and replace the 'segmented' structures and practices most frequently present in functionally differentiated companies.

A number of models have been proposed that enable the change process within organizations to be analyzed and evaluated. Of particular significance in this respect is the dynamic model of innovation proposed by Utterbeck and Abernathy (1975). This model is founded on the premise that innovations and manufacturing systems design should be evaluated and categorized using a three-part classification: product, process and work organization. By understanding the type of innovation and its process, it is argued that managers are better able to understand the full extent of their implications. A number of studies have adopted the Utterbeck and Abernathy model and provide an interesting insight into organizational development and design innovation within manufacturing companies (Abernathy, 1978; Whipp and Clark, 1986).

Organizational culture

The third area of organization studies to be reviewed here is that of corporate culture. Interest in the culture of different organizations is not a new phenomenon (Jacques, 1951). However, the prescription of cultural cures for organization and process design has struck a popular chord in the 1980s and 1990s, with writers quoting and advocating the lessons learned from successful US, Japanese and British companies (see respectively Peters and Waterman, 1982; Pascale and Athos, 1981; Goldsmith and Clutterbuck, 1985).

The work of Peters and Waterman deserves particular mention here because this, more than any other contemporary management research, has been seized upon by senior executives and instilled within their style of management (Peters and Waterman, 1982; Peters, 1988). They claim to have identified those policies that successful companies adopt and which lead to superior performance. These are:

1. A bias towards action rather than analysis.
2. Simple organization with the minimum of staff.
3. Autonomy to encourage innovation.
4. A climate of trust enabling the development of consensus building.

The central argument of Peters and Waterman is that successful companies strive towards achieving excellence in their decision making and activities. This goal of perfection fits well with the Japanese-inspired goals of total quality management and just-in-time manufacture (Schonberger, 1986). The evidence from these and other texts suggests that companies with unified, strong and quality-oriented cultures tend to be those that exhibit flexible and adaptable organization structure (see, for example, Crosby, 1979).

Having reviewed the literature on organization structure, innovation theory and organizational culture this chapter will now concentrate upon presenting the DRAMA methodology for organization: its key parameters, design options and the overall process of design.

7.2 Key parameters

Identification of key parameters

Theory and current practice show there to be a myriad of determinants, influences and issues in organization design. However, this book concerns the development of market-oriented manufacturing activities and so a strongly market- and strategy-deterministic approach is advocated to the development of the organization. This position is explained by the proposition that to be successful a production system must align closely and respond to the market and technological changes in the environment, competitive pressures and the needs of customers. However, the approach within DRAMA acknowledges the need to identify and consider other influences upon the organization including power, control and politics.

The key parameters shaping organizational choices have been identified as the market and manufacturing strategies of the company developed in previous DRAMA components, plus the existing status of organization design, which acts as a restraint upon decision making. These key parameters are illustrated in Figure 7.1. The outcome of the organizational design process is the implementation of suitable corporate and manufacturing organizations. Each of the key parameters will now be considered in turn.

Market strategy

The type of market strategy adopted by a company will have an influence upon the type of organization structure employed and the style of management and processes within the business. A market-focused company will normally invoke those sort of structures and mechanisms best suited to customer satisfaction and flexibility, whereas those that stress the need for internal control will use those best suited to stability.

Manufacturing strategy

The arguments underpinning strategic determinism lead one to conclude that the strategy adopted for manufacturing operations will affect and impinge upon the decisions made on organization design.

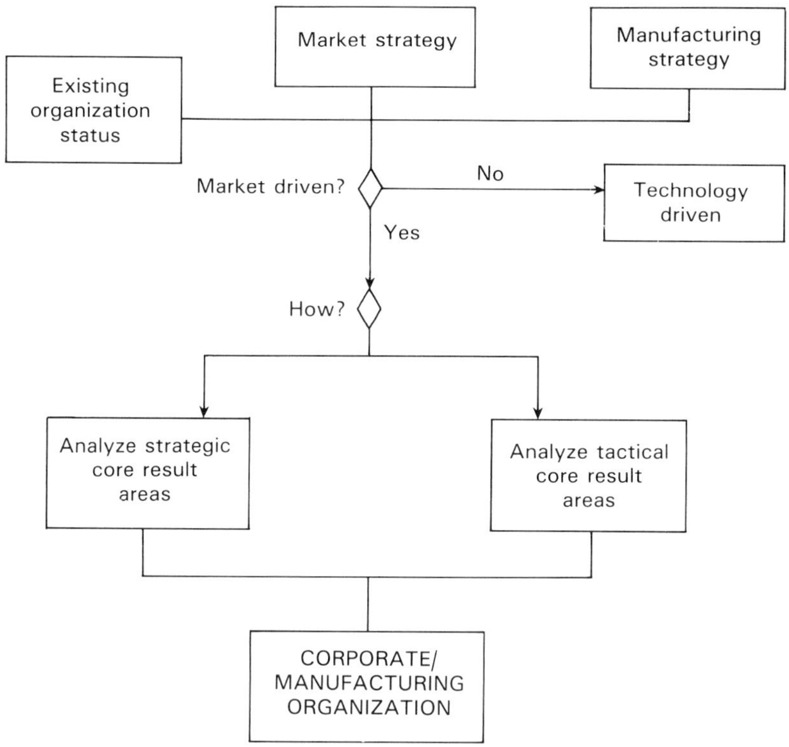

Figure 7.1 Key parameters for organization.

Existing organization status

In developing organization designs the current structure and state of the organization act as important determinants of future designs. The inertia and resistance to change provided by existing conditions can be seen as a negative influence upon the process of organizational development, but decision-makers should be aware of these current conditions and beware should they seek to radically alter the organization in case the changes infringe upon an unacceptably high number of corporate norms. The most appropriate way ahead is frequently one of evolving the organization into a form required through a series of incremental and justifiable steps.

Application 7.1: Inter-functional relationships at Rank Xerox in the early 1990s

Rank Xerox is a European Partner of the Xerox Corporation and has its main UK manufacturing site for photocopier and printing products and electronics manufacturing at Mitcheldean in Gloucestershire. The company operates in markets where there has developed intense competition from Far Eastern and particularly Japanese companies. The main competitive factor appears to be the ability to supply advanced high quality products at low prices. Throughout the 1980s and 1990s competition has intensified to the point where product life cycles, and those of their production systems, are of the order of eighteen to twenty-four months. In addition, industry-wide improvements in productivity and process technology have dramatically reduced assembly times in terms of build hours per machine.

Manufacturing within the Mitcheldean plant is therefore geared towards achieving flexibility and responsiveness to market requirements by achieving short product introduction times. There is, as a consequence, a strong emphasis placed upon the development of product-focused organizational structures, and notably a close link between Xerox factories across the world manufacturing similar products. These inter-plant relationships ensure commonality of manufacturing plant layout, processes and tooling. To complement this, a programme of quality improvement has been in operation for a number of years where importance has been placed upon developing market and customer awareness amongst all employees. Total quality at Rank Xerox covers product quality, costs and delivery but also incorporates such activities as employee involvement, continual improvement and benchmarking. The company instigated a drive for quality at every stage in the life cycle of a product and through every facet of the business: from design, through materials conformance, production processes, to employee awareness and training.

The organization structure of the Mitcheldean plant, in common with the wider Rank Xerox organization, has displayed an increased market focus and a reduced number of management levels. Also the number of managers who directly report to the Plant Director for the core business has been reduced from as many as eight to five, thus enabling Business Centre managers to have inter-disciplinary responsibilities across a number of the more traditional organizational functions. For example, the Mitcheldean Business Centre managers have their own Product Quality and Technical support staff, and in some case Materials staff, whereas previously a centralized site service function provided this capability.

For the 1990s a vertical integration strategy has seen a new Electronic and Interconnect Business Centre added to the Mitcheldean Manufacturing organization, again a self contained unit with its own dedicated Quality, Materials and Technical support staff.

Closer links between the Design and Manufacturing functions have also been addressed with multi-disciplinary teams involved in the simultaneous development of products and manufacturing processes. This initiative involves the secondment of manufacturing engineers and purchasing staff from the Mitcheldean plant to Xerox design centres across the world. The results of these changes to organizational structures and inter-functional processes and relationships have enabled Rank Xerox to remain competitive in world markets with reduced times to markets for new products, a more thorough testing and proving of product designs and build strategies prior to production, and the engendering of a right-first-time attitude among all personnel. Moreover, aligned with the total quality management programmes now deployed within the company, the creation of market-focused Business Centres enhances the relationship, via the Marketing and Field Service organizations, with the end customer – the product user. Understanding the customer requirements more fully at the design stage has enabled Rank Xerox to establish a more stable product specification. This, coupled with using a structured product development and delivery process, with agreed targets, check-points at key stages, and rigid transition criteria, means that engineering changes are minimized and products reach their maturity to plan.

Senior managers now believe that the level of customer awareness amongst employees at all levels in the business enables them to maintain their position at the forefront of the world photocopier and office systems market.

7.3 Selection and configuration of design options

Design options for organization have been categorized into two areas: organization structure and organization state. Organization structure relates to the composition of the company in terms of functional differentiation, lines of responsibility and hierarchy, whereas organization state relates to the prevailing conditions within a company, including such notions as corporate culture, values, beliefs and prevailing attitudes.

Organization structure

The DOG for organization structure is illustrated in Figure 7.2. The choice of structure needs a number of factors to be taken into account. With this in mind, the result criteria within Figure 7.2 include the need for effective lateral and hierarchical interaction of functions and a structure whereby links to the market and environment are fostered and maintained. The result will be an organization which lends itself to management control, but with minimal bureaucracy and which operates at an acceptable overall cost to the organization as a whole. Six design options are offered and the organization designer must opt for the option or configuration of options that best suit their needs. The choices are as follows:

Hierarchical structure

The hierarchical structure represents the traditional form of organizational

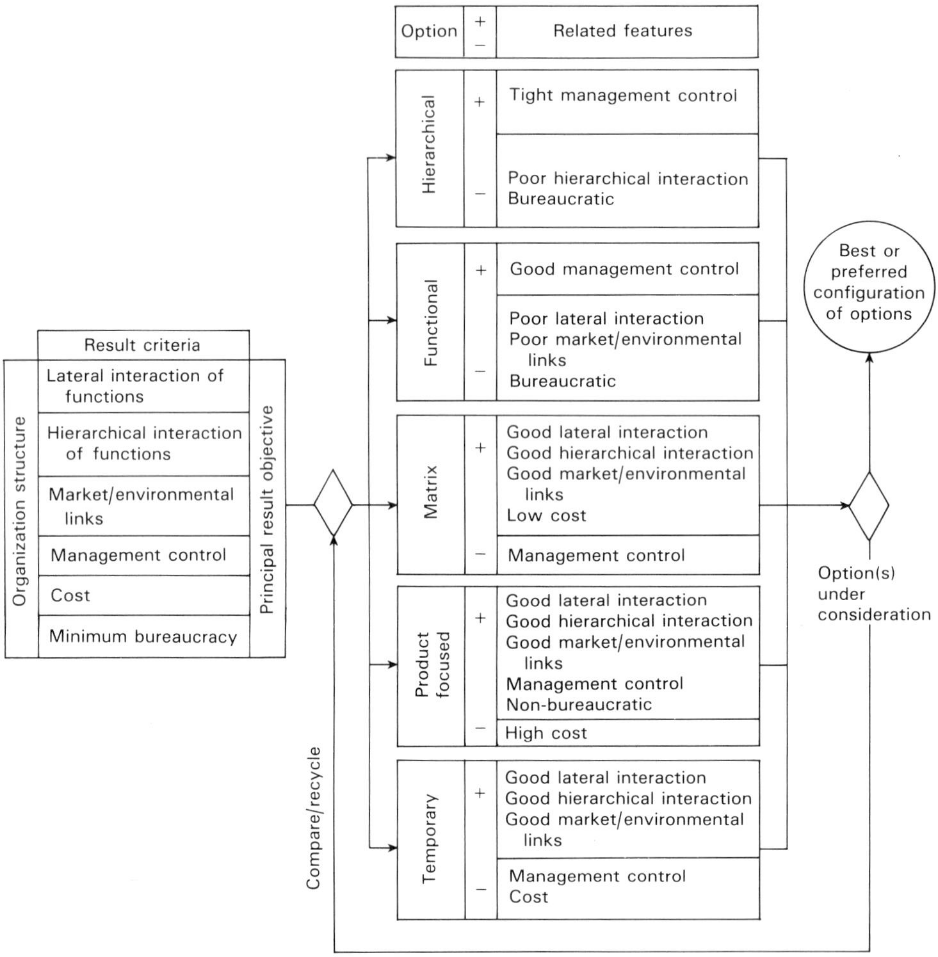

Figure 7.2 Design option guide for organization structure.

structure where the reporting system comprises a number of tiers with information and discussion usually passing from one tier to the next before being passed on once more up or down the hierarchy. The hierarchical structure can be seen to provide tight management control in certain circumstances, but can result in poor hierarchical interaction should the number of levels of management become excessive and is liable, as a result, to be bureaucratic in nature.

Functional structure

Whereas hierarchy represents a vertical segregation of the business, functionality represents a horizontal differentiation in different areas of expertise, responsibilities or operations. The main benefit is seen to be that it provides the opportunity for the effective management control and enables specialization of personnel and makes the most effective use of expertise within the organization with people concentrating upon accomplishing those tasks where they are seen as proficient. The disadvantages include those of poor organizational integration between different parts of the business and the potential for politicking and bureaucracy within the business.

Matrix structure

The matrix structure is where individuals report to two or more superiors or groups for separate parts of their overall activities. The advantages widely cited for this option are those of improving hierarchical and lateral integration through the organization as people from different functions of the business work together for part of their work time and report to a wider range of departmental heads. However, when implemented, matrix structures can result in strains upon the business in terms of management control and a potential for increased interdepartmental friction, as individuals may retain their functional loyalties and cosuperiors may argue as to who has authority in particular aspects of the individual's work.

Process-orientated structure

Different products may be best processed using different facilities, machinery, technologies, control systems and human skills. It can therefore make sense to differentiate parts of the organization on the basis of process-focus. The advantages of so doing include minimizing the duplication of facilities and tooling, the ability to concentrate technical and management skill where it is needed, and that it is easier to balance loads across plants and maintain high utilization of facilities. The drawbacks include the danger of slowing a company's response to new product development and changes to market requirements, it can result in long manufacturing throughput times and high work-in-progress, and the risk of poor co-ordination of processes and product routing.

Product-focused structure

Where a highly market-focused organization is required, a product-focused structure would seem to offer significant advantages. Usually such a structure will manifest itself in the form of a cellular organization where discrete parts of the business are targeted to deal with particular customers or with a certain family of products (see Appendix I). This would be expected to provide good links to the market through more direct relations with an identifiable number of customers. It also overcomes many of the causes for poor lateral communication that are to be found in functionally organized companies. A major caveat, however, is the cost of implementing product-focused structures in terms of duplication of resources and processes and the resultant deterioration in economies of scale in the business as a whole.

Temporary structures

Temporary groups are those that are brought together for a limited time period for a particular objective. The most common form of temporary structure is that of the project team whose members work together for the duration of a particular activity, some on a full-time and others on a part-time matrix basis, and disband upon the completion of the project. Other temporary structures include 'task forces' and 'kaizen teams'. Such structures will normally encourage interdisciplinary working and good lateral interaction, but there are caveats including the effectiveness of management control during the duration of existence and the cost of providing staff to such teams.

Organization state

Figure 7.3 shows the design option guide for organization state. When developing the conditions within which an organization operates, it is imperative that a number of key criteria are met and an indication of these are given in Figure 7.3 in the result criteria. Each company has its unique blend and emphases of desirable properties for the organization, but will normally include items such as a high degree of market orientation and an appropriate work culture, which includes flexibility to change, adaptability of people within the organization and conducive to effective communication throughout the business.

Harmonization

Harmonization is the employment of all employees on similar terms and conditions throughout the organization. This gives the appearance of democracy and can provide flexibility in people for the benefit of the business as a whole. A major question in harmonization is that of the cost of such an employment policy.

Training and development

Training and development in quality and skills to accomplish tasks is increasingly being seen as an effective means to induce a greater understanding of market and business needs within manufacturing organizations. As such, it is often used as a way of reorienting the views and attitudes of people within the company towards a culture that is compatible with its business needs. Once again, the benefits of investing in quality training must be balanced against the costs of such a programme to ensure that the right balance is achieved.

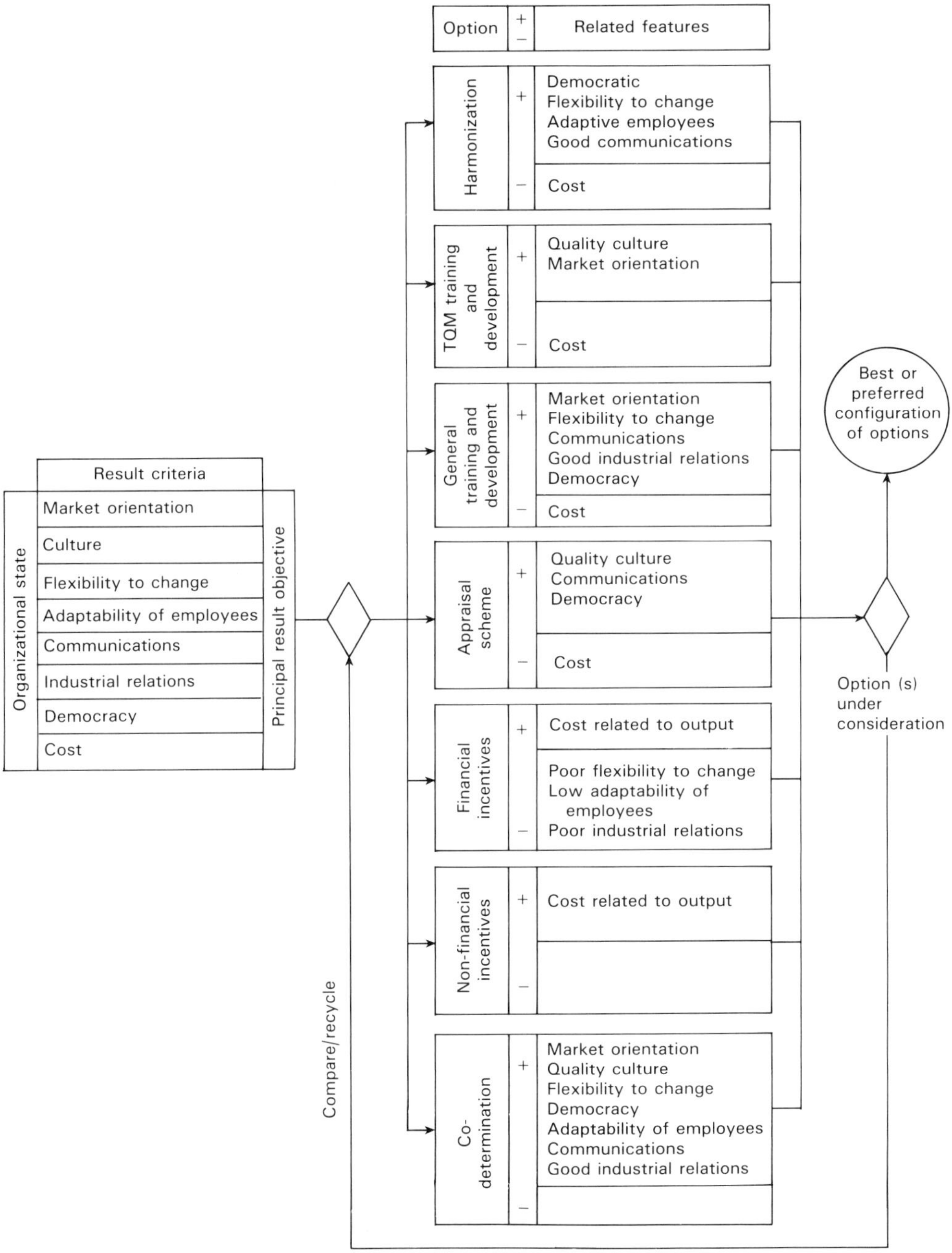

Figure 7.3 Design option guide for organization state.

Appraisal scheme

An appraisal scheme is where employees are formally assessed on an annual or more frequent basis to evaluate performance in their work and to set objectives for the next period. In addition the employee is allowed to raise issues at the appraisal meeting that may affect their current or future performance. The benefits cited from such a scheme is that it enables direct interaction between managers and all employees of the organization, improves communication and facilitates a more democratic style of leadership, but once more the cost of running appraisals is a major concern for adopting companies.

Financial versus non-financial incentives

There are many different forms of incentives, the effectiveness of which is open to debate. Financial incentives include payment by results schemes, where a direct relationship between performance and pay is realized (e.g. piecework, production bonuses) and daywork and merit rating, where there is no direct link between performance and pay (e.g. hourly-paid schemes, measured daywork, job evaluation and appraisal) but where some other form of performance assessment comes into play.

Despite its continued widespread use, pay as a means for motivation has been criticized for many years for the reported effect it has upon the workforce. It is argued that financial incentives can result in poor flexibility and adaptability of people at work to change and is often the main issue in industrial disputes. The use of non-financial incentives is increasingly being used as an alternative. These are often intangible and difficult to identify (e.g. a nice environment to work in), but non-monetary rewards can also include extra holiday entitlement, greater security, improved status, special working conditions and a company car. However, these must be seen to be fair and so need to be used with care.

Co-determination

In developing the organizational state of the business it is often thought necessary to demonstrate to employees that they do have a say in the running and decision-making processes in the company. Co-determination represents the various ways in which the workforce as a whole can be involved in policy matters. It materializes in many forms, from joint committees through to employee and trade union representation on the corporate board of directors. The benefits of democracy and communications that are seen to derive from co-determination are seen to be so important in countries such as Sweden that a minimal level of co-determination has been written into law and the statute books. However, many companies remain sceptical of such high degrees of involvement, as they feel it would constrain the effectiveness and pace of their decision-making processes.

7.4 Generalized methodology

The detail of organization design varies considerably from company to company. The flowchart shown in Figure 7.4, however, provides a generalized and straightforward method for designing and developing organizations for the advent of market-focused production systems. Although not providing precise guidance on the exact forms of organization design that should be adopted for given situations, the process of evaluation outlined should raise a number of important organizational issues and considerations for the manufacturing company:

1. **Develop objectives for organization designs.** The key parameters of market and manufacturing strategies (see Chapters 5 and 6) provide the initial inputs to determining the objectives required from the organization design.
2. **Design organization 'structure'.** In market-oriented operations one might expect the eventual structure to have a large degree of product-focus. Given this, the structure must be designed so that the degree of centralization does not hinder the need for effective functional links, particularly between marketing, design and manufacturing. Communications should be allowed to run laterally and hierarchically to support the functional

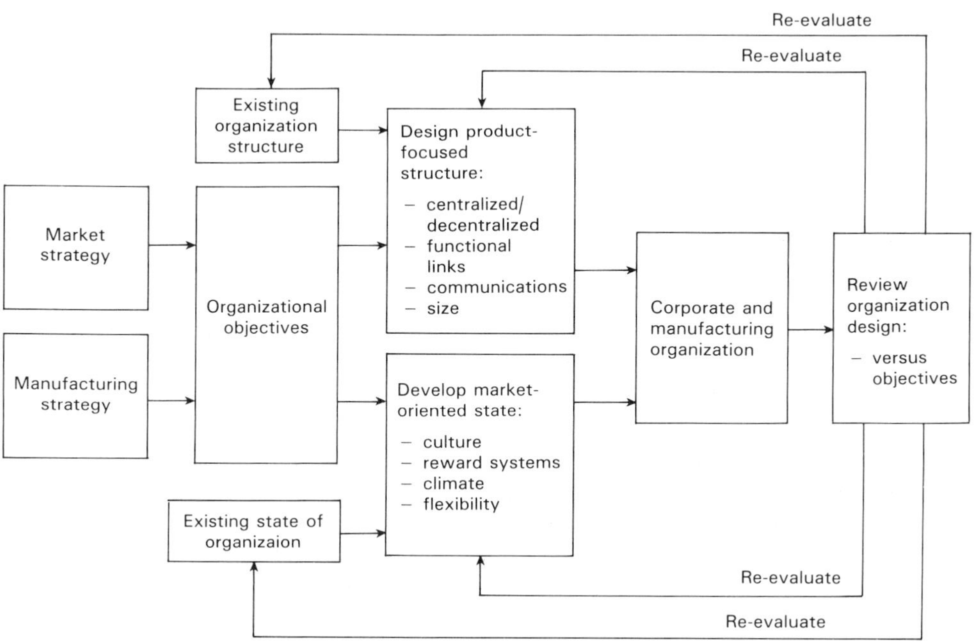

Figure 7.4 Generalized design methodology for organization.

links and overall corporate and unit sizes should be optimized to prevent the potential for bureaucracy and wasted resources.

3. **Develop organization 'state'.** Flexibility, appropriate reward systems and the wider climate within the company should reflect the need for market focus. If deficiencies occur in any of these areas attempts should be made to adjust the state of the organization through the use of training and awareness for employees and, possibly, a review of incentive schemes, away from those geared upon productivity and output to ones which more directly reflect objectives for customer service and product quality.

4. **Integrate into overall corporate and manufacturing organization.** Using the results of your evaluations in 2. and 3., develop an appropriate organization design.

5. **Review and refine organization design.** Compare the existing organization structures and state, compare with the objectives set out in 1., refining these over time if necessary, and evaluate periodically the effectiveness of the prevailing design, adjusting when mismatches are identified.

Thinking of organization design in the two areas of structure and state is a useful means of analyzing and developing both strategic (i.e. corporate) and tactical (manufacturing) organizations. Organizational development should always be ongoing and particularly so in the case of market-focused companies. The organization must be able to adapt and change frequently and at short notice in response to changes in the market and environment. A continual analysis of structures and state should pervade the organization to make this occur in practice.

Application 7.2 ICL: Organization

As has been explained in the previous sections, the 1980s were a period of turbulent change and reorientation for ICL in terms of strategic direction and technological developments. However, changing statements of strategy is comparatively easy compared with making the requisite changes in organizational structure, processes and culture. As a result, such change in ICL, in common with most larger organizations, has been more incremental in nature and tended to lag somewhat behind statements of intent and technological developments. However, taken together over a period of years, there have been significant changes in the organization of people, functions, relationships and attitudes within the company. Two developments deserve special mention here, namely the development of product-focused structures within ICL and the introduction of harmonized single status terms and conditions for employees.

At the corporate level, the move towards increased market orientation through a niche approach resulted in the establishment of product-focused 'business centres', which cut across the traditional functional boundaries of design, marketing and manufacturing. The purpose of the business centres is to interact with customers, help them determine their distinctive IT needs and mobilize the necessary design and manufacturing activities to achieve these requirements. A key feature of business centres is their desire to build up long-term and lasting relationships with ICL customers. At the manufacturing level, business centres manifested in the formation of product-focused structures where 'product centre managers' replaced the former functionally differentiated production managers.

Harmonization had been in the minds of senior ICL executives for some years, but was finally launched and gained widespread recognition in late 1987. The main thrust of the harmonization process was two-fold. First, all employees were placed on the same terms and conditions of work, which involved the abolition of hourly paid contracts on the shop floor in favour of monthly salaries for all, standard entitlements for holiday and allowances for absence through sickness, and the introduction of appraisals for all employees. Second, changes were made to the work environment and cosmetics of the factory, including open-plan offices, the abolition of reserved parking for managers, single status catering facilities and, wherever possible, attention was paid to the decor and conditions of work. The objective of these measures was to create a place of work where employees feel that they have an equal footing and purpose wherever their position on the organizational chart and, thereby, break down the traditional hierarchical and functional boundaries that one normally expects to find in the traditional manufacturing company.

The key observations in the development of ICL's organization since 1980, with particular reference to manufacturing, are as follows:

1. Business centres have been established to deal with ICL's chosen niches in its marketing strategies with the aim of improving coordination of the design, marketing and manufacturing functions.
2. Wherever possible, functions have been devolved to the lowest possible level in the organization structure. Thus responsibility for finance, personnel and quality have largely been delegated to individual plants and divisions within ICL, and a sense of ownership of these activities has been nurtured at the factory and individual levels.
3. The development of networked computer systems and the emergence of common databases have assisted in cross-functional collaboration and the sharing and interchange of information within ICL. Integrated manufacturing is seen as a key goal, both in a technical and organizational sense.
4. ICL initiated a process of harmonization as a means of developing the single status organization. Implementation has been top-down in nature and reached the shop floor level in 1988 with the introduction of monthly remuneration for all employees.

5. Modularization of the assembly processes at Ashton with the introduction of the 'cascade' system has brought with it a corresponding introduction of 'cellular' work organizations within the plant.
6. Whereas manufacturing and systems engineering personnel have traditionally been the main players in process design projects, there is an increasing recognition of the need for multidisciplinary and cross-functional approaches to systems development. The 'user' of the eventual system is now encouraged to play a key role in design.
7. A 'product introduction' function was introduced to ICL factory organization structures to improve and better co-ordinate the design to manufacturing links within the company.

Source: by kind permission of International Computers Limited.

7.5 Conclusions

Organization design is a complex and pervasive subject to come to grips with, not merely at a theoretical level but more especially at the practical level. The difficulty is that no two organizations are precisely alike and the variables underpinning structures, processes and cultures within companies range considerably between organizations. However, a number of views have developed from theory and practice which have been reviewed within this chapter and then incorporated within the methodology's approach.

Theory points to the need for organizations to sense and interpret intelligently market and environmental factors, and for their structures to be flexible enough to adapt to customer requirements and competitive opportunities and threats. However, one must recognize that there are other determinants upon structure which must be acknowledged when designing organizations ranging from technology to corporate size and politics and power. Latterly the need for a top-down and market-focused approach to manufacturing organization design has been promoted in order to generate strategic edge within markets. TQM literature points to the need to develop a strong and customer-oriented corporate culture to support the structure within manufacturing and this book as a whole certainly adheres to this conviction.

In practice, however, whilst some companies exhibit product-focused organization structures and manufacturing processes, others remain overtly functional and hierarchical with excessive management levels resulting in a lack of market orientation and adaptability. Many companies are now recognizing, however, the need to improve interdepartmental and functional links, particularly as a result of the need to cut product introduction times through the closer integration of the design and manufacturing functions of the business.

The DRAMA methodology for organization is that design should be derived from the earlier analyses of market and environment (Chapter 5) and the statements concerning market and manufacturing strategies (Chapters 5 and 6, respectively). Design should be seen as two separate but entwined parts – structure and state – both of which are strongly influenced by the historical development of the organization, political dimensions within the business and the various human aspects that impinge when people are brought together to work within groups. Implementing change in organizations is, given the above, often a difficult and perplexing activity. Despite the complexity, the designer should still aim to evolve the organization in a logical and structured manner and the DRAMA methodology provides a structured approach for this. Once the market-focused organization has been implemented, which might in itself be a radical process, the design should then be developed incrementally in the light of changes both from within and external to the organization and periodic reviews of effectiveness should be conducted.

The organization component of DRAMA represents the interface between the strategic domains of decision making and the tactical considerations of developing a case for justifying investment in new systems and project management of the system design process. It is to these components that we will now turn.

7.6 Further reading

Carnell, C. A. (1990) *Managing Change in Organizations*, Prentice Hall, Hemel Hempstead.

Mangham, I. L. (1987) *Organization Analysis and Development: A Social Construction of Organizational Behaviour*, Wiley, Chichester.

Pearson, G. J. (1992) *The Competitive Organization: Managing for Organizational Effectiveness*, McGraw-Hill, Maidenhead.

Perrow, C. (1970) *Organizational Analysis: A Sociological View*, Tavistock, London.

Wilson, D. C. (1992) *A Strategy of Change: Concepts and Controversies in the Management of Change*, Routledge, London.

8 Justification

Objectives of the chapter

On completion of this chapter you should:

1. Appreciate the current debates and concerns in the use of financial appraisal techniques to justify investment in new production systems.
2. Understand how a business-based and market-oriented justification of new systems can be performed.
3. Be able to use the justification statements as terms of reference in systems development, implementation and evaluation.

Chapter outline

New production systems requiring substantial investment need to be justified in the light of strategic objectives in order that capital can be released for their development. The justification process greatly influences subsequent developments because it is here that decisions are made whether to accept or reject proposed system developments. Proposals for new manufacturing systems therefore need to be presented in a favourable light if they are to be sanctioned by senior management, and the more a case for investment is directed toward the achievement of corporate objectives the greater its chances of success. This chapter explores the above issues by presenting a review of widely used investment appraisal techniques and some of the contemporary views that now exist concerning the justification process. The DRAMA methodology is then used to provide options and general guidance for systems designers in the justification of new production systems in relation to business and market requirements. Issues such as who should be involved in this process and the tools and techniques available are covered in this section. Finally the generalized design methodology for the process of justification is presented, which illustrates the process of moving from general and over-arching corporate and manufacturing objectives to developing a case for investment in new systems.

■ 8.1 Theory and current practice

Investment appraisal techniques

Production systems designers and managers may see the need for investment in new manufacturing facilities, but a case for this investment will still need to be expressed in financial terms within most organizations. Companies usually adopt some form of investment appraisal technique for this purpose. The basic methods for economic justification have not really changed significantly over recent years. Appraisal techniques are normally based upon comparing the initial and subsequent outlay for a capital project against the estimated financial benefits of the project in the future. Five such methods are reviewed here, these being the 'payback period', the calculation of 'return on investment', the discounted cash flow techniques of 'net present value' and 'internal rate of return', and finally the 'life-cycle costing' method. Each of these will be described and their relative merits discussed. Arnold and Hope, and Lumby, provide a detailed and comprehensive review of capital investment procedures (Lumby, 1988; Arnold and Hope, 1990), whilst Harrison gives a review of financial appraisal in relation to advanced manufacturing technology including a useful section on risk and sensitivity analysis (Harrison, 1990).

Payback period

The payback period technique is based on the time taken to recover the initial capital outlay through the cash flows generated. Essentially, the payback period is calculated according to the time into the future when the predicted income resulting from the investment pays back the initial and any subsequent capital expenditure. Companies using this method will normally have a target payback period within which time all investments will be expected to have been repaid. It is the simplest and, historically, the most commonly used method, and is useful as a first rough check to judge whether a new project is likely to be financially sound. The method has its limitations, however. First, it ignores income accruing from the investment after the payback (i.e. breakeven point) has been reached. Second, it is inadequate as a rigorous and systematic means of comparing alternative project investments. Finally, it tends to be biased against high capital investments, such as advanced manufacturing technology, where the benefits are viewed in terms of long-term flexibility and responsiveness, rather than as a quick financial return.

Return on investment

An alternative to payback period is the 'return on investment' (ROI) method of capital appraisal (Bennett, 1981). ROI is more normally used in the analysis of the overall business, but it would seem to make sense to adopt this approach in the assessment of individual investments, linking the decision of whether to approve expenditure with the wider business objectives. ROI involves calculating the ratio of average annual net benefit against the capital employed, and is usually expressed as a percentage. However, it is probably better as a yardstick for judging the success and return of past projects rather than assessing future projects. ROI does not take into account the life of the project nor the expected useful life of the investment. The timings of when benefits are expected to accrue are also excluded.

Net present value

In contrast to those techniques described above, discounted cash flow (DCF) methods take into account the life time of the investment and the times of expected benefits. One technique using DCF is that of 'net present value' (NPV). The NPV is the value of an investment expressed in present-day terms, taking into account items such as inflation and the opportunity cost of capital, which could have been employed elsewhere. The method acknowledges, therefore, that an amount of money spent or received in the future is worth less than the same amount of money today because money can be invested and used to generate interest in the future. In calculating NPV, negative values (i.e. capital outlay and costs) must be included to offset the income expected (revenues and returns). So the basic rule is that if NPV is greater than zero it is a worthwhile investment as it yields contribution for the organization, provided there are no projects competing for the same funding that have a greater NPV.

Internal rate of return

Another calculation often used within DCF analysis is the 'internal rate of return' (IRR). This is rather similar in principle to the ROI method, but takes into account the discounting factor over the expected life of the new system. The company will normally select a rate of return that is expressed in percentages which equate to the minimum rate of return it will accept from its capital investments. Any projects with an IRR lower than this will not normally be accepted. The IRR method is now widely favoured and is more commonly adopted than NPV. It gives a concise figure for an investment, which can be compared to other projects or investment options within the business. However, it only gives an overall figure which, whilst taking into account the rate of inflation and costs of capital, does not show the timings of such returns.

Life-cycle costing

Based upon the concepts of 'terotechnology', life-cycle costing is an alternative means of investment appraisal. Terotechnology involves a combination of management, financial, engineering and other practices applied in pursuit of economic gain. Life-cycle costing investigates the 'total costs' connected with an investment, including those associated with feasibility studies, research and design, maintenance, recruitment, training, and the like. These items are not normally accounted for separately in other methods and are assumed to be part of general overhead costs. Using this method, the preferred investments will be those that provide the greatest NPV once all costs included have been taken into account (Sizer, 1979). This method aims to more realistically represent the full costs and financial implications associated with an individual project. However, it is in reality often difficult to obtain realistic figures for some of these other tasks, such as the proportion of systems designers' time spent on developing the proposal. Additionally there is the argument that many such costs are already spent and 'sunk' into the expenditure of the organization, and so should not be taken into account when assessing future values and worth.

Contemporary views of justification

With the advent of advanced manufacturing technologies there has been a growing awareness that the established financial justification methods outlined above do not represent the full benefits liable to accrue from the development of manufacturing systems (see, for example, Cooke, 1986; Meredith, 1986). Considering projects merely in direct financial terms may not be the best way to judge the wider business benefits to the company. For example, increasing response to customers in terms of reducing design or manufacturing lead-times may produce benefits in terms of increased sales. Likewise, the enhanced organizational image, favourable publicity and improved product quality likely to result from the development of an advanced manufacturing system may also yield unforeseen benefits. It appears, not surprisingly, that measuring capital projects merely in financial terms may not be the best or only way to determine the likely success of a market-focused production system (Bower, 1985; Marsh *et al.*, 1988). Hayes *et al.* (1988) point out that conventional methods of justification are now quite commonly seen not as the most appropriate or sole means of assessing the worth of large projects to the organization. Harrison, however, adds the proviso that we should be careful that the procedures and rationale of financial appraisal are not undermined by a collusion between design engineers and accountants and advocates a balanced approach to justification (Harrison, 1990).

Quantitative calculations of, for example, labour savings, inventory cost savings and increased customer satisfaction are important. However, the intangible benefits offered by these, though difficult to quantify, also need to be assessed because they are likely to be significant in the longer term. A number of alternative ideas on capital appraisal and justification of advanced manufacturing systems have therefore been developed with the objective of providing improved procedures for justification. Conventional methods tend to overlook some of the intangible or non-quantifiable benefits liable to accrue from an investment. The nature of these intangible benefits depends on the manufacturing environment and the type of technology adopted, but can include factors such as quality improvements, improved manufacturing control, reduction in delivery lead-times, opportunity costs, flexibility for accommodating new product designs, different mixes and volumes, and the marketing advantages of having a 'showcase' factory (see the discussion on these points in Kaplan, 1984; Bhattacharyya and Gibbons, 1992).

The contribution of Primrose and Leonard at UMIST in Manchester is worth mentioning here. They have advocated supplementing traditional cost-saving measures with a quantification of the perceived intangible benefits (see, for example, Primrose and Leonard, 1986; Primrose *et al.*, 1985). The intangibles they considered include:

1. Improvements in product quality and the reduction in rework and scrap.
2. Shorter lead-times causing a better response to customer demand and a consequential increase in sales.
3. Faster introduction of new products giving a technical lead in the market and better economies of scale through an increase in market share.

To assist the systems designer in justifying new production facilities for which a number of intangible benefits have been identified, Primrose and Leonard have developed a computer-based tool which enables a well-structured case for investment in appropriate systems to be developed.

Kaplan also points to the need to justify new manufacturing facilities on more than an 'act of faith' basis (Kaplan, 1986). He argues that many managers have abandoned the use of conventional accounting and justification procedures because they are not seen as applicable in the case of advanced manufacturing systems where the business benefits are apparent, but difficult to precisely quantify. However, to make critical decisions on investments without financial information seems to Kaplan to be a very risky approach. He puts the case for the intelligent use of financial appraisal techniques in the justification of new systems. Inherent in this is the need for common sense to prevail in the mind of the decision-maker in considering both financial analyses and the business issues such as increased flexibility and shorter throughput times.

From a practical perspective, it now appears as though most companies use only an outline design concept for a new system, together with an analysis of the reasons for choice, when running through the justification procedure. These capital approval papers normally contain estimates of the times to perform activities and the relevant budgetary plans for such development. In addition, some companies are coming to a realization that there are virtues in spending more time and effort prior to the submission of a capital approval paper in carrying out detailed design work, so obtaining better estimates of the time-scales and benefits (for further discussion on some of these issues see Slagmulder and Bruggeman, 1991).

Standard justification techniques based on payback periods and DCF methods are widely employed by industry in the justification of new manufacturing systems. They provide a sound foundation on which to quantify the expected savings by means of financial cost–benefit analysis. However, in industries where labour costs are relatively low in comparison to material cost, greater emphasis is placed on working capital savings from reduced inventory levels. There is also a growing awareness that the benefits from adopting advanced manufacturing systems are not confined to the short term (Cooke, 1986; McLean, 1988). Account is also being taken of the intangible benefits likely to result from the proposed development, although difficulty still exists in their quantification. It is nevertheless realized that the inclusion of intangible benefits means there is a better chance of the proposal gaining approval. Close co-operation between engineers and accountants is also considered to be essential in order that the proposal can provide a concise description and analysis of both the financial and engineering benefits expected.

The degree of systems design detail included within the justification proposal normally depends on the level of investment involved. Alternatives are first screened and analyzed and, in order to give weight to the case, reasons are given for their selection. In some cases, due to the requirement to submit proposals against tight deadlines, the degree of design detail can be low, but the proposed development can still be presented concisely, with a projection of time-scales and budgets. Experience shows that, although approval may be gained by this method, the project may then suffer from the development team being pressurized to conform to a projection which may not be realistic. This can have adverse consequences on the design and implementation phases. Some companies are beginning to realize the virtues of spending extra time in carrying out detailed design work, allowing more realistic projections of the time and budgetary plans to be estimated. This may minimize the degree of pressure encountered at the postapproval stage.

So, the justification process not only involves seeking approval to proceed with systems design, but also provides a 'capital paper' to serve as a 'specification' for design. Justification within DRAMA is seen as occurring with the tactical domain of organizational decision-making: it involves a high degree of interaction between middle and senior management. The procedure requires acknowledgement of corporate objectives, together with a determination of both tangible and intangible benefits.

Application 8.1 ICL: Justification

The process of justifying new production systems within ICL tends to follow a set procedure of completing a capital paper and then submitting this to senior managers, who take the decision to accept or reject the proposed project. The amount requested determines the level at which approval must be sought with large project expenditure proposals requiring acceptance from

ICL's main board. Capital papers within ICL are normally developed jointly by manufacturing engineers and financial specialists, the latter having the task of quantifying the perceived benefits offered by the system and converting this figure into payback periods and rates of return.

The details of design are usually relatively limited at this stage and provide an outline as opposed to design specifics. Engineers in ICL are told to justify new systems using a business rather than a technological case to ensure that the degree of fit between proposed system designs and corporate objectives can be assessed. This results in capital papers not merely identifying the more immediate quantifiable benefits in terms of reduced manufacturing costs, headcount savings and inventory reduction, but also a statement of the harder to quantify benefits including improved service levels, product quality, increased flexibility, etc.

In the case of separate cells within the modular assembly cascade at Ashton, implementation was expected to contribute to the achievement of manufacturing and market plans within ICL in terms of providing products at specified levels of quality, reliability, cost and delivery performance. The main benefits quoted for investment, which were accepted by the ICL main board as proving an adequate business case, included clear quantified expectations in terms of:

1. Reduced inventory holding.
2. Improved labour efficiency.
3. Improved space utilization.
4. Reduced product introduction times.
5. Improved quality through tighter control of processes.
6. Manufacturing cost reduction as a result of 1. to 5. above.

More detailed comments on the process of justification within ICL are now given:

1. The justification of early modules within the cascade were based largely upon inventory and indirect labour cost savings. The capital papers of later modules laid more emphasis upon intangible, but important, business benefits such as collapsed product introduction times, greater flexibility and improved levels of service.
2. The change from being technology-led to market-driven in corporate approach influenced and changed the parameters upon which investment proposals were developed and evaluated. There resulted increased uniformity in the cases made for process investments with the benefits quoted frequently reflecting the identified business drivers of productivity, cost, quality, inventory and service.
3. From the situation where only a broad outline of proposed developments was presented, ICL now realize the value of a more comprehensive description of systems design and functionality, which can act as a

reference model both for systems development and post-project investment appraisal purposes.

4. The phased introduction of new process designs is a feature of ICL capital papers. Project phasing is based upon various criteria, including minimal outlay at the initial stage to allow manual proving of systems concepts and basic operation, manual operation as a step prior to investment in automation, and the incremental and lower risk spend of capital with potential step-off points during the introduction of new systems.

Source: by kind permission of International Computers Limited.

8.2 Key parameters

Identification of key parameters

The key parameters seen as shaping this procedure are shown in Figure 8.1. They comprise corporate policy, market strategy, manufacturing strategy, financial strategy and product/process technology. The influence that each of these has upon the justification process in the production system design process is now considered.

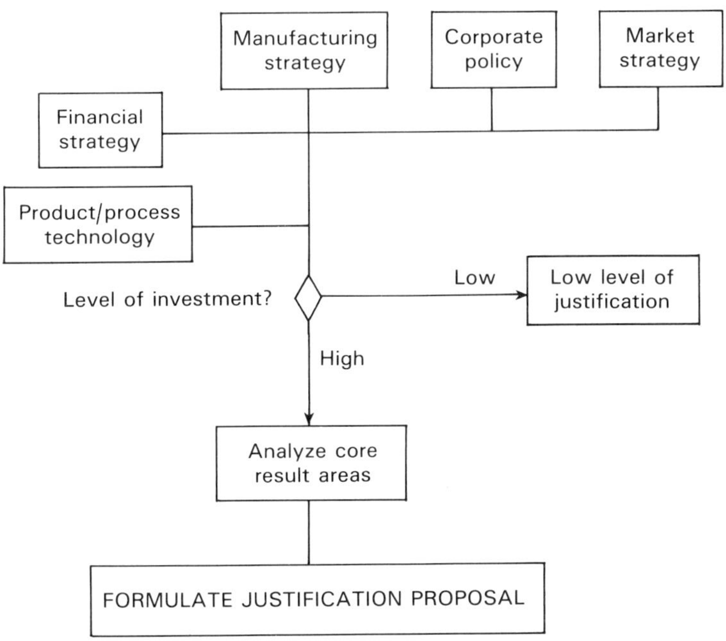

Figure 8.1 Key parameters for justification.

Corporate policy

The justification of a major investment in new production facilities must necessarily take into account the corporate policy developed in the market and environment component of the DRAMA methodology (Chapter 5). The responsibility here is two-fold. First, the systems designer, in developing a case for the investment, must consider the objectives of the business in the longer term and should develop a 'capital paper' in line with goals relating to profitability, growth, customer service and any other intentions that the business as a whole may have. Second, the senior managers responsible for accepting or rejecting the project, must ensure that any expenditure sanctioned does not conflict with the long-term objectives and that the money is to be invested in the most beneficial manner.

Market strategy

As with corporate policy, above, the case for investing in new processes must be in line with the organization's aspirations and strategies in the market. Thus, if the market strategy is one of competing on the basis of price, then the production facility should be capable of producing at lowest manufacturing cost. Likewise if the strategy is one of responsiveness to customer orders or flexibility to customer requirements, the manufacturing system should enable short manufacturing lead-times or be adaptable enough to cope with variety.

Manufacturing strategy

The manufacturing strategy developed in Chapter 6 provides the context within which operations and all decisions regarding production systems design should take place. Being a distillation of corporate policy and market strategy, the manufacturing strategy is in effect a translation of corporate objectives and market needs into the language of production. Therefore, any system design proposals that do not reflect manufacturing strategy should normally be rejected, unless it can be proved that the system might present areas for competitive advantage not previously envisaged in the strategic plans of the company.

Financial strategy

The financial strategy, if one exists within the organization, should act as an important determinant in the choice of investment and must be adhered to at the stage when systems designers are developing a case for approval. The financial strategy may contain guidelines on what types of investment projects are most favoured, the maximum payback period allowed, corporate internal rates of return and returns on investment expected, together with a corporate view on the treatment of such items as intangible benefits. The tactical managers of an organization, in preparing capital papers, should be conversant with the financial strategy prevailing at that time as it is often a key determinant of whether a proposed project is given approval.

Product and process technology

When formulating capital papers, some thought must be given to the available technology in the design of processes and products. In forwarding the case for justification, managers and systems designers should be confident that, should the funds for their project be sanctioned, they will be able to deliver what has been promised. The figures detailing costs, benefits and advantages to the business will have been based and probably accepted by senior decision-makers on the assumption that the project is technically feasible. Should this not be the case, the projected figures will be undermined.

8.3 Selection and configuration of design options

The justification component of DRAMA includes one design option guide, which is illustrated in Figure 8.2. As shown in the result criteria, the justification process should seek to align any solutions that emerge with the objectives of the business. It should also seek to minimize risk, time and cost wherever possible and be able to provide analyses of the potential benefits of emergent system designs including aspects such as efficiency, quality, control, product introduction times and flexibility. Bearing in mind these result criteria, the design options shown in Figure 8.2 will now be discussed.

Multidisciplinary development

Using people from different departments and functional backgrounds within the organization enables emergent systems to be justified from a number of perspectives and so adds to the depth of argument for any new investment. This also facilitates a high degree of consensus from different areas within the business due to widespread participation in the process compared with the situation where investment proposals are developed by a limited number of people in the company. The main drawbacks of the multidisciplinary approach

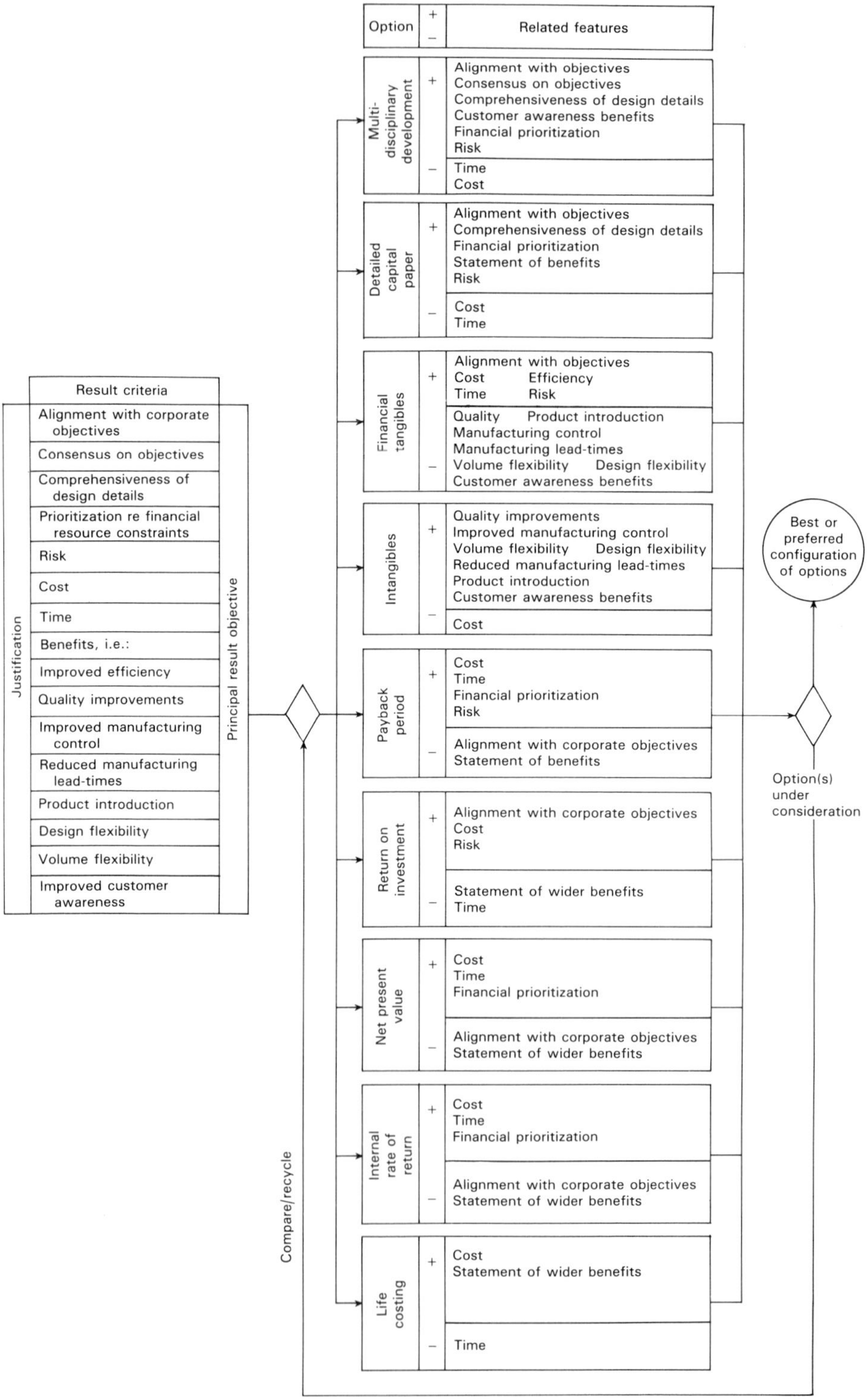

Figure 8.2 Design option guide for justification.

are the cost and time of involvement and, when delays occur due to fundamental differences, the time for the justification paper to develop.

Detailed capital paper

A detailed capital paper will provide comprehensive coverage of the different aspects and considerations of the new system. Likewise it will provide an initial benchmark and statement on expected benefits against which the performance of the new system, once in operation, can be compared. Once more these benefits must be balanced against the time and cost of development.

'Financial tangibles' and 'intangibles'

Any justification paper should contain quantitative analyses of the new system in terms of the expected reduction of costs of manufacture, efficiency improvements and the like. However, not all benefits are best quoted in financial terms and, sometimes, this is very difficult to do in any case. Therefore the statement of financial benefits must normally be complemented with statements on the intangible and non-financial benefits of new systems including the potential for increased flexibility, faster response times and improved quality and service.

Method of investment appraisal

Five methods of investment appraisal were presented earlier in this chapter and these represent options available for use in the justification process. Each has its advantages, but also drawbacks. Payback period offers good coverage on aspects such as cost, time, prioritization of alternative investments and, because it is geared to return on the initial outlay, is a measure of risk of the investment (generally speaking the shorter the payback period the less the risk). However, the other options, such as ROI, NPV and IRR discounting methods, and life costing, take into account the level of total return for the investment and so provide a more accurate representation of the true value of a proposed new system.

■ 8.4 Generalized methodology

The generalized methodology for justification is a multiple step process, which results in a capital paper for senior management consideration. Illustrated in Figure 8.3, the methodology is as follows.

1. **Define problem/requirements from new systems.** Consider any existing manufacturing facilities and relate them to the statements made in the

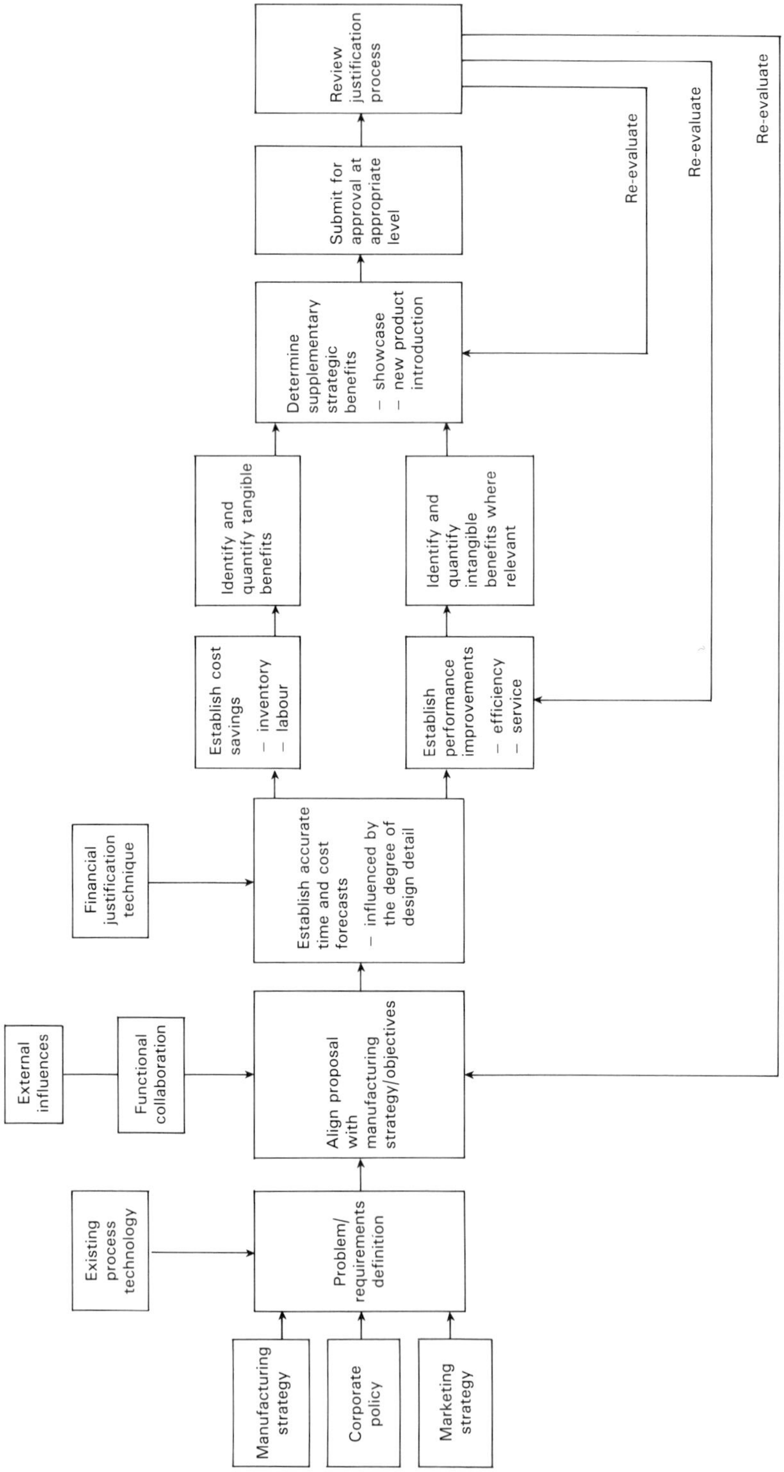

Figure 8.3 Generalized design methodology for justification.

corporate policy, market strategy and manufacturing strategy. Evaluate the gap between current system capabilities and performance and the business requirements of the system. If there is no existing system, compare the business requirements with the expected results of the new operation.

2. **Align capital proposal with strategic objectives.** Using a process of functional collaboration throughout the organization, develop the basic proposal in line with the business needs identified in 1. above.
3. **Establish time and cost estimates.** Establish time and cost forecasts both for the development of the new system and for its operation once implemented. The way in which this is done, and the information required, will vary depending upon which investment appraisal technique has been adopted.
4. **Develop tangible benefits.** Establish the associated financial costs and returns over the duration of the project and of the proposed system once in operation, and thereby establish the tangible benefits expected to derive from the investment.
5. **Develop intangible benefits.** Determine any potential benefits which cannot be readily quantified. Analyze these and develop a schedule of advantages and disadvantages expected to result from the new system. Review these once more in an attempt to quantify the benefits if at all possible (cost being a common denominator or the 'bottom line' [Kaplan, 1986]) and move quantified items into the statement developed in 4.
6. **Determine supplementary strategic benefits.** Identify such items as the marketing benefits of having a 'showcase factory', greater efficiency of new product introduction, etc.
7. **Submit capital paper for approval.** Using the data collected and stated in 4., 5. and 6. develop a comprehensive business case for investment in the new production process and submit to senior decision-makers for approval. If accepted, the project can now proceed.
8. **Review justification process.** In the light of experience when moving from 1. to 7. there should be some views on the part of all those participating on the efficiency and appropriateness of the justification process within the company. The opportunity should therefore be taken to establish the views of participants and so improve the standard and efficiency of internal practices within the organization.

Application 8.2 Accounting for CIM at Allen-Bradley

Manufacturing and accounting experts have contributed much to the knowledge and theories on justifying advanced production technologies, yet still the process remains intimidating. Traditional accounting methods lose relevance when applied to manufacturing systems that are flexible and reusable. Some of the most valued benefits of Computer Integrated Manufacturing (CIM) are not quantifiable by existing accounting methods, nor are there proven new ways of evaluating them. At this point the justification of CIM remains as much a strategic issue as a financial one.

Traditional justification models were not designed for today's CIM systems, which can promise benefits such as improved staff productivity, better capital resource productivity, higher quality, economies of scale, and a rapid response to the marketplace. It is a major challenge, therefore, to measure those benefits, which are difficult to quantify. Existing models, like return on investment and internal rate of return, focus on the time value of money and give little thought to the strategic opportunities and threats that technological advances present.

It has become clear that management should rethink its guidelines for justification procedures. CIM requires new ways of looking at investments. For years management made long-term financial projections. Successful investments must yield returns greater than the cost of the capital invested. It is important, however, for CIM to be evaluated in the light of long-term objectives. In other words management's viewpoint must change: it should start with a strategy, not an investment. Unfortunately the IRR and NPV models are project oriented, not strategy oriented. To survive, firms have to look beyond tactics.

Many companies have fought their way through the justification barriers and implemented CIM with surprising results. One example of successful CIM implementation is at Allen-Bradley's world contactor assembly plant in Milwaukee. The move was prompted by the threat of merging international standards for motor controls to one of Allen-Bradley's core businesses. To remain competitive it not only needed a new product, but also one that could be sold profitably. Off-shore sourcing was one option, but this was quickly discarded, as all the engineering skills were already in place at Milwaukee. The solution was a new product and a new production line, developed by design and production engineers working as a team. Other experts were assigned to the project from quality assurance, MIS, marketing and purchasing.

The objectives of the project were to develop a world-class product and production facility with the following goals:

1. High volume of up to 600 products manufactured each hour.
2. 125 different product variations in batch sizes of one.
3. Products that could be built-to-order in 24 hours.
4. To locate the new CIM production facility within the realms of the existing manufacturing plant.
5. To keep labour costs to a level that was competitively insignificant.

The result was that not only did Allen-Bradley meet all its objectives, but today it now has an integrated CIM facility whose relative cost per unit is 60 per cent that of machine-assisted labour. The system has already demonstrated that it can respond to market demands. Instead of the original 125 product variations it now offers 837 variations, in batch sizes as low as one. Relative return on assets is five times what it would have been with traditional manufacturing. And the plant is consistently producing higher quality contactors at a lower unit cost than any other supplier.

Source: adapted from Tracy O'Rourke (1990). Accounting for CIM, *Industrial Computing*, **March**, 21–27.

8.5 Conclusions

This chapter has argued the need for a business-oriented justification prior to the development of any new production system. The virtue of going through this process is:

1. To clarify the worth of the new system to the business.
2. To provide information for senior decision-makers upon which they can make a decision whether to accept or reject proposals for systems development and new investment.
3. To provide 'terms of reference' for the anticipated benefits against which to judge the eventual success of implemented systems.

A number of conventional investment appraisal techniques, frequently used in the justification process, have been reviewed, including the 'payback period', 'return on investment', 'net present value', 'internal rate of return' and 'life-cycle costing'. However, in many cases, these established financial methods do not fully allow for the scale of benefits likely to accrue from large investments in advanced manufacturing technologies. Their main drawback is that they consider the value of investments merely in direct financial terms and ignore the intangible and indirect benefits likely to accrue from such effects as increased quality, customer satisfaction and reduced design and manufacturing lead-times. It now appears that many companies, rather than merely accepting investment in advanced new systems on an 'act of faith' basis, are seeking to develop justification or capital investment papers, which outline both the quantitative and intangible benefits and so enable the senior decision-maker to make a more balanced and informed judgement on whether to accept or reject proposals for new systems development.

The process of justification is carried out within the tactical domain of the organization and should involve interaction between strategic decision-makers and systems designers, co-ordinated and channelled by senior to middle level manufacturing and engineering managers amongst others. Close collaboration between managers, engineers and financial specialists should take place in the development of any capital proposal in order to establish the financial, technical and market consequences likely to arise from the design and operation of the new system. This cost–benefit information should then provide senior managers with an adequate foundation upon which they can make valid and informed decisions and judgements.

8.6 Further reading

Lumby, S. (1988) *Investment Appraisal and Financing Decisions*, VNR, Wokingham.
Saaty, T. L. (1980) *The Analytic Hierarchy Process*, McGraw-Hill, New York.

9 Project Management

Objectives of the chapter

On completion of this chapter you should:

1. Have an understanding of the basic process and techniques available for the management of systems design projects.
2. Appreciate the major management and organizational issues involved in the planning and control of systems design projects.
3. Have available a series of alternative approaches and options for use in developing new systems.

■ Chapter outline

Effective project management is the key to success or failure in any production systems design programme. It represents the translation of strategies and plans into practice and operation. Although project management is not only the organizational interface between higher levels of management decision making and the operational domain, it determines to what extent the technological design is effectively converted into a working system. This chapter reviews the process of project management. After giving an insight into what comprises the process of project management, the techniques available to assist in the planning, scheduling and control of projects are presented. These include 'Gantt charts' and the network analysis techniques of 'critical path method' (CPM) and the 'programme evaluation and review technique' (PERT). The role and responsibilities of the project manager within the organization are discussed, and then project management in practice and the concept of the simultaneous engineering of product and process designs are covered. The second half of the chapter presents the DRAMA methodology for the project management of new market-focused production systems. It is shown that rather than being proficient with the widely publicized tools and techniques of project management, the most important skill for the manager is the ability to draw together the many different functions of the business for the implementation and support of new systems. Design options for the management and process of project management are given in two areas, namely the management of the project and, second, in relation to the available tools and techniques. The chapter concludes with the generalized methodology for the planning, scheduling and control of production systems design projects.

9.1 Theory and current practice

Perspectives on project management

Project management is critical in establishing whether a project will be a success. Areas of management responsibility have been defined as planning, control and implementation. The initial stage of project management consists of a feasibility study (Morris, 1983), where a clear definition of the goals and ultimate benefits needs to be established. Top management support for project management is important for giving authority and direction, and to ensure that the goals of the organization are achieved at the end of the project. The form of support can also influence the degree of resistance the project encounters (Manley, 1975).

Effective project planning and the adoption of adequate monitoring and feedback mechanisms are critical in establishing whether the implementation of a new production system will be a success. The literature advocates the need for user consultation during the design and implementation process, where the degree of user involvement can influence the ensuing support for the project. Knowledge, skills, goals and personalities are all characteristics that must be considered in assessing the environment of the organization (Boer and During, 1987). The project team should possess the necessary technical skills, as well as having access to adequate technology to allow them to perform their tasks.

Effective project plans, which detail the stages of implementation are important for successful development. Nutt has proposed that the stages of project management should be seen as those of formulation, conceptualization, detailing and evaluation (Nutt, 1983). More commonly, however, the three stages of project planning, scheduling and control are seen as comprising the process of project management (see, for example, Cleland and King, 1983; Moder *et al.*, 1983; Heizer and Render, 1988). Each of these three stages will now be considered in turn, together with the tools and techniques appropriate for use at each stage.

Project planning

Project planning is the set of 'front-end' activities that precede commencement of a project. Once the need for a new production system has been identified, the planning of the design project should start. This involves the setting of project objectives, organizing the project team, the development of a clear project brief, which defines the project's scope and the responsibilities of team members, and a rough estimate of the time and cost of anticipated project activities. The project planner makes use of a number of sources of information during this process, including budgetary and cost data, strategy and justification documents, personnel records and engineering diagrams. Often the first estimate of project duration and activities is done using a Gantt chart (commonly known as a milestone plan), which is in effect a series of horizontal bar charts that provide an easy-to-develop means of representing the project in a form that can be readily understood by everyone involved in the organization.

Project scheduling

Once general agreement on the overall plan has been agreed, the project manager can proceed to develop more accurate schedules for those activities that need to be carried out. Project scheduling involves the allocation of resources in the form of people, finance and equipment to individual activities. As the project commences a more detailed relationship between activities will develop. This will involve setting more accurate and realistic times for activities and a regular updating of the schedule as the project itself commences.

The tools appropriate at this stage include cash flow estimation and, once more, Gantt charting. However, these techniques are somewhat limited for more complex projects of the type one might expect for the introduction of new production systems. For this reason the network analysis tools such as the 'critical path method' (CPM) and 'programme evaluation and review technique' (PERT) were developed in the 1950s to provide project managers with a more powerful and flexible means of first scheduling and later controlling projects. The development of a project network diagram of the type shown in Figure 9.2 is common for both the CPM and PERT methods whereby events (points in time during the project) are represented by circular nodes and activities are shown as arrows. Within the nodes the 'earliest start' and 'latest start times' for the following activities are given and so the route of critical activities through the project can be identified.

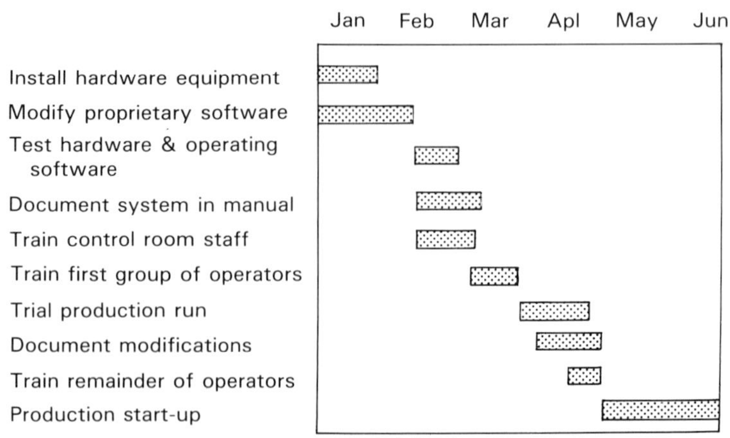

Figure 9.1 A simple Gantt chart for use in project planning.

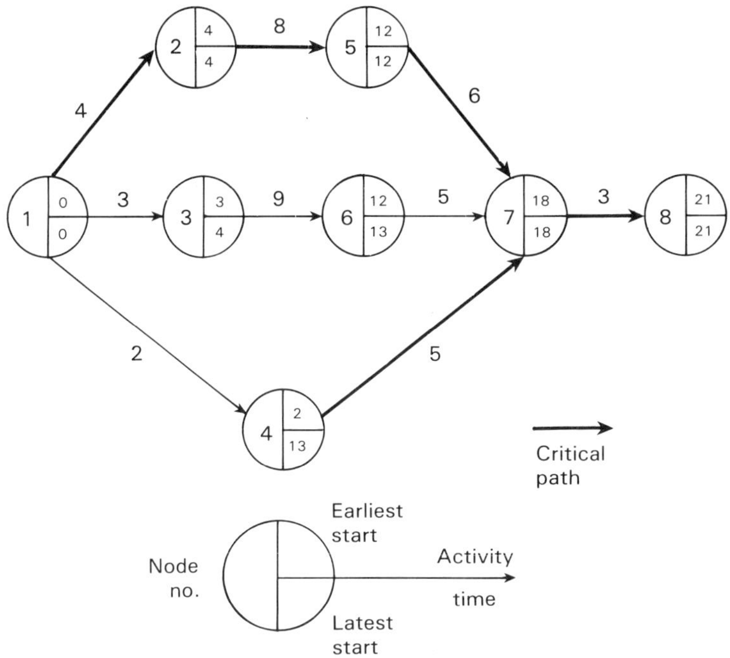

Figure 9.2 A project network for use in CPM and PERT analysis.

Heizer and Render (1988) have identified six common steps in CPM and PERT:

1. Define the project and its significant activities and tasks.
2. Develop relationships among activities and decide upon the order and precedence of these.
3. Draw the network connecting activities.
4. Assign time and/or cost estimates to each activity.
5. Compute the longest path through the network by performing by calculating the earliest and latest start times: the so-called 'critical path' is where these are equal.
6. To use the network to help schedule and control the project.

The fundamental concepts and techniques of project management have now been presented. A more detailed appraisal of the use and benefits of CPM and PERT can be found in those references quoted so far in this chapter plus others (including Kerzner, 1984; Meredith and Mantel, 1985). It is, therefore, to the third phase of project management that we now turn – project control.

Project control

The aim of project control is to ensure that the project proceeds according to the overall plan and the more detailed schedules that have been developed. It involves monitoring costs, time achievements, resource usage and human performance, revising and changing schedules as and when necessary. It may also concern shifting resource allocations over time to meet the time, cost and design objectives set for the project. Project control requires adequate monitoring and feedback mechanisms at each stage of the project, by these means key personnel can compare progress with initial projections. This enables the project manager to anticipate problems, take corrective action, and ensures that no deficiencies are overlooked (Pinto and Slevin, 1987; Toft, 1988).

For less complex projects use of Gantt charts and a review of milestone attainment may be sufficient for control, but increasing use is now being made of computerized CPM and PERT packages. PERT in particular offers a powerful means of controlling projects and can be used to give:

1. Probability-based estimates of completion time as the project proceeds.
2. Details of costs of the project on an ongoing basis and information regarding the additional costs of completing activities in reduced times.
3. A post-audit of projects both as a means of evaluating individual design processes and for improving project management techniques within the organization in the future.

The project manager

The responsibilities of the project manager have been identified as follows (Bowenkamp and Kleiner, 1987):

1. To plan thoroughly all aspects of the project, soliciting the active involvement of all functional areas involved, in order to obtain and maintain a realistic plan that satisfies their commitment for performance.
2. To control the organization of manpower needed by the project.
3. To control the basic technical definition of the project, ensuring that technical versus cost trade-offs determine the specific areas where optimization is necessary.
4. To lead the people and organization assigned to the project at any given point in time. Strong positive leadership must be exercised in order to keep the many disparate elements moving in the same direction in a co-operative manner.
5. To monitor performance, costs and efficiency of all elements of the project and the project as a whole, exercising judgement and leadership in determining the causes of problems and facilitating solutions.
6. To complete the project on schedule and within cost, where this is the overall standard by which the performance of the project manager is evaluated.

The role of the project manager is therefore a key determinant in the outcome of a project. The manager must be carefully selected and, once in post, must act as a champion and facilitator for the rest of the project team and the eventual users of the new system. The project manager must co-ordinate designers and users to ensure that the eventual system design matches those objectives explicitly stated at the justification stage.

Project management practice

Approaches adopted in the project management of new manufacturing systems vary considerably, but can be categorized broadly into either 'autocratic' or 'democratic' in style. Most companies surveyed during the course of the authors' research in developing this book adopted a fairly autocratic style of project management. An autocratic style usually influences the traditional, and most popular, approach where a project team is established comprising system design specialists who bring in expertise when necessary from other parts of the business, or who come from external sources such as consultants or equipment vendors. User requirements are usually articulated at the outset and checked on an on-going basis, but users themselves often do not participate in the design process. Once the new production system is designed, the autocratic style will normally lead to it being 'imposed' on the production area, although this is often preceded by presentations of the system and its operation and purpose to the end users.

In contrast, there has been a growing awareness of the importance of user involvement in the design process, especially for the introduction of advanced manufacturing systems, due to their potentially wide ranging impact. Thus, efforts towards a 'democratic' style of project management have been attempted through the involvement of users on a regular basis during the design phase. This can increase the development time but experience has shown that imposition of a solution can result in a reluctance for it to be accepted; imposition may also lead to complaints and suggestions for change thereafter. A more democratic style, through participation, does not necessarily mean that everyone will always get their own way. However, it does allow users to have a say in the design, as well as having a better understanding of what is going on. This type of approach is useful in breaking down any barriers that may exist. Thus, although a participatory approach may mean that the design process takes longer, it can result in a shorter implementation time, being more readily acceptable by production personnel. The result is also likely to be closer to the 'optimum' design. So the design of an 'optimal' manufacturing system involves not only the use of functional specialists, but also complete consideration of user requirements. Although a total participative approach is desirable in project management, a balance must be achieved between obtaining a consensus and a degree of design imposition. The main factors in establishing the style of project management to be adopted include organization and justification, and so these form the key parameters for the project management component of DRAMA.

Application 9.1 ICL: Project management

The development of the modular assembly cascade at ICL Ashton placed challenging demands upon the company's personnel to ensure its successful design and implementation. A two-tier project management structure was adopted, which can be seen as representing the tactical and operational domains of decision making. The top tier was that of the 'project board', which was, in effect, the steering committee comprising senior plant and system development managers. The second tier was that of 'working parties', comprising system design engineers and the users of systems from various functions within the plant.

Project board meetings were held on a regular basis, typically on a fortnightly cycle, and were formally chaired by a senior user, usually a Product Centre Manager or the like. The terms of reference for project boards were as follows:

1. To monitor the performance and progress against agreed milestones and report to the Plant Director on these.
2. To review vendor performance against the contractual agreement.
3. To review actual capital expenditure on projects.
4. To resolve issues referred to the Board by working parties.

The working parties were involved with the detailed design and implementation of specific elements in the overall production system. Membership of the working parties differed from one group to the next, but would usually include a core from the manufacturing engineering and manufacturing systems functions, plus personnel drawn from various user groups, including materials control, production management and supervision, production planning and control, test engineers, quality control and so on. The terms of reference for working parties were:

1. To formulate detailed plans relating to the specification, development and implementation of all aspects of the cascade.
2. To ensure that all necessary actions were initiated so as to enable the project plan to proceed according to plan.
3. To monitor performance against those plans developed in 1. and to identify and resolve any problems that hinder progress.
4. To report progress against major agreed milestones and to communicate any problematic issues that the working party may find difficult to resolve to the project board.

The key points concerning project management and its structure within ICL Ashton are therefore as follows:

1. The working parties were primarily responsible for detailed system design, but any major decisions regarding physical layout changes, system control, and generic modes of operation needed authorization from the project board.
2. The initiative of forming working parties was seen by senior managers as encouraging a wider participation in the system design process and, as a consequence, increasing the chances of the final system design being accepted by the users, thus reducing implementation time.
3. During the development of the cascade system the intention was to allow periods of reflection between stages of development to allow each module to be tested and debugged and for project team members to evaluate the project management procedures adopted. However, pressures for completion of the cascade project as a whole tended to reduce such reflection time.
4. The design of the various modules was effectively an iterative process and the fine details of design only began to take shape once the overall concept of layout and operation was agreed.
5. Representatives from user parties were responsible for ensuring their departments' requirements were recognized during the specification of design so as to facilitate the early resolution of potential problems.
6. Factory-wide presentations were given to shopfloor personnel and all other users who did not participate in the project teams in an attempt to communicate design proposals prior to implementation and allow for debate on points of issue.
7. The ICL approach could be seen as 'design by committee' and one feature of this is that it tended to increase the design and development times. However, this was offset by a reduction in the implementation and debugging time for new systems.

Source: by kind permission of International Computers Limited.

Simultaneous design engineering

A key issue in the product design process is the management of the interface between the design and manufacturing function. All too frequently these work in isolation, with product design engineers designing the product from the marketing department's specification and then passing this on to the manufacturing and process engineers, who will incorporate it into existing systems or develop new processes as appropriate. This approach, 'sequential engineering', can often be very time consuming, with a high number of design iterations between the design and production functions. 'Simultaneous engineering' (otherwise known as 'concurrent engineering'), however, involves the parallel development of products, processes and production organization (Riedel and Pawar, 1991; and Sweeney, 1992; see also Chapter 1). This, it is argued, can not only dramatically cut time to market for new products but also develops a 'right-first-time' (quality by design) attitude in the product design process.

Simultaneous engineering of product and process is becoming increasingly common in manufacturing companies. This is especially so where early and continuing involvement in new product design by process engineers and production and materials management is seen as essential in ensuring products are properly managed throughout their life-cycle and that the most effective systems designs are evolved and implemented. Simultaneous design implies continual interaction and parallel developments and decisions throughout the product and production system design processes.

9.2 Key parameters

Identification of key parameters

As shown in Figure 9.3, project management is strongly influenced by two factors, the organization of the company and the outcome of the justification process. Each of these will now be discussed.

Organization

A key concern of the project manager is to bring together the appropriate people and skills from within the organization to facilitate the introduction and smooth operation of new systems according to plan. Clearly, decisions taken at higher levels (see Chapter 7 on Organization) will determine the degree of ease or difficulty the project manager will experience in taking a multidisciplinary approach to systems introduction. The structure of the organization will provide lines of command and, most probably, functional distinctions between various parts of the business. The historical development of the organization in terms of culture, norms and values will also have a large influence upon the degree to which individuals around the organization are willing to participate in detailed design and development. So, before embarking upon the management of any system design and introduction programme, project managers must have a clear understanding of the structure and state of the organization within which they are operating.

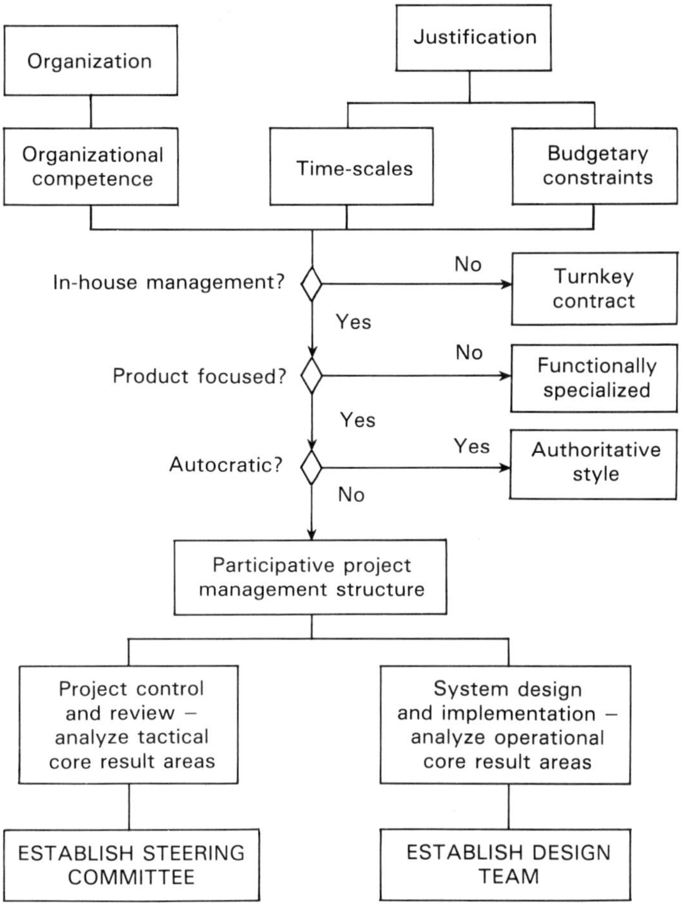

Figure 9.3 Key parameters for project management.

Justification

The outcome of the justification process should be an agreed 'capital paper' or similar document, which details the aims, objectives and anticipated benefits of the system design project (see Chapter 8). This represents a refinement of the broader company objectives and manufacturing strategy into a focused statement on the context and role the new system will play within the organization. As such, the justification paper should serve as the terms of reference in the development of the new system and provide the basis for evaluating the success of the system and its design process, both during the design and implementation project and following the introduction of the system.

9.3 Selection and configuration of design options

Project management process

As illustrated under the result criteria in Figure 9.4, the key issues in deciding upon the structure and process of project management for new production systems include the need for tight project and cost planning and control, communication and participation across functions in the business, a high degree of commitment from design teams and system users, and effective utilization of resources. The options available are now identified.

Steering committee

Production system design projects will require guidance and direction from the strategic domain of the organization and the use of steering committees has proved to be an effective means of providing this. The steering committee will normally comprise senior managers from those functional areas affected by the changes to result from the design project plus the project managers. The benefits of using a steering committee is that it encourages top management awareness and support for projects, but a drawback can often be disillusionment at the level of project teams, who see the important project and design decisions being taken in an arena to which they have little or no access.

Project team structure and composition

There are many options concerning the structure and composition of project teams. A working party method will involve drawing together team members on specific aspects of design on a part-time basis. User groups will usually represent a significant proportion of members on working parties. The task force approach involves the use of full-time design specialists to develop solutions. This provides a fast means of arriving at new designs, but often results in very limited participation on the part of users and, consequently, can result in their lack of commitment. The time saved at the design stage, therefore, can be more than balanced out through excessive time and difficulties in implementation. Finally, a decision must be made as to whether project team members report within the context of the normal hierarchical organizational structure or whether matrix reporting and organization should be introduced. The benefits in terms of lateral communication and team working are apparent for matrices, but there still remain caveats in their use in terms of the effectiveness of management control and the political tensions caused through dual reporting roles.

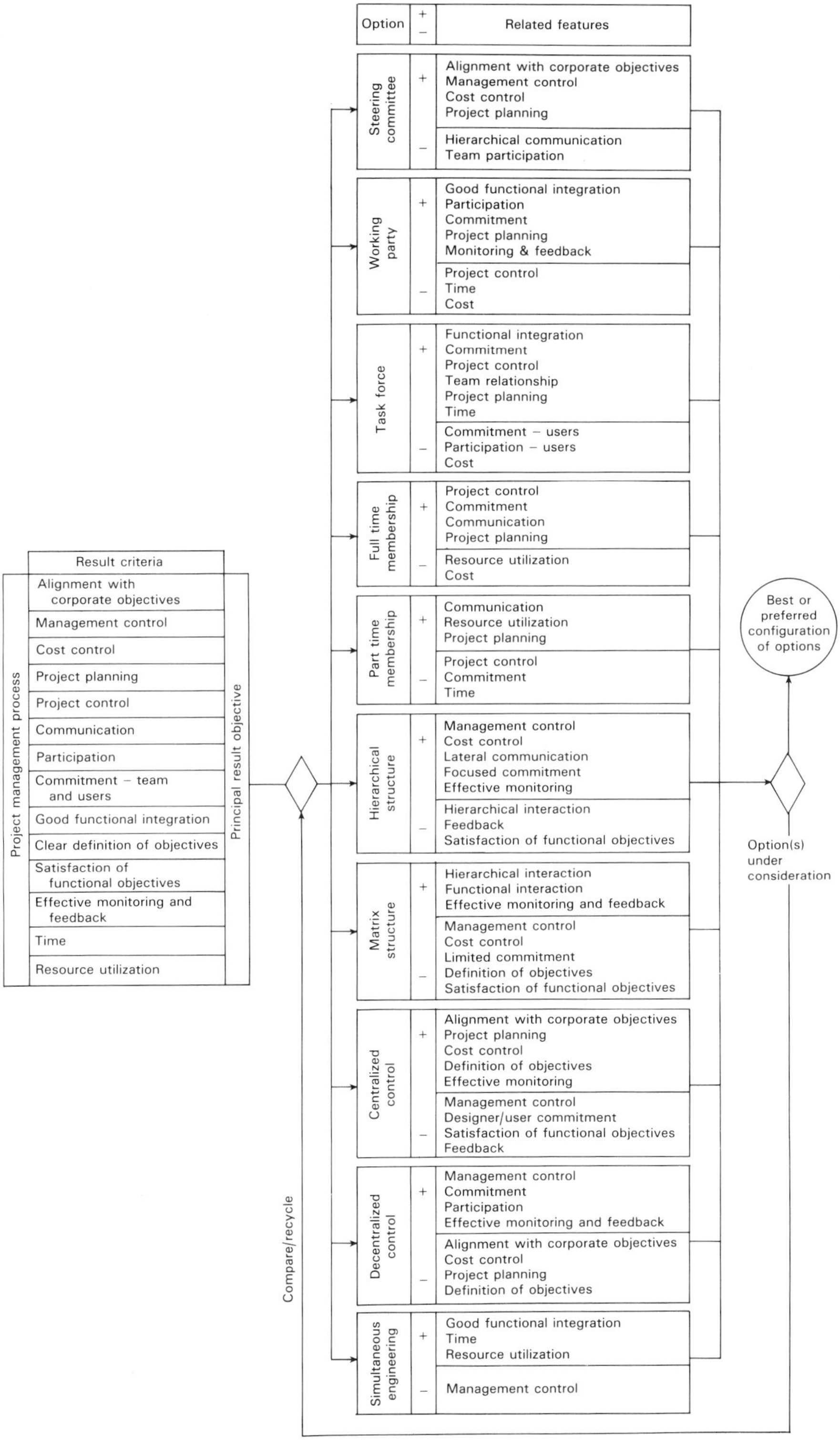

Figure 9.4 Design option guide for project management.

Project management control

Linked to the above decisions on structure and composition, a fundamental choice in the management of projects must be made between centralized and decentralized control. Whereas centralized control offers advantages in terms of reflecting corporate objectives and placing the project developments into context of the overall organization, it suffers with respect to its 'distance' from the project activities causing a lack of management control in day-to-day activities and, with limited control in their own destiny, low levels of team and user commitment. A decentralized control, where authority is passed to project managers, counteracts many of the caveats associated with centralized control of projects, but can result in the project developing with little relation to overall corporate objectives.

Simultaneous engineering

The most striking advantage of simultaneous product and process design is that of collapsing the time taken from the identification of the need for a new or modified product and its production and availability to the customer. Simultaneous engineering also results in improved functional integration and, if managed correctly, a more effective use of corporate resources. The main problem, particularly immediately following the adoption of simultaneous engineering, is that of management control for functional managers who are now expected to share resources and people in a matrix structure.

Application 9.2 Concurrent engineering in the aerospace industry

Although perceived as a military initiative, the computer-aided acquisition and logistic support system will have repercussions throughout manufacturing.

Down in South Carolina, USA, there is a view of the future, a navy funded initiative, RAMP, which integrates CAD-CAM and process planning into a just-in-time concept. Rapid acquisition of manufactured parts (RAMP), based at the South Carolina Research Authority, is a pilot spare parts manufacturing facility and part of the US Department of Defense's (DoD) computer-aided acquisition and logistic support (CALS) project. This is seen as a model for manufacturing facilities, which may be located in factories, forces bases or even in the field. What makes the demonstration even more interesting is the use of design and manufacturing data based on the still to be released product data exchange specifications (PDES).

RAMP brings with it two important messages, CALS is real not just an academic discussion, and its effect on manufacturing practices will be far-reaching.

Over recent months it has become increasingly clear that the Department of Defense's CALS programme will be a major factor in shaping industry through the 1990s, and not just for the defence giants. But far from being the big stick it is often perceived to be, CALS is turning out to reflect sound commercial practices.

Many companies, including Boeing in the US and Rolls-Royce in the UK, are on record as saying that there is little difference between their company goals and strategies and the aims and expectations of CALS. CALS aim as regards the DoD is a reduction in the cost of weapons systems, and is seen as growing in importance as defence cuts start to bite in the USA. In meeting this aim, which involves the capture of wide-ranging business, design, analysis and manufacturing information, CALS is focusing attention on the use of concurrent engineering practices. And what company is not heading towards concurrent engineering (CE), DoD or no DoD.

Robin Blake, head of Rolls-Royce's computer-aided engineering and manufacturing, explains that the commercial need for reduced lead-time is driving CE and would be happening without CALS.

Rolls-Royce's first move towards CE, in common with many other companies, is centred on the establishment of integrated co-located engineering teams, getting data and people into a common environment. This is often impractical, and data has to be shared between different sites or even companies separated by hundreds or thousands of miles. Data has to be transferred electronically, but the current options, IGES and proprietary translators, cannot exchange the depth of information required to make CE work. This is where CALS comes in. CALS is focusing attention on the development of techniques to exchange product definition data, not just geometry and drawings. The term 'product data' denotes the totality of data elements, which completely define a product, including all the geometry, topology, tolerances, relationships, attributes and features necessary to design, analyze, manufacture, assemble and support a part or product. This requires more than the transfer of geometry possible with the initial graphics exchange specification (IGES), although IGES is specified in the early phases of CALS. However, all that this requires is that illustrations and engineering drawings can be extracted from the design process.

More than geometry

Because companies moving towards concurrent engineering need to exchange more than just geometry, a more far-reaching standard is needed.

According to Blake, much of the work Rolls-Royce contracts out needs a full understanding of the data. This forces RR to insist that some contractors use identical systems and exchange only native data. Obviously, this has limitations, and the company has joined PDES Inc, the group working on the transfer of product definition data. PDES, and its European counterpart, the standard for exchange of product model data (STEP), will be the key to the success of CALS. Although unlikely to be a requirement until 1995, the impact it will have reaches far beyond the military field.

It is conceivable that PDES will be an integral part of the design of future CAD systems. The long-term aim, says Les Shaw of the CADCAM data exchange centre, is to create a concurrent engineering environment in which new systems are designed, developed and supported by multiple industry and government agencies through the use of standardized product definitions and logistics data. The enabling technology for this is the integrated weapon system databases, which store this data across distributed heterogeneous systems. Increasingly, the trend is towards integration of software tools and technologies. The MAP/TOP standards, and ESPRIT projects such as CIM-OSA, have defined application environments in which CAD and CAM are seen as just two elements in a more complex picture.

Emphasis on product data, says Shaw, reflects a change in the perception of CADCAM by users and developers, from ever-increasing capabilities of systems, to increased emphasis on the value data and the ability to reuse this data.

For CALS to be truly effective, advances are needed on several fronts. Electronic data interchange (EDI) is now common for commercial information, but limited for technical data by the existing communication infrastructure. A telephone network is quite capable of handling 1 Kb order files but not 5 Mb CAD files, so CALS is stuck with exchange on magnetic media until adoption of improved standards such as ISDN.

Source: Ian Bowman, 'From CALS to CADCAM', *Manufacturing Systems*, April 1990, 33–35.

Project management tools and techniques

The design options above concerned the process of project management. Once these wide-ranging decisions have been taken, however, choices need to be exercised between the tools and techniques that are available for the planning, scheduling and control of projects. As well as providing effective planning and control tools, the type of technique adopted must also be capable of measuring time achievements, utilizing resources efficiently and be relatively straightforward in use. Figure 9.5 illustrates the options discussed in this chapter, those of Gantt charts, critical path method and PERT.

Gantt charts

Gantt charts are a simple and low-cost means of ensuring that all activities are planned for, their order of precedence are accurate, activity time estimates are approximately correct and that the overall time can be evaluated. The strength of Gantt charting is in its planning of time, but it is limited in the way in which it can plan or control costs or resource inputs.

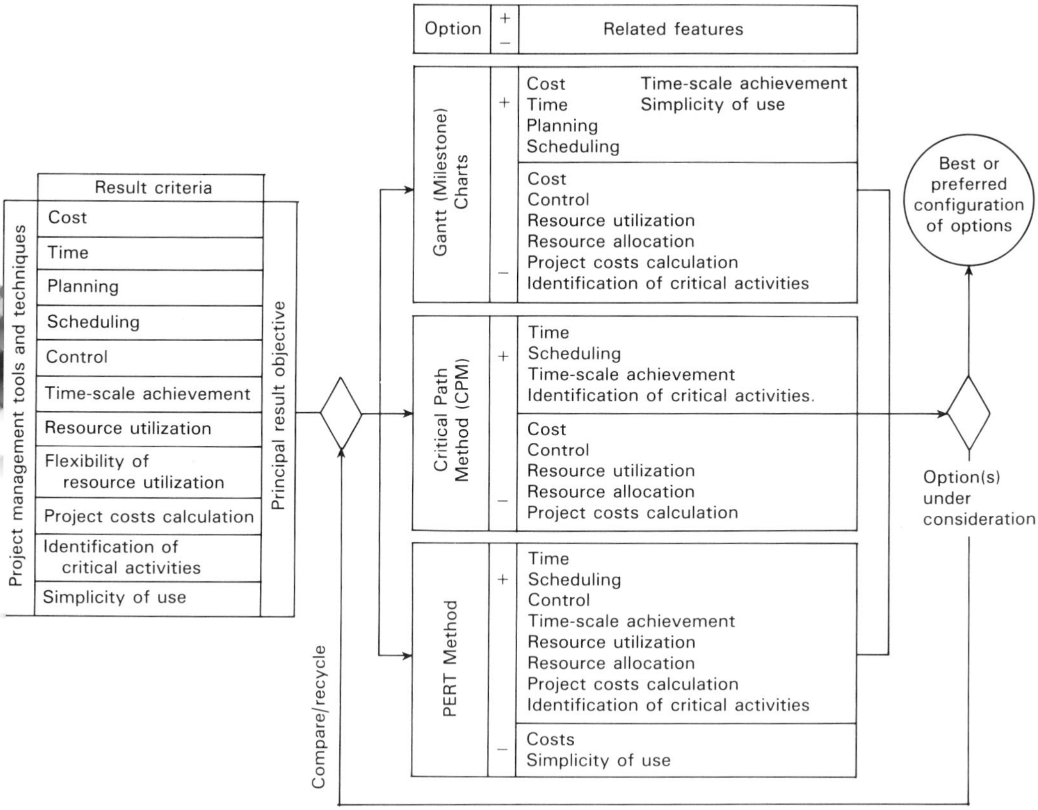

Figure 9.5 Design option guide for project management tools and techniques.

Critical path method (CPM)

CPM provides benefits over and above those given by Gantt charts. In particular it identifies critical activities that need to be accomplished on time within the overall project and provides a means for evaluating the degree of flexibility available for the completion of tasks. However, it still makes little impact on the management of resources and involves more work up front to develop a network than is involved in the construction of Gantt charts.

PERT method

The PERT method offers the advantage over CPM of providing the means for resource management and utilization. However, the costs of gathering the requisite information to run PERT effectively can be high and can also be time consuming. The amount and complexity of work involved in running and updating the PERT programme are also barriers to its more widespread adoption.

■ 9.4 Generalized methodology

The generalized design methodology for project management is illustrated in Figure 9.6 and the suggested process for this is now described.

1. **Determine the established constraints on the project.** The approved justification proposal will necessarily contain details relating to the time for development and the sanctioned cost of development. This will provide the time-scales for the overall project and the budget available for the scheme.
2. **Establish the structure for tactical management of the project.** Plan and prepare the function and operation of a high level project review group or 'steering committee'. Evaluate the time and financial constraints and ensure that the project will continue to align with manufacturing strategy and corporate objectives.
3. **Create a user-oriented design team.** Select project team members to fit within the structures determined in 2. These individuals could be on a steering committee or work within more focused design groups. Wherever possible, try to recruit design team members who will be responsible for operating the new system or will be served by it (the 'users').
4. **Establish degree of participants' involvement.** Despite the attractions of full-time design team members, the constraints on manpower will normally lead to widespread use of part-time participants, who will continue to perform their usual functions and responsibilities during the design process. Where resources allow, users may be seconded on to the project on a full-time basis, but user participants more frequently tend to be part-time and work closely with specialists such as manufacturing engineers, information systems personnel and the like.
5. **Identify project leader.** Where no project leader has been determined at the outset, it is now important to choose a leader. Use the information gathered in steps (2. to 4.) to act as a reference in selecting the most appropriate person. Where a leader was selected from the outset, this choice should now be reviewed on the basis of this information before proceeding any further. An important decision relates to whether the project leader should be a user or functional specialist. Many organizations advocate a user as leader, but given the constraints highlighted in 4. and the likely constraints on users' time for participating, this may not be feasible.

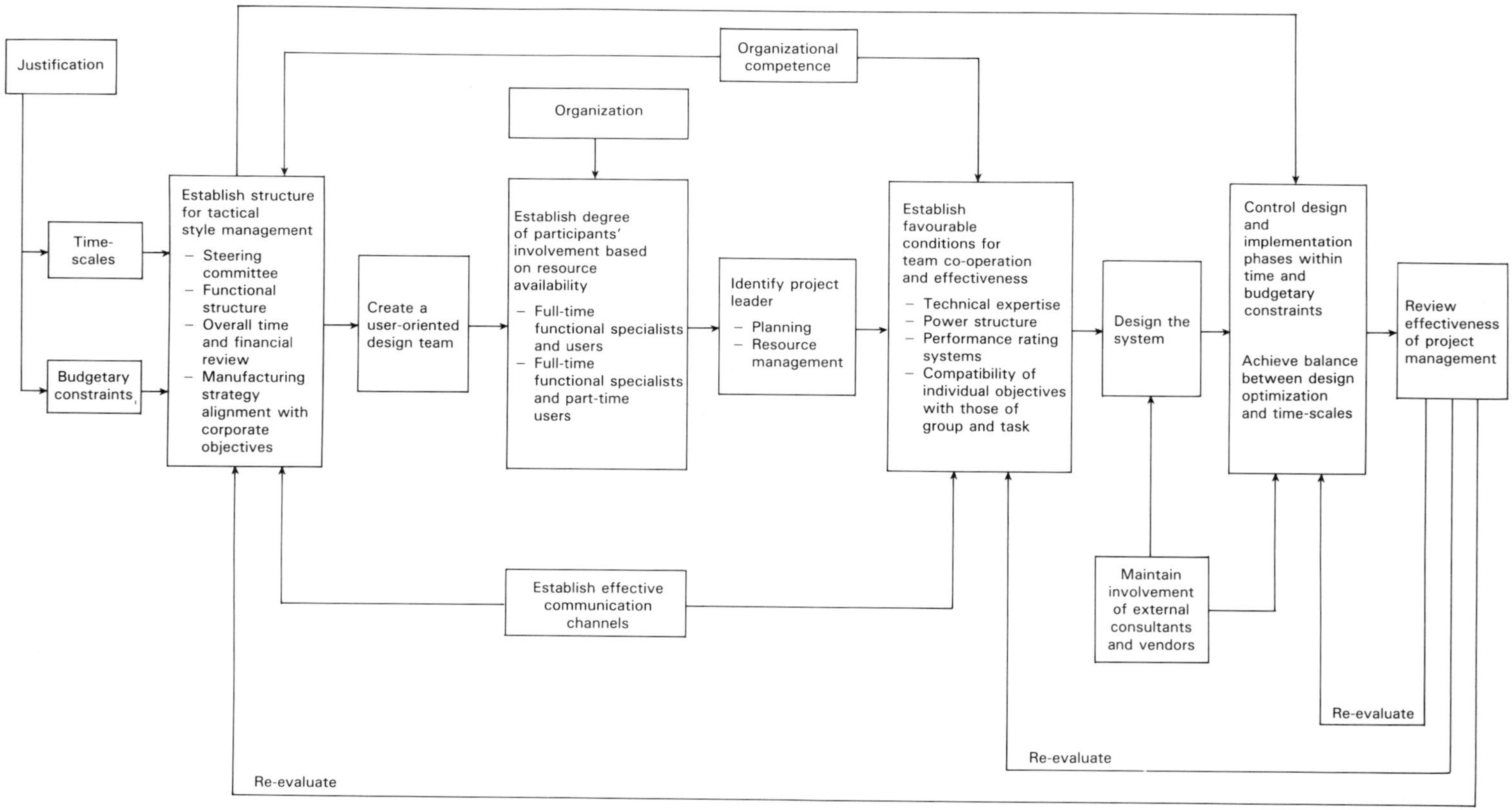

Figure 9.6 Generalized design methodology for project management.

6. **Establish co-operation and co-ordination.** The engendering of favourable conditions for team co-operation and co-ordination requires a number of aspects to be taken into account. The competencies prevailing within the organization need to be considered so that technical expertise and knowledge of the system can be best utilized and balanced. It is important to establish effective communication channels both between the steering committee and working parties and laterally between groups. Here it is important to understand the power structure within the formal organization and seek to ensure that this does not hinder progress in the design project. Finally, design team members must have mutually agreed performance rating measures upon which to judge their work and there should be a high degree of compatibility fostered between individual, group and corporate objectives.
7. **Design the system.** Using the methodology components detailed in Chapters 10, 11 and 12, design the physical system, control systems and work organization.
8. **Control design and implementation phases.** Manage and control the design and introduction of the new system ensuring that they adhere to the agreed time-scales for development and are contained within the financial constraints. Whilst seeking to develop the best system possible, there should always be a recognition of a trade-off between design optimization and the time and budgetary constraints impinging upon the project.
9. **Evaluate the process of project management.** Upon completion of subprojects and of the overall project, review the effectiveness of project management and the tools and techniques adopted in order to improve the process in the future.

9.5 Conclusions

Effective project management and the adoption of adequate monitoring and feedback mechanisms are considered necessary in ensuring successful project development and implementation. Theory and practice provide a wide range of tried and tested tools and techniques for use in planning, scheduling and controlling projects and these have been reviewed within this chapter. Furthermore, the literature advocates the need for a high degree of user consultation during the design and implementation of new production systems. This is mainly because the degree of user involvement during the early stages of a project can influence the ensuing support for the project by developing a feeling of 'ownership'. However, the reality is often some way detached from this ideal. Many companies practise a relatively autocratic style of project management, but argue this is necessary to ensure the terms of reference laid down in the capital justification paper are strictly adhered to in the design of new systems. User requirements, as a result, tend to be checked on an ongoing basis with the degree and scope of user participation in design somewhat restricted. Thus project teams are frequently structured around a number of functional specialists with expertise brought in as and when required from external vendors and consultants.

Given that the tools and techniques for project management are well developed and documented in the literature, the DRAMA methodology for project management has concentrated upon the process of managing a project and has provided practical guidelines for effective planning and control. A participative style of management is necessary in order for the design and implementation of market-focused production systems to be successful and readily accepted. The structure of any project should recognize the distinction between the tactical domain, where a steering committee may have responsibility for overall financial control and project reviews, and the operational domain, where individual design teams are responsible for the fine detail of systems design and implementation. Such a division of tasks results in more effective control and stimulates the alignment of project developments with corporate and business objectives.

Finally it is stressed that team co-operation, communication and effectiveness are achieved if certain favourable conditions are developed within the team structure. This calls for a balance between technical and organizational expertise, a power structure conducive to creativity, and the continued commitment of all to the development of successive systems. It is therefore essential that the effectiveness of project management practices and procedures are continually assessed and developed in order to improve the process and to ensure that future developments can be tackled with greater confidence and assurance in the mechanisms for development.

9.6 Further reading

Goodman, L. J. and R. N. Love (1980) *Project Planning and Management: An Integrated Approach*, Pergamon, New York.

Harrison, F. L. (1981) *Advanced Project Management*, Halstead, New York.

Weist, J. D. and F. K. Levy (1977) *A Management Guide to PERT/CPM*, Prentice Hall, Englewood Cliffs, New Jersey.

10 Physical System Design

Objectives of the chapter

On completion of this chapter you should:

1. Be familiar with the main theoretical and practical issues in physical system design.
2. Understand the principles and the main options to consider when designing and installing a new production facility.
3. Comprehend the stages in the process of designing a new physical system.

■ Chapter outline

This chapter covers the decisions and options available for the design of physical facilities in the production system. Issues of concern to the designer such as layout, automation and flexibility are covered in the discussion on theory and current practice. The key parameters upon the process of physical system design are then identified and their influence discussed. These range from strategic considerations of market and manufacturing strategy to the available process technology prevailing in the environment at the time of design and the concurrent development of control systems and work design. The options for physical systems are wide and varied, as one would expect at this detailed level of design, and the chapter contains ten design option guides to assist in the process of choice. The DOGs are categorized into three types, those at the 'factory level' of decision making, and those at the 'module' and 'utility' levels. DOGs for consideration include storage and transportation (factory, module and utility levels), layout configuration (factory and module levels), physical integration (module level) and tooling and work instructions (utility level). The DOGs provide guidance, from factory concepts of design to specific hardware choices at the utility level and give the related positive and negative features for each option Finally the process of physical system design is summarized in the generalized methodology flowchart.

10.1 Theory and current practice

Theory points to the need for process simplification and rationalization prior to introducing automation, with facilities being laid out along coherent routes of material flow to facilitate easier manufacturing control. In a high product variety environment the literature advocates the development of autonomous units based on make-complete production combined with the principles of flexible manufacturing and, where feasible, the use of appropriate technology for materials handling and storage. Three distinct areas will now be considered: production facility layout, manufacturing automation, and flexible manufacturing systems.

Production facility layout

The layout of manufacturing facilities can be categorized into three main areas: (1) fixed position, where the material or major components remain fixed while the equipment, labour, tools, etc., are brought to the place of work; (2) functional (or process), where equipment or processes of the same type are grouped together and products move between them, usually in batches; and (3) product (or operation sequence) layout, where products or components are sequentially moved within one area or along a flowline (see Wild, 1984; Bennett *et al.*, 1988; Hill, 1991 for more discussion).

The shift from manufacturing standardized products in high volumes to providing a wide variety of products in low volumes has necessitated a dramatic rethink of manufacturing strategies (Frazelle, 1986). There has been a move away from the traditional approaches, which seek to exploit economies of scale, often to the detriment of other factors such as quality and service (Blois, 1986; Talysum *et al.*, 1986). This has influenced the design of production systems and has seen the emergence of 'flexible manufacturing' (Ranky, 1983; Randhawa and Bedworth, 1985), 'just-in-time' production (Hannah, 1987) and 'total quality management' (Schonberger, 1982), which are all now widely recognized as effective means for enabling corporate objectives to be met at the operational level of the business.

Manufacturing automation

Technological advance through the introduction of automation is widely seen as a fundamental means to improve production efficiency and product quality (Meredith, 1987). Automation of the physical system occurs in three areas (Keogh, 1987; Moinet, 1987; Wheaton, 1987). These include:

1. Materials processing, including the introduction of computer numerical control (CNC) machine centres.

2. Storage systems, such as automated storage and retrieval systems (ASRS) using robotic cranes and the like.
3. Transportation using conveyors and, more recently, automated guided vehicles (AGVs).

The development of integrated systems rather than the evolution of so-called 'islands' of automation, which has occurred in the past (Bolwijn and Kumpe, 1986), is increasingly attractive, some might say essential. This integration can be achieved not only through physical system design, but also through linkages in the control and information systems to be discussed in Chapter 11 (Sweeney, 1986; Maji and Banerjee, 1988). In the integration of systems, the simplification and rationalization of processes prior to automation is widely recommended in order to reduce the complexity of system development and also to avoid the risk of automating any inherent deficiencies and inefficiencies that may be present in the current system (Kumpe and Bolwijn, 1983; Vuzelov, 1983).

Flexible manufacturing systems

Manufacturing facilities with finite capacity need to be flexible within the modern operations systems where product variety is high (Buzacott, 1982; Bessant and Haywood, 1986; Zelenovic, 1986; Slack, 1986). The techniques of FMS to respond to this demand are now relatively well developed (for discussion see Gustavsson, 1984; Vonderembse and Wobser, 1987), but its application has mainly been directed towards component processing industries where production systems invariably incorporate advanced machining equipment operating in conjunction with automated handling devices (Hartley, 1984; Bennett, 1986; Lim, 1986, 1987).

Flexible manufacturing for assembly is less well developed than for component processing as automation is frequently not feasible or cost-effective where product variety is high (Foyer and Drazon, 1986). An additional factor is that automation cannot respond to major design changes for products whose life-cycles are inherently short and the volumes per design type are never great enough to justify robotics and the like (Bennett and Forrester, 1988). Mixed and multimodel lines (to be discussed more fully in Chapter 12) are a step towards increasing the variety of products assembled (Aneke and Carrie, 1986) but there are a number of drawbacks associated with them. Most notably the periodic resetting of lines for new models causes disruption and, when running, there is often low efficiency in operation due to balancing losses (Bennett, 1986). The development of autonomous working based upon the principles of 'make-complete' production, can enable a greater variety of products to be assembled in parallel systems (Sandberg, 1982; Zelenovic *et al.*, 1987). However, the desire to attain high levels of manufacturing flexibility should be balanced against the fundamental requirement in manufacturing of maintaining productivity levels (Beaton, 1983; Galonek, 1987).

The design process for physical systems

Traditionally, the activity of designing a physical system for production was considered to only be the concern of engineering function within manufacturing companies (Forrester and Hassard, 1992b). On two counts the approach was, typically, of a 'closed systems' nature:

1. There was little interpretation of the business needs of the organization and few efforts to link these with system specification and design.
2. There were only limited attempts to properly learn from the experience of other organizations in the area of process technology and design, and little collaboration with external systems vendors.

Current theory and best practice indicates a shift in this type of approach to one where the specification and design of a production system is linked very closely with the strategies developed by senior managers and the demands of the customers. Additionally, technical systems designers are actively encouraged to survey the environment to ensure that the best available or most appropriate process technologies are utilized in the design of any new production system. System design and development work is also increasingly being subcontracted to external suppliers and vendors, who are often regarded as having greater technical expertise or more experience in the field of designing new production facilities.

■ 10.2 Key parameters

Identification of key parameters

Physical system design has a number of internal and external key parameters which influence the eventual solution and should be taken into account when starting the process of design. These are shown in Figure 10.1 and have been identified as:

1. External:
   (a) the available process technology.
2. Internal:
   (a) market strategy;
   (b) manufacturing strategy for the production unit;
   (c) organization design;
   (d) the outcome and specification from the justification process;

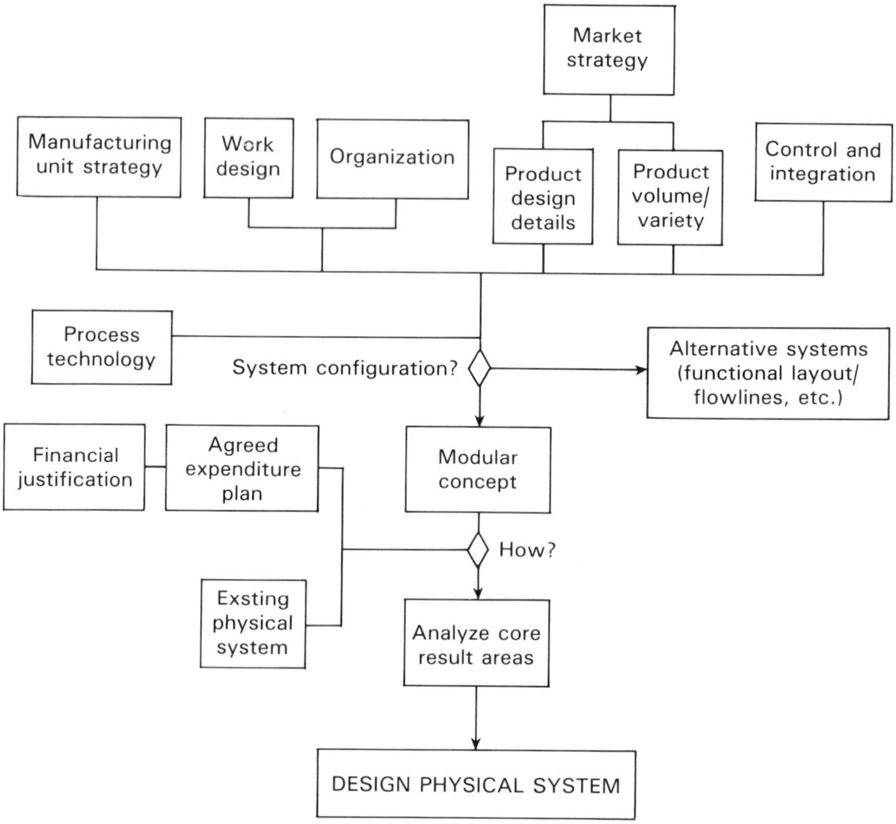

Figure 10.1 Key parameters for physical system design.

(e) the existing physical system design;
(f) the concurrent design of the other components comprising the operating system, namely control and integration, and work design.

Available process technology

All personnel involved in the design of the physical system must be aware of both the possibilities and limitations of process technology; this applies not only to senior engineering managers. A successful manufacturing company will have a whole host of 'sensing mechanisms' in the operational domain keeping it in touch with developments in the 'environment' (Kanter, 1984). So conferences must be attended, magazines read, other manufacturers consulted, new employees from other organizations debriefed and consultants and systems vendors used where necessary. In so doing, the organization ensures that all designers are kept abreast of technological advances and recent experiences to guarantee that the new production system takes advantage of 'state of the art' hardware and software opportunities.

Market strategy

The market strategy is an explicit statement relating to the markets and customer policy to be adopted by the organization. It is an interpretation of the current state of the market, customers and competitors and, as such, should be used in physical system design as a blueprint against which the eventual production system must match-up. Market strategy will directly influence the details of product designs and the volumes and varieties, which need to be manufactured within the operating system. If the organization has no explicit or coherent market strategy, then one must be defined prior to embarking upon the design of the physical production system. The market and environment component of DRAMA can be used as a guide to developing a new market strategy.

Application 10.1 ICL: Physical system design

The rationale behind the modular assembly cascade system concept was to ensure that manufacturing would be operated in a more integrated, cost-effective manner and market-oriented mode. Production systems were completely revised by system designers and users in the project teams with manufacturing flexibility based upon cellular manufacturing principles being a prime consideration. The key aspects of the physical system design process and the detailed configuration of facilities within ICL Ashton were as follows:

1. The modular assembly cascade concept links the ideas of FMS and JIT by providing a flexible infrastructure for the physical system. The basic configuration is one of a factory-wide hierarchy of integrated autonomous sub- and final-assembly modules. The modules themselves derive flexibility from generalized, rather than dedicated, work operations and materials distribution enabling a wide range of products and assemblies to be handled.
2. Most modules comprise an input store for incoming kits of parts, flexible assembly workstations and an output store for completed subassemblies. The inclusion of the 'satellite' input and output stores may appear to be in conflict with the principles of JIT, but their inclusion is necessary in order to cope with the problems of high variety manufacturing prevalent within the plant. They can in fact be viewed as similar in concept to the buffering provided by 'kanban squares', which are often found painted on the floor in production areas running under JIT.

3. Various designs based upon conveyorized lines were considered during the development of the initial 'Mercury' (1 m^3) module for final assembly, but the limited availability of floor space led to the adoption of an integrated crane and racking facility. This also gave advantages in terms of speed, reliability and the flexibility of parts delivery to workstations.
4. The Mercury module is an integrated design: the workstation, storage and handling system being self contained in one facility unit. However 'Apollo' (0.5 m^3) and the subsequent modules developed are of a segmented design whereby the workstation, storage and handling utilities are physically separate. The latter offers a high degree of visibility and also a greater resilience to hardware and software failure than the integrated design as separate utilities within the facility can be run in manual mode.
5. Subassembly modules were divided into two separate modules. The Apollo module is concerned with the assembly of small subassembly items that do not require special tooling and handling. The 'Dedicated Subs' module caters for larger, more unwieldy subassemblies and contains dedicated workstations with specialized handling and processing equipment. Thus efficiency is increased by separating 'complex' and 'routine' operations.
6. An input store was incorporated into the Apollo module holding a ready supply of kits for assembly. This was in the form of an automatic carousel system, which was used to enable robotic handling of kits into and out of the storage system and also to enable the fast and efficient auditing of stock.
7. Other storage areas, where a greater range and complexity of parts and kits are held, employ simple 'live racking' systems and manual picking of parts.
8. Automated materials handling systems are used for transportation between and, in some cases, within modules. Conveyorized options were originally considered, but AGVs were eventually chosen because of their advantages of flexibility of routing and adaptability to design layout changes.
9. AGVs used within Apollo are of a 'non-intelligent' type, which works under the control of a local PC. The AGVs perform a complex array of transactions with and between the adjacent main stores, the carousel input store, workstations and test operations. AGV routing is controlled by a signal-emitting wire track embedded in the floor. At the time this was thought to be less risky than using the relatively new, but more flexible 'painted-line' routing technology.
10. Various assembly workstation designs have been used in the different modules in the cascade. For Mercury, the stations are on the outside of the crane and racking system with the crane servicing assembly operations. The Apollo, dedicated subs and cableform (no longer in existence) modules incorporate manual assembly work benches whilst the 2 m^3 final assembly module uses a bay or docking system offering flexibility in the size and dimensions of products that can be assembled within it.

11. Integration in a physical sense is achieved by the factory AGVs, which procure partially kitted containers from the main stores and build up the kits cumulatively by visiting the output stores of modules around the factory in a circuit before delivering completed kits to their final input store destination. The finished products are also transferred by AGV from the output ports of final assembly modules to the packing and dispatch area.
12. A prime factor throughout design was that of safety and systems, equipment and tooling have been designed with features that ensure safety for both people and products.

Source: by kind permission of International Computers Limited.

Manufacturing strategy

A manufacturing strategy must be in place and referred to in the design of a new production facility. Preferably this will have been devolved to the level of the operating unit (e.g. the factory or process) so that it can be linked directly to the requirements of the physical system. The manufacturing strategy component of DRAMA can be used for developing a new strategy.

Organization

The prevailing organization structure and state (refer to the Organization component of methodology in Chapter 7) impose a direct influence over the type of system that is adopted. The characteristics of the organization effect both the process and, hence, the outcome from the production system design project. Organization must also be taken into account when considering how the system will be operated and managed when it is up and running, which can lead to the choice of one option over another.

The existing physical design

It is rare for a manufacturer to ignore totally the current production facilities when designing a new system. Even if the new design is to be radically different or for the production of a brand new product range, there will always be comparisons drawn with existing facilities as these serve as experiential benchmarks against which to predict or evaluate future performance.

Control and integration and work design

Physical system design is only one component of three in the design of operating systems. Therefore, design of physical systems must be conducted concurrently and in conjunction with the control and integration and work design components (see Chapters 11 and 12 for these parallel design activities).

Application 10.2 Computer integrated paint and press facilities at Vauxhall Motors

Vauxhall has commissioned a new paint shop costing £60 million at its Ellesmere Port plant, which features a European 'first' – application of a sophisticated, computer integrated control system called Manufacturing Automation Protocol (MAP 3.0), which is essentially a means of bridging the gap between different computerized control systems in a manufacturing plant with a fast, reliable and efficient communications system.

The new paint shop currently runs at 35 car bodies per hour (net) but will have the capacity to handle 45 bodies an hour (net) by 1992, to match forecast volume requirements. The latest paint process, materials and 'clean room' concepts have been adopted to enhance paint quality, durability and appearance – this includes the first use in the UK of water-borne primer paint technology. The new technology will help reduce manufacturing costs, as paint processes represent a significant factor in the total cost of building a vehicle.

Environmental benefits result from: (1) the use of water-borne primer, with solvent emissions almost eliminated; (2) efficient management of overspray paint (including pioneering research methods of recycling waste material); and (3) energy savings through heat recuperation. Working conditions are improved by: (1) glass sidewalls in the spray booths; (2) ovens isolated from the working areas; (3) temperature-controlled working areas; (4) extensive use of automated equipment; (5) the introduction of team working concepts; and (6) improved rest areas.

MAP 3.0

Vauxhall's new paint shop represents the first application in Europe of the Manufacturing Automation Protocol (MAP 3.0) standard to provide a full computer-integrated communications network for rapid monitoring and control of all painting operations.

Tri-Axis press

Major advances in Ellesmere Port's manufacturing facilities also include a massive 1600 tonne Tri-Axis Transfer Press and new High Speed Coil Cut Line, together involving a total investment of £30 million.

VAUXHALL
Ellesmere Port Paint Unit

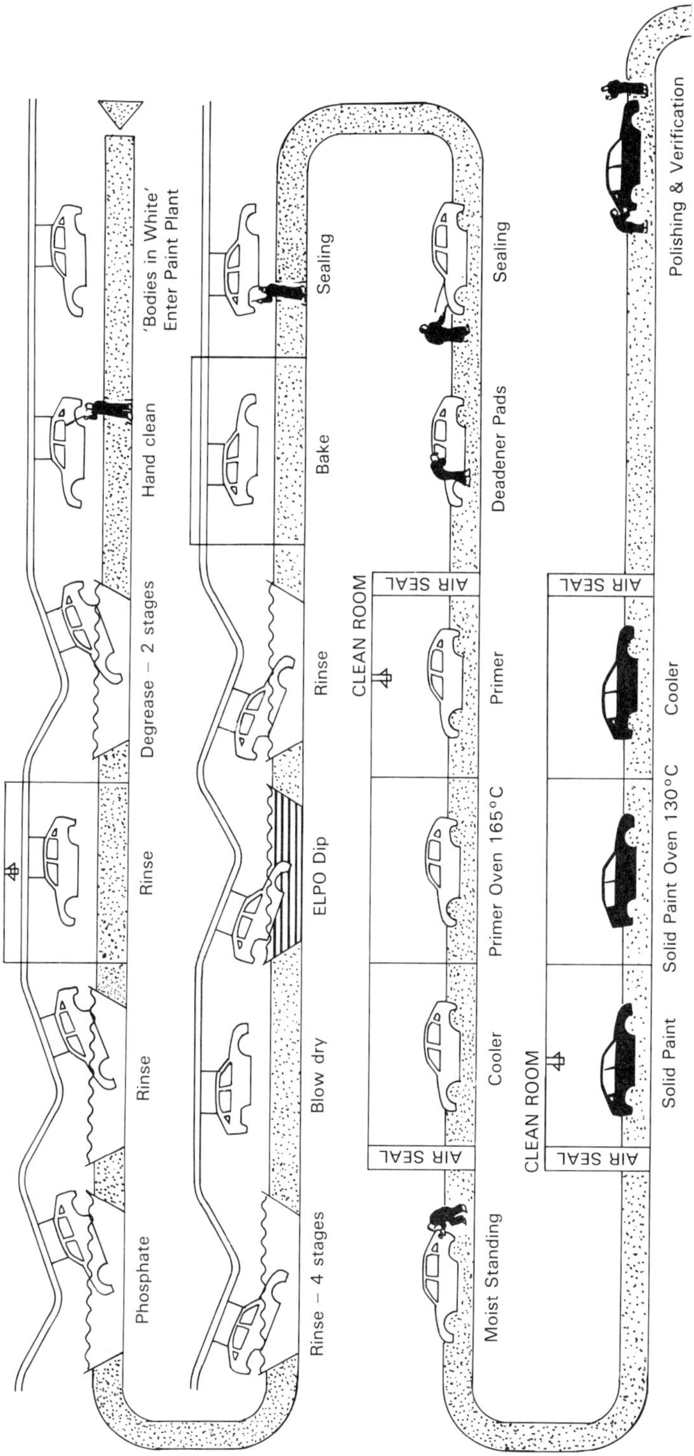

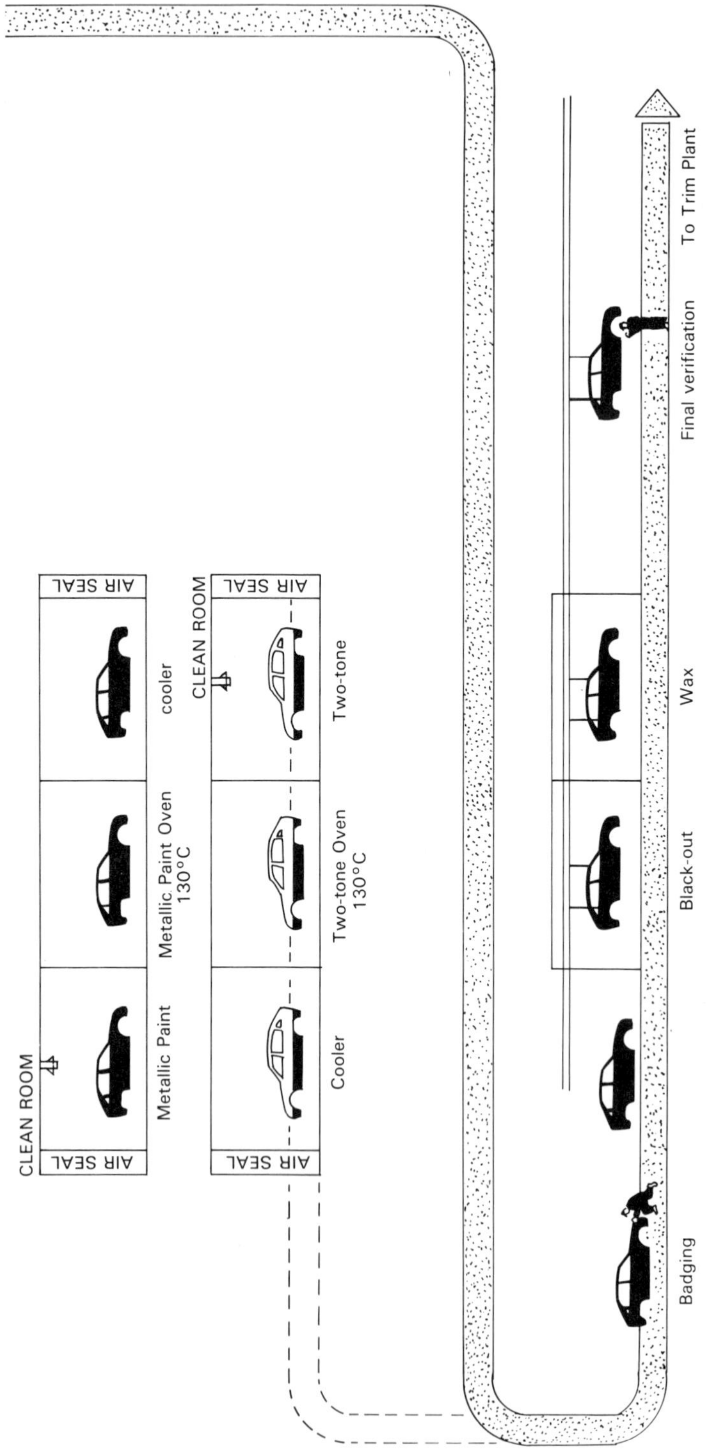

CLEAN ROOM
AIR SEAL
Metallic Paint
Metallic Paint Oven 130°C
cooler
AIR SEAL
CLEAN ROOM
AIR SEAL
Cooler
Two-tone Oven 130°C
Two-tone
AIR SEAL
Badging
Black-out
Wax
Final verification
To Trim Plant

Considered to be the most technologically advanced of its kind, the fully automated Large Transfer Press can produce body panels with a performance equivalent to four conventional press lines. The main benefits to Vauxhall from the new press will be: (1) negligible inventory; (2) a near continuous manufacturing capability; (3) lower tooling costs; and (4) greater flexibility of manufacture.

Apart from enhanced productivity, other advantages include: (1) further improved body quality from more consistent pressings; (2) safer and quieter operation; and (3) a more pleasant working environment.

European integration

A significant benefit to Vauxhall from these investments will be improved harmonization between General Motors' European manufacturing operations, particularly important in the context of export programmes and overall efficiency.

The Large Transfer Press, for example, is not only the latest of its type but is one of three similar machines currently being supplied to other General Motors' plants in Europe.

Source: by kind permission of Vauxhall Motors Limited.

10.3 Selection and configuration of design options

Design of the physical production system may be considered at three levels:

1. **Factory level.** Factory-wide concepts for layout configuration, storage and transportation. Specific hardware options are not conceived at this level, but global decisions regarding production configuration, degree of centralization and the scope for automation are taken. Three design option guides are reproduced to assist design at this level (see Figures 10.2, 10.3 and 10.4).

2. **Module level.** The decisions for the design for individual modules within the operating system. Here decisions are made on the conceptual operation of the module and include layout configuration, materials storage, intramodule transportation, physical integration and tooling/work instructions. These, respectively, are represented by the DOGs in Figures 10.5 to 10.9.

3. **Utility level.** This is where specific choices of hardware equipment are taken. Specifically, the DOGs shown in Figures 10.10 and 10.11 relate to choice of materials storage system and type of transportation for the production system.

Each of the DOGs for Physical System Design will now be considered in turn.

Layout configuration (factory level)

The layout of the factory as a whole will have consequences for more detailed decisions on physical systems. Figure 10.2 shows some of the criteria by which any factory layout is judged, of which space utilization and a rational flow of materials around the system are increasingly being seen as key indicators of a 'good' layout. The options for layout configuration given by DRAMA are as follows.

Group technology

An introduction to the concept of group technology was given in Chapter 2 and group technology (GT) is detailed in Appendix I. At the factory level the choice of whether to opt for a GT concept hinges upon the desirability of dividing the factory into smaller product-centred work cells instead of the functional layout usually employed by process-oriented factories. The benefits include those of rationalizing material flows and increasing the autonomy and customer focus of individual units, although there have been questions regarding the effectiveness of GT in terms of space utilization, machine productivity and levels of work in progress (due to the possible duplication of certain processes and constraints on volume and mix flexibility).

Functional

Functional layout is traditionally the most common form of factory organization. The idea is one of grouping together all similar component manufacturing or assembly processes and, wherever possible, producing in batches to generate economies of scale and operator/supervisor expertise in particular parts of the total manufacturing process. However, functional layouts have some widely accepted shortcomings, particularly with regard to high levels of work-in-progress as a result of excessive queuing in the system, long throughput times from when raw material enters the factory to when it leaves as finished goods and a lack of customer focus, particularly for 'upstream' operations, where there is often a lack of product identification. A fuller description of functional layouts is given in Chapter 2.

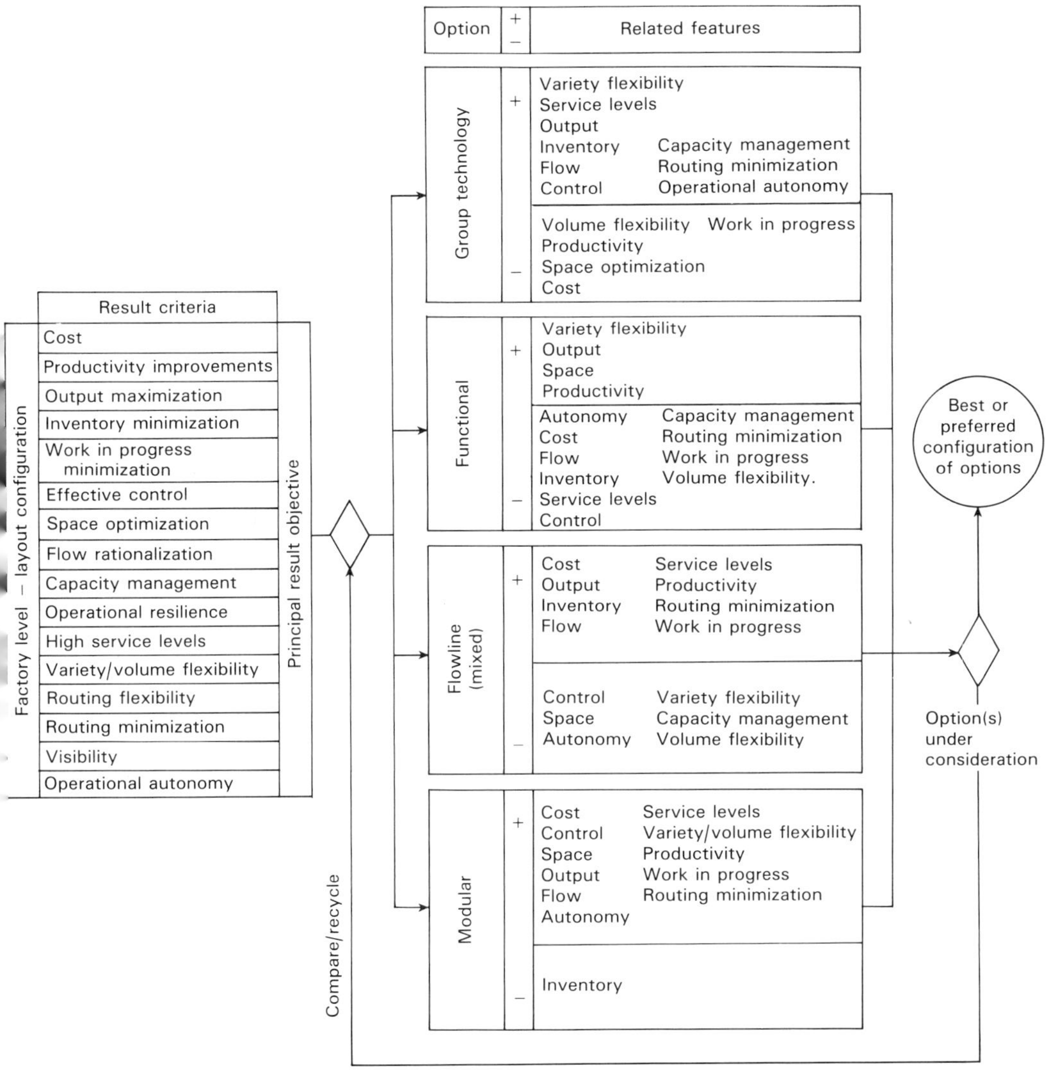

Figure 10.2 Design option guide for layout configuration (factory level).

Flowline (mixed)

The advantages and drawbacks of flowline working, particularly for assembly, are well publicized and have been considered in some detail in this book (see Chapter 2). However, the view of flowline working as producing high volumes of absolutely identical products (the 'any colour so long as it is black' notion of Model 'T' Ford production) is dated. Customers now demand variety, products for specific purposes and differentiated offerings from manufacturing companies. Where flowline working is adopted, it is usually in the 'mixed' model variant, where a restricted range of similar products or different models are produced using a sequential layout.

Modular (autonomous assembly cells)

The modular approach is based on similar ideas to those that give rise to group technology cells. In essence it comprises small, group-based working, where restricted ranges of similar assembly operations are carried out within modules throughout the factory. The layout of ICL at Ashton-under-Lyne, which uses the modular assembly cascade concept (see Chapter 4), demonstrates this. The key advantage of this configuration is that it provides more flexibility than flowlines and greater product focus than functional layouts. It also offers increased opportunities in terms of the ability to add or remove certain cells as product designs and the mix of product demand alters. The main caveat is that, unless carefully planned and managed in operation, modular layouts can cause a large increase in the levels of inventory by virtue of their autonomy. To counteract this, the introduction of modular systems is often accompanied by the parallel introduction of improved manufacturing control using just-in-time pull techniques (see Chapter 11).

Storage (factory level)

Within the factory layout configuration, a decision on the type and form of storage facilities within the operation must be taken. Major considerations shown in Figure 10.3 include the minimization, wherever possible, of inventory holdings, effective control and the ability to provide high service levels. Four options have been identified within DRAMA.

Centralized storage

One option is to operate with all stores within one central area, the factory store. There are advantages in terms of cost, space and (in theory) low inventory holding due to the avoidance of duplication of stores around the factory. However, in practice, a number of caveats and disadvantages have been identified, including lack of control (poor visibility for operations managers), poor response and lower service levels with no specific focus to storeroom activities, and an interruption to the smooth flow of materials and products throughout the factory.

Decentralized (subassembly-independent) storage

The decentralized option is one where a large proportion of stocks are held

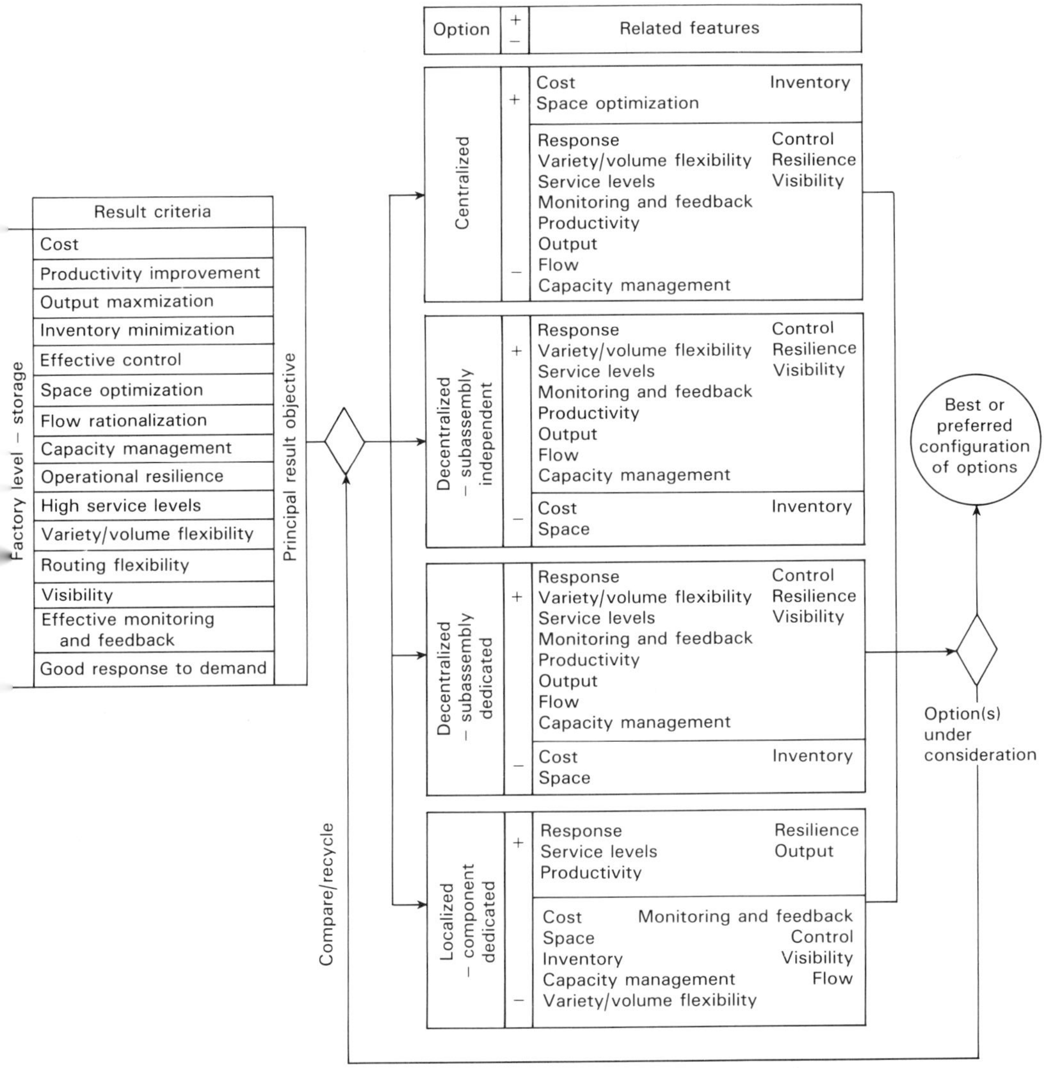

Figure 10.3 Design option guide for storage (factory level).

away from a main stores in a number of satellite locations around the factory. This may enable different types of stock (on the basis of size or special storing requirements) to be separated and stored in different locations according to their properties. This increases the focus on storeroom activities and can help in rationalizing the flow of materials around the factory. However a major danger is that inventory can proliferate with the increase in storage locations which will adversely affect cost and space requirements.

Decentralized (subassembly-dedicated) storage

The subassembly-dedicated decentralized storage option involves storing and locating inventory on the basis of the subassembly, or perhaps the final product it will go into. Thus the attempt here is to separate categories of inventory on a product-type basis, not merely by virtue of its properties. The advantages and disadvantages of this option are similar to those for subassembly-independent decentralization outlined above. Most usually this option is chosen with the aim of introducing parts kitting, an option to be explored in Chapter 11.

Localized (component-dedicated) storage

This is a form of decentralized store, but one where storage of parts and materials is governed by its point of use in the manufacturing process. Stores are therefore located around the factory and close to where the materials are manufactured into parts or products.

Factory-wide transportation (factory level)

In addition to deciding upon layout and storage choices at the factory level, guidelines for factory-wide transportation to link the various operations and storage locations must be developed. Result criteria for transportation, shown in Figure 10.4, include cost minimization, control, flexibility and ease of reconfiguration. The over-riding options are to have a fully automated system, a partially automated system or manual transportation.

Automation

The main reason for deciding upon a high level of automation for factory transportation will be to increase control and response, and thereby improve efficiency and output within the factory. A major concern for this choice, however, remains that of cost for the purchase, development and operation of such a system. Some forms of automation can also constrain systems operation and design change through a lack of routing flexibility and the problem of reprogramming.

Semiautomation

The semiautomated transportation system is one where the movement of materials is done by machine, and may be supplemented by a computerized control system of the type to be described and evaluated in Chapter 11. Although likely to be of lower cost, involving less risk and liable to have

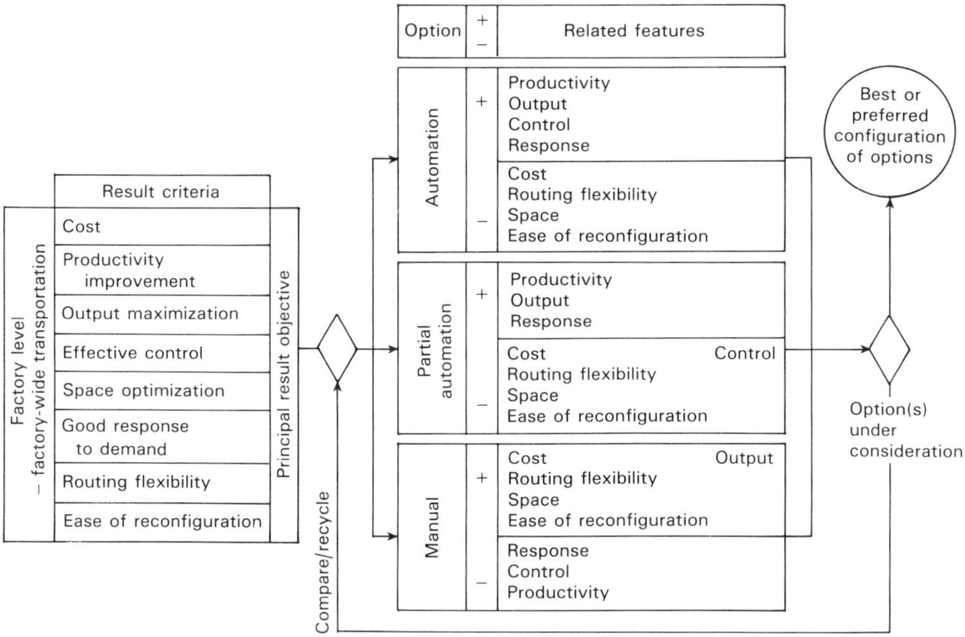

Figure 10.4 Design option guide for transportation (factory level).

greater inherent adaptability, the related positive and negative features are much the same as for the fully automated option above.

Manual

Manual, non-automated, transportation systems avoid many of the problems associated with the options described above, particularly the cost and risks involved in development. They also benefit from the flexibility and adaptability offered by the human operator. However, the cost of operating manual systems is becoming increasingly high in relation to automated transportation, and so there remain questions regarding their productivity for factory-wide transportation as materials handling systems develop into the future.

Layout configuration (module level)

Decisions regarding layouts at the module level are obviously largely dependent on the direction taken at the factory level for overall layout configuration (see Figure 10.2). However, there are instances where choices at the module level can be different to those offered at the factory level, but are more appropriate for the part of the overall manufacturing process in question. For example, one department may adopt flowline working whereas at the factory level a functional layout has been agreed. This example is indeed very common for factories which are largely functionally organized and make their components in batches within one area, but assemble the final product in another part of the factory. At the module level, the design options have been identified (Figure 10.5).

Single- or multistage build

Should the operation be performed in one step or should it be broken down into its constituent elements and produced in a number of steps? The benefits and associated negative aspects of each of these choices mirror the wider debate in the literature, which considers whether it is more effective in the longer term to produce in a make-complete manner, or whether the segregation of tasks benefits overall productivity and economies of scale (see Adam Smith's example of the pin manufacturing factory (Smith, 1776) and all ensuing debates, for example Gyllenhammer, 1977).

Group technology

If a cellular approach has been agreed at the factory level, specific layout decisions still need to be made at the module level. The group technology option at the level of the module refers to the decision whether to opt for a product-oriented work layout using machines and people positioned appropriately to perform the sequence of tasks required by a predetermined family of parts or products. Benefits include those of improved work flow and reduced work-in-progress, whereas the caveats to be avoided are restricted volume flexibility, high use of space and cost of manufacture.

Functional

This option is one whereby the functional configuration of the factory is taken down to the next level in terms of similar operations within the area in question being grouped together and machines or people involved with batch production in their own specialist area. The main drawbacks to this approach is the inherent queuing, high work-in-progress and long throughput times which may result.

Flowline (mixed)

In common with the group technology option at the module level, the flowline option involves the use of a product or sequential layout. It differs in that it is dealing with a more restricted range of parts or products (variants of one design) as opposed to group technology, where the range is greater, being a family of parts. The flowline option is liable, then, to be more dedicated and

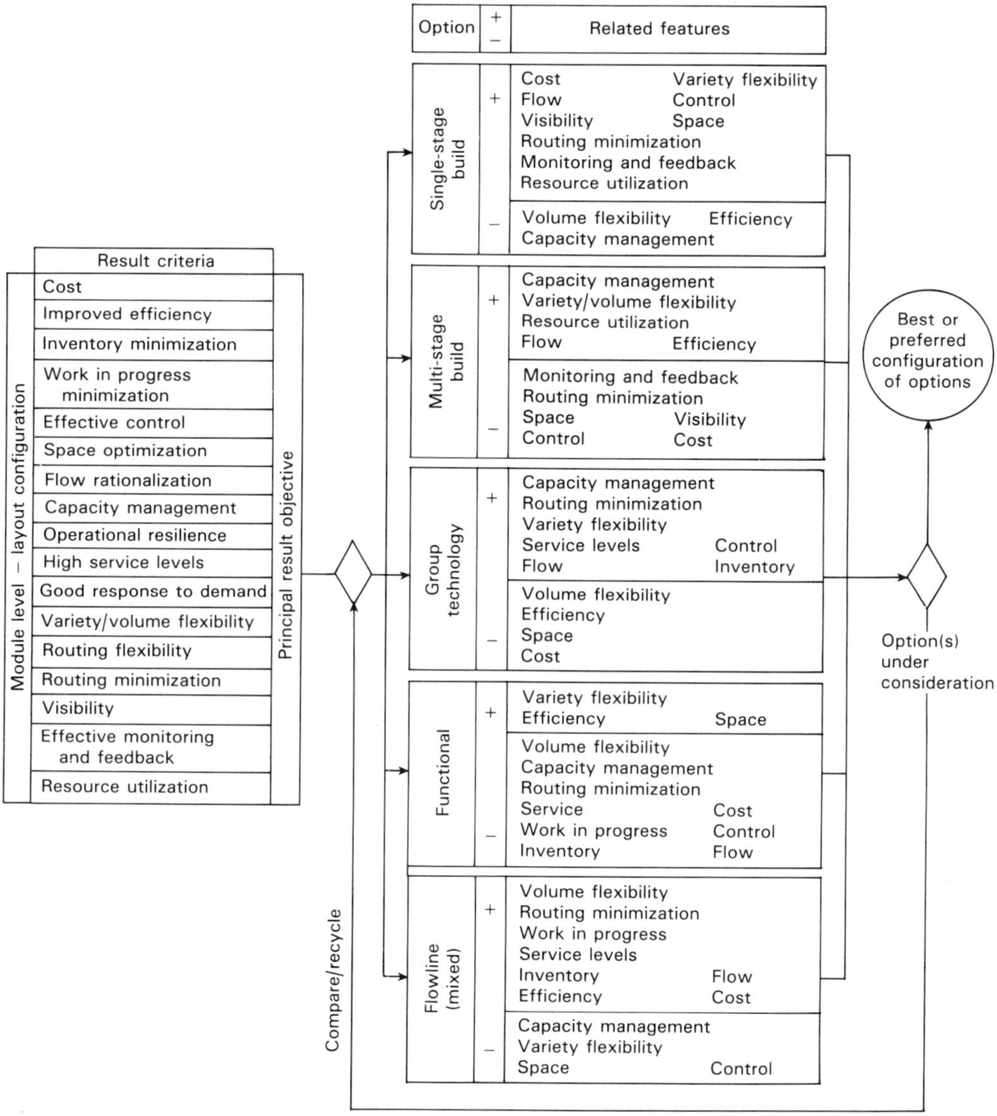

Figure 10.5 Design option guide for layout configuration (module level).

less flexible in terms of its ability to accept new product designs and is more commonly used for assembly rather than component production.

Storage (module level)

At the level of the module the options and their combinations are illustrated in Figure 10.6. Once more, specific choices on module storage are frequently

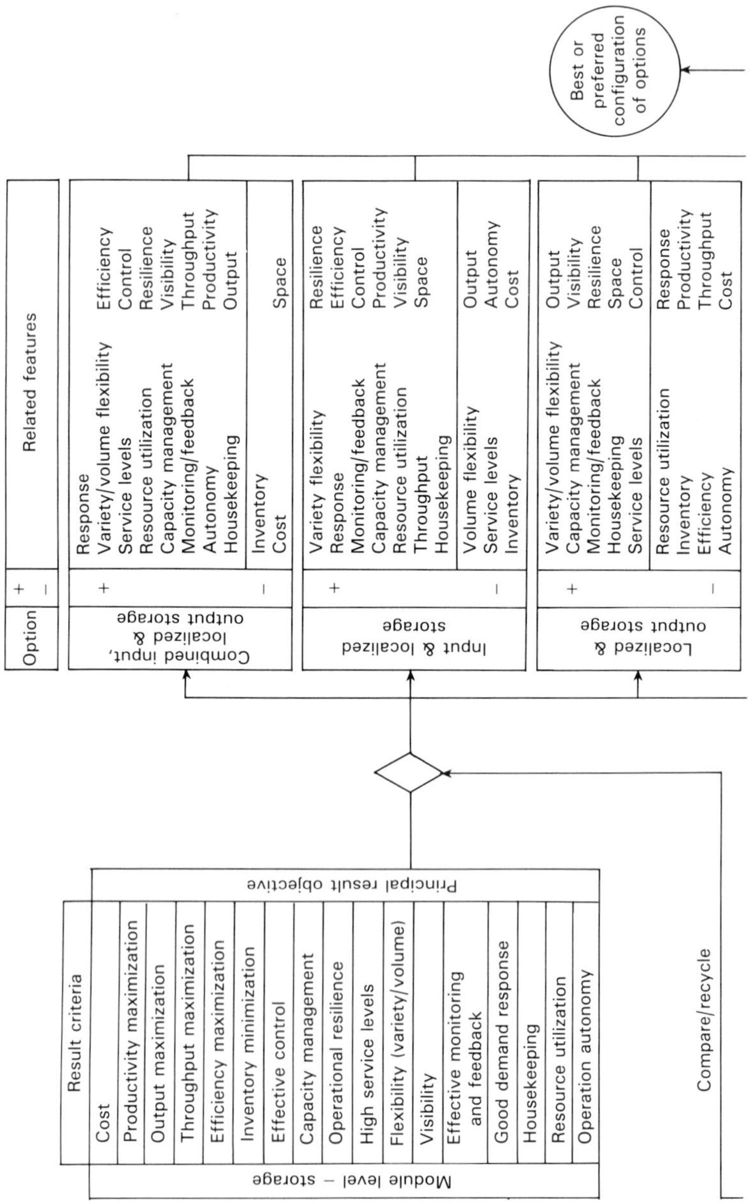
Module level – storage
Result criteria
Cost
Productivity maximization
Output maximization
Throughput maximization
Efficiency maximization
Inventory minimization
Effective control
Capacity management
Operational resilience
High service levels
Flexibility (variety/volume)
Visibility
Effective monitoring and feedback
Good demand response
Housekeeping
Resource utilization
Operation autonomy
Principal result objective
Compare/recycle
Option
+
–
Related features
Combined input, localized & output storage
+ Response, Variety/volume flexibility, Service levels, Resource utilization, Capacity management, Monitoring/feedback, Autonomy, Housekeeping, Efficiency, Control, Resilience, Visibility, Throughput, Productivity, Output
– Inventory, Cost, Space
Input & localized storage
+ Variety flexibility, Response, Monitoring/feedback, Capacity management, Resource utilization, Throughput, Housekeeping, Resilience, Efficiency, Control, Productivity, Visibility, Space
– Volume flexibility, Service levels, Inventory, Output, Autonomy, Cost
Localized & output storage
+ Variety/volume flexibility, Capacity management, Monitoring/feedback, Housekeeping, Service levels, Output, Visibility, Resilience, Space, Control
– Resource utilization, Inventory, Efficiency, Autonomy, Response, Productivity, Throughput, Cost
Best or preferred configuration of options

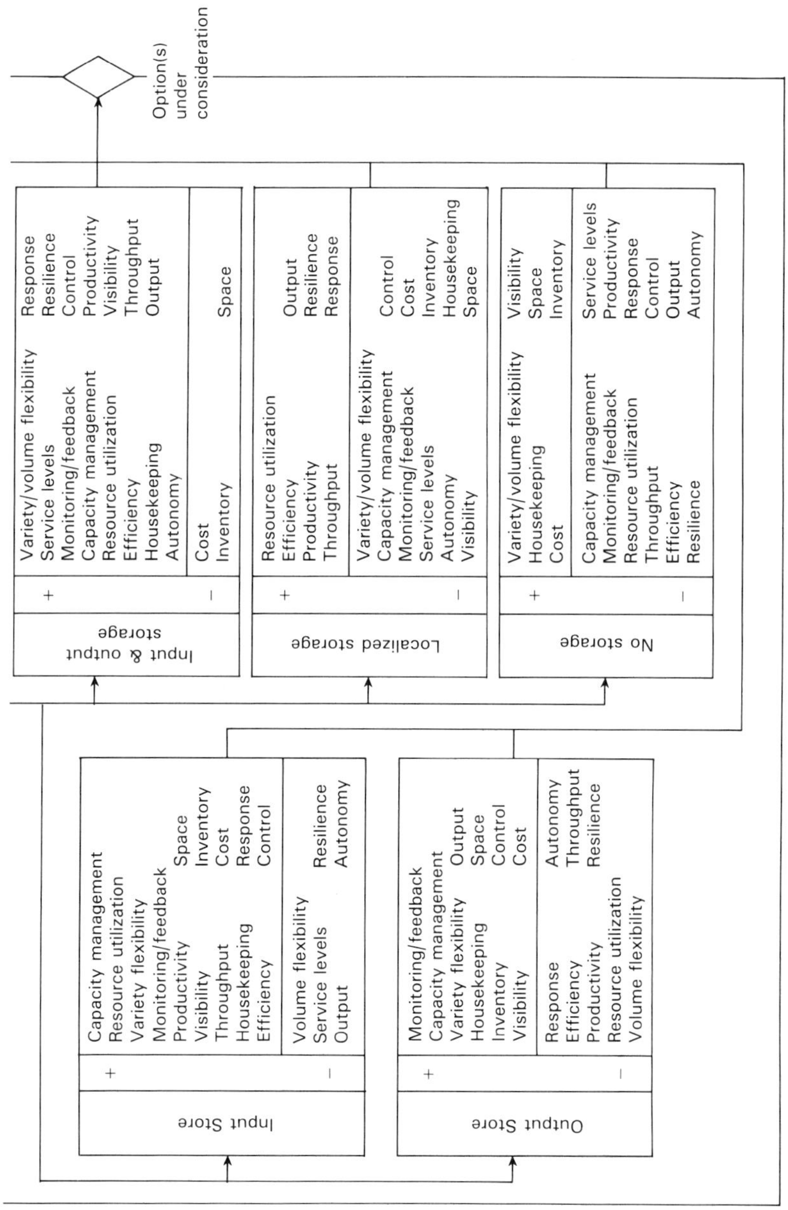

Figure 10.6 Design option guide for storage (module level).

influenced and constrained by choices made at the factory level. The design options are now explored.

Input store

An input store is a store of materials and parts waiting to be processed within the module. The use of input stores enables module managers to have immediate stocks of incoming items within their control. The reason for maintaining an input store is normally to ensure that the work area is supplied with a continuous flow of components and material items. Input stores are becoming more frequently used in production operations when parts kitting is adopted, and so in these instances will contain the kits of parts to be assembled in the module. When used in operations under 'just-in-time' (JIT) control, the input store can also be seen as analogous to a kanban square, an area marked off on the factory floor for a certain volume of storage which, when nearly empty, signals the need for replenishment from upstream operations (the conveyance kanban). The chief advantage of using input stores is that of response to demand and control through visibility for the module leader. However, there is a counter argument to say that the input store adds yet another opportunity to keep stocks in the system, thus it is debatable whether input storage at the module level actually reduces inventory.

Output store

An output store will hold completed components or products that have been produced within a production module and services the downstream operations with a supply of these items. As with input storage, the output store can be used in production operations in a parts-kitting environment, providing a supply for kitting operations in succeeding modules. Likewise, it can also be likened to a kanban square in a JIT operation, being used to trigger the production of more parts or products for replenishment purposes (the 'production' kanban). Advantages and drawbacks are similar to those for input stores, but maintaining stocks of finished items throughout the factory will cause a reduction in manufacturing throughput times from the point material enters the factory to the time it leaves as part of the finished product.

Localized storage

Localized storage is, in effect, a less sophisticated form of decentralized storage than the input and output stores outlined above. The main purpose is to provide a supply of readily available parts and materials in close proximity to the production operations rather than merely in an often distant central stores area. The main benefit of localized storage is the expected improvement in response and throughput times as a result of having materials on hand. The main drawback lies in the probable increase in inventory levels due to the potential for parts storage duplication, the demands upon space on the factory floor and the resultant cost consequences of these.

Combinations of the above options

Figure 10.6 shows a number of combinations of input, output and localized storage at the module level. Of particular note is the dual adoption of an input and output store within the context of a kanban control system. It can be seen from the design option guide that this is expected to be the most advantageous combination from those offered.

No storage

In the design of an individual module the decision may be made to have no storage areas at all, even if the organization as a whole has decided upon some form of decentralized storage at the factory level. The advantage of merely drawing requirements from the central store include potentially lower costs, reduced space requirements, faster throughput times and improved housekeeping in the area concerned. The main drawbacks are those of increased throughput, reduced responsiveness in satisfying demand and a lack of control at the module level.

Intramodule transportation (module level)

Whereas the design option guide in Figure 10.4 considered transportation between different production and storage areas within the factory, Figure 10.7 addresses transportation within individual modules. Once more the options are automation, partial automation and manual systems. Each is now considered in turn.

Automation

Full automation of materials movement within a production module can provide benefits for improved control, response and productivity increases, but these may be offset by higher cost and other factors, both in the design and operation of the new system.

Partial automation

Partial automation may be more attractive from the point of view of cost of development, risk and maintenance than full automation, but may not provide the same benefits once implemented. Partial automation may also represent the first phase of a move towards fully automated materials handling, thereby

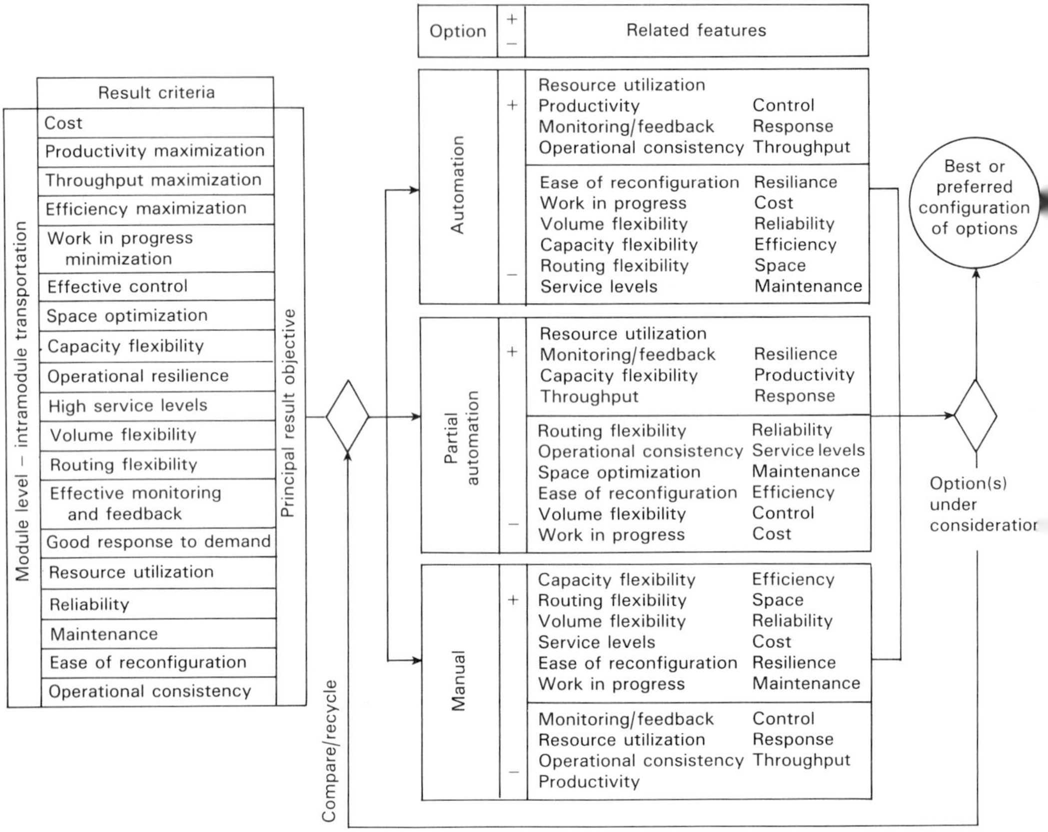

Figure 10.7 Design option guide for transportation (module level).

providing a 'stepping-off' point if full automation appears initially to be too expensive or difficult to implement.

Manual

The use of manual transportation may be seen as appropriate should the costs of automation be seen to outweigh the benefits in terms of reduced operational costs and increased productivity and responsiveness. Indeed, in most organizations it can be viewed as the 'safe' option, requiring least change to the status quo and established methods of working. However, the continuing and escalating cost of labour needs to be taken into account when considering the use of manual transportation, particularly as automated material handling is becoming more efficient and costs less.

Physical integration (module level)

Integration in terms of control systems will be covered in Chapter 11. However, the system designer must also have regard to the way in which the physical parts of the production module (the workstations, the storage points and the transportation system) will link together. Figure 10.8 gives a list of result criteria upon which the success of integration should be judged and also provides two basic options, namely the 'integrated' design and the 'segmented' design.

Integrated design

An integrated module design is one where the workstations, storage and materials handling system are contained within one single facility. Examples

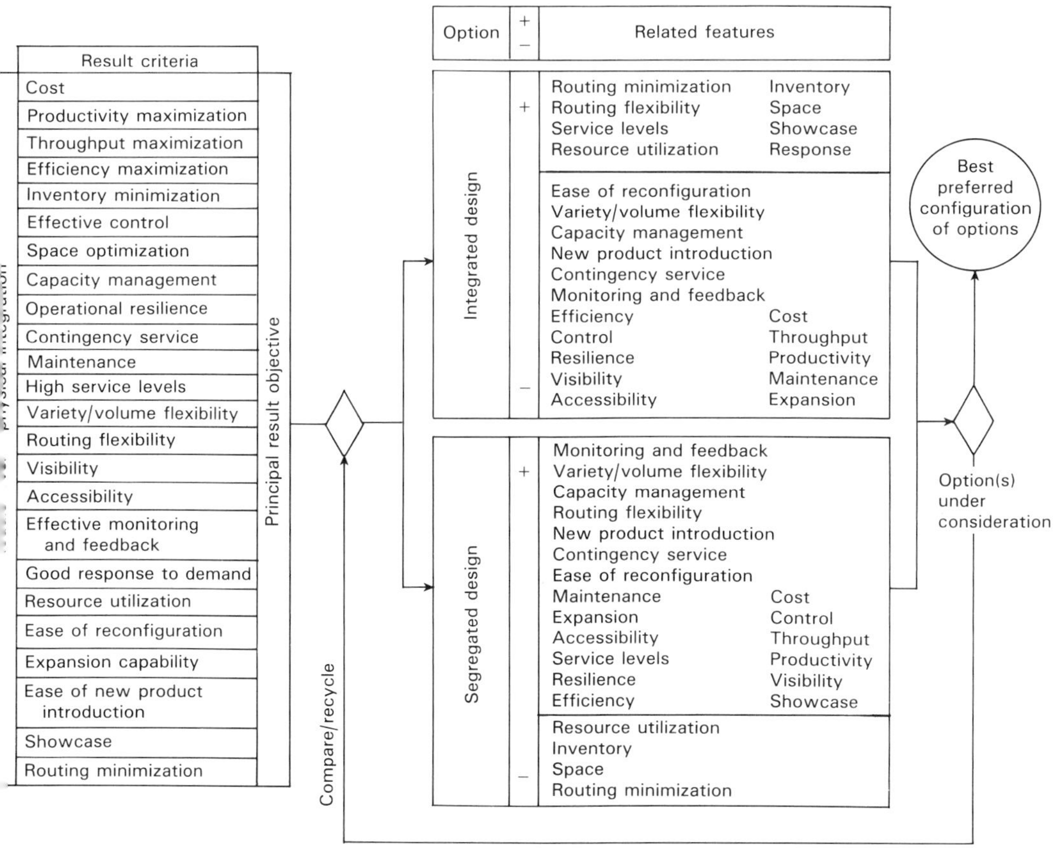

Figure 10.8 Design option guide for physical integration (module level).

include a conveyorized layout, where materials are stored on the conveyors with workstations located alongside, or a crane racking system, where storage is in racks served by a high-level robotic crane, which supplies workstations up and down its run. Benefits of an integrated physical system include simplified routing, constrained levels of inventory and space saving. The drawbacks include inflexibility for longer-term expansion and adaptation for increased volumes or new products as a result of the system's holistic property.

Segmented design

The alternative to an integrated physical design is that of a segmented design. Here workstations, storage points and the transportation system are physically separate, being integrated instead by the control system. Whereas this usually results in increased space requirements and higher levels of inventory, the segmented design offers important advantages in terms of resilience to change, as smaller discrete parts of the module can be replaced or adapted in the light of new products or changed volumes. The segmented design also offers increased contingencies to cope with disruptions such as downtime, as parts of the system can continue to work even if there is a failure of one station or of one element in the transportation system. If part of an integrated module fails it is likely that the whole system will stop.

Tooling and work instructions (module level)

An important consideration within the design of any module is the location and availability of tooling for machines and operators and, in a human-centred module, the provision of work instructions. The principal objective here is normally the effective utilization of resources available, the need to control costs and to optimize efficiency and response. The design options, with their related positive and negative features are shown in Figure 10.9.

Tooling

The key decision in the provision of tooling, whether this is simple hand tooling for operators or sophisticated dies and cutting equipment for machines, is the amount of duplication the factory is willing or able to bear. Centralized tooling is frequently chosen where costs of equipment are high and resource utilization is a chief concern, although it may also be the preferred option where special storage conditions are required or tooling utilization is very low, so one tool can serve the requirements of several or all the workstations. Centralization has the added advantages of good housekeeping and space utilization, but can adversely affect efficiency and productivity compared with the situation where tools are readily to hand at the workstation. Distributed tooling is where tools are located close to or at each of the

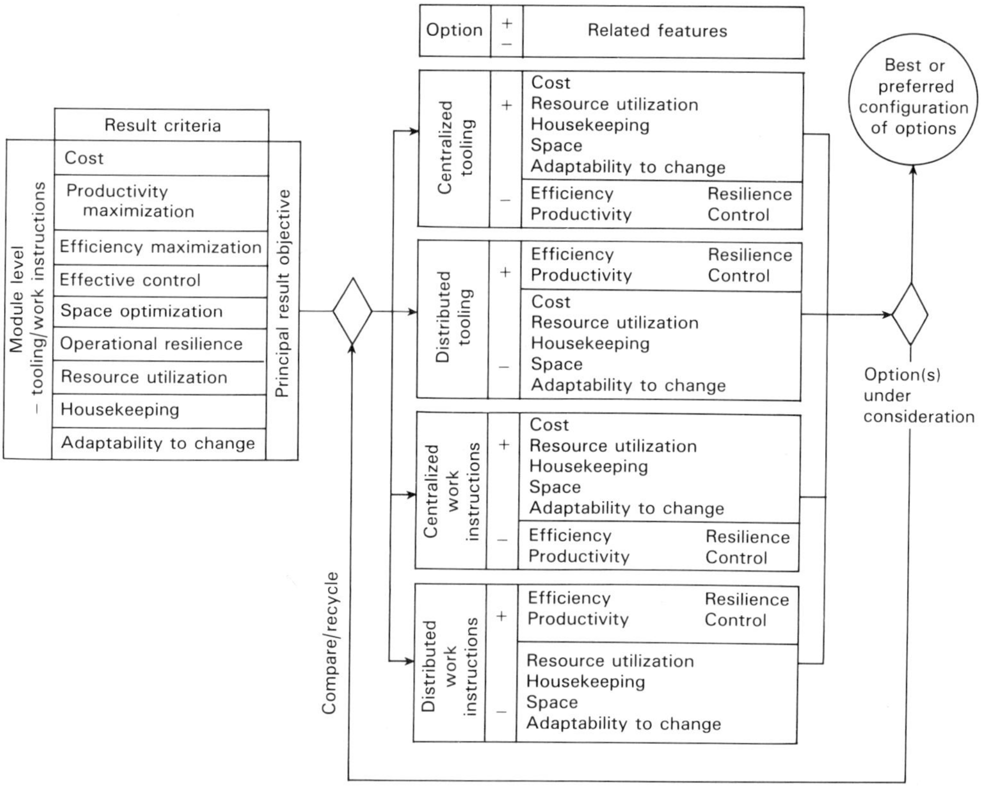

Figure 10.9 Design option guide for tooling and work instructions (module level).

workstations. Costs of possible duplication can be high and resource utilization low, but for cheaper and frequently used tooling items it offers the advantage of eliminating lost time in the procurement of equipment and potentially reducing set-up times during machine or operator facilitated changeovers.

Work instructions

In human-centred production modules the provision of instructions detailing methods of work are very important, both in the training of operators to do tasks and in the maintenance of agreed and established procedures. However, the question of whether these instructions should be centralized or distributed is again an issue. The rule of thumb developed above for tooling should again come into play, specifically that expensive or low usage work instructions should be centralized whereas inexpensive or high usage instructions should be distributed to the operators.

Storage (utility level)

The principles of storage and the generalized options at the factory and module levels were shown in Figures 10.3 and 10.6. At the utility level these general decisions are translated into specific hardware and equipment choices. Figure 10.10 provides the design option guide for storage choices at the utility level and the following text describes these various options.

Storage bins

Storage bins represent a very simple form of storage and have the advantages of very low cost to install, easy to maintain and adapt, and simple in design.

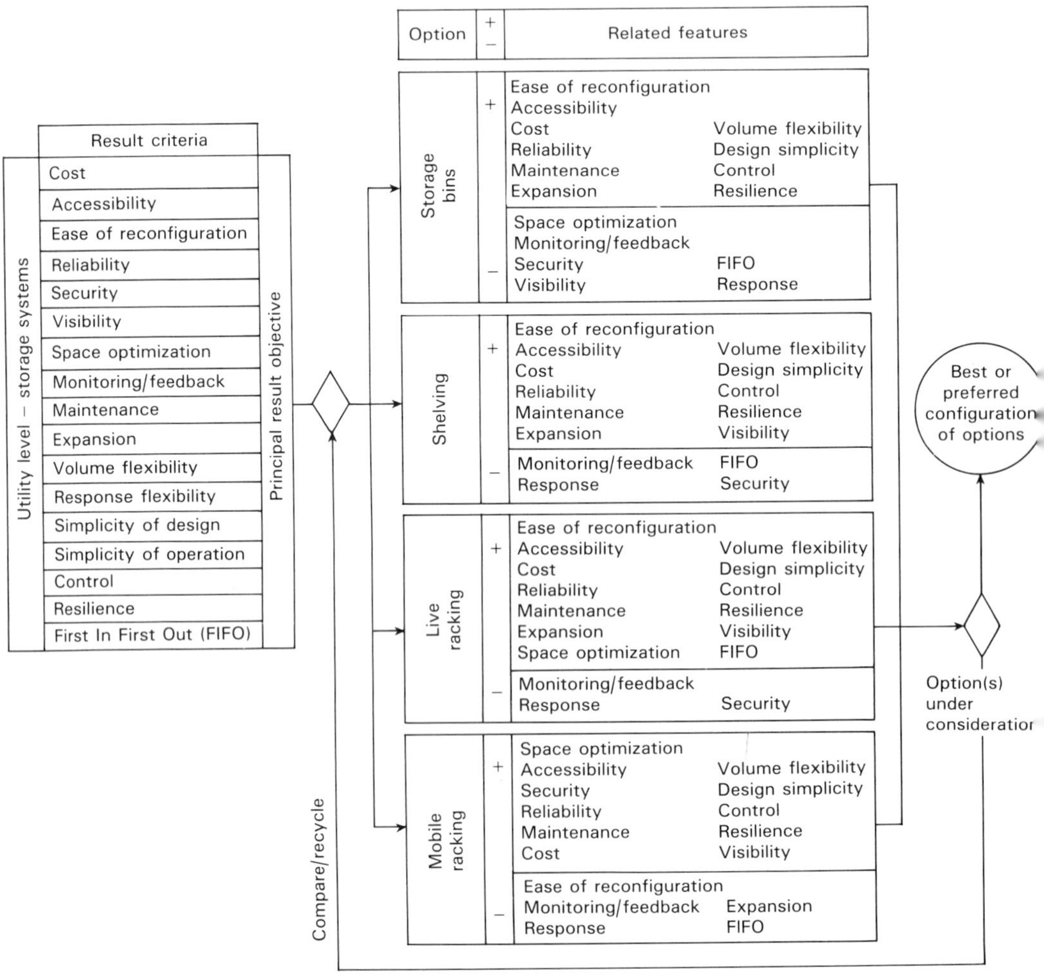

Figure 10.10 Design option guide for storage (utility level).

However, they have the disadvantages of being potentially insecure and difficult to operate on a 'first-in, first-out' (FIFO) basis. For low value items, however, they can provide a highly cost-effective means of controlling stock, particularly when used in a 'two bin' control operation (see Chapter 11).

Shelving

Shelving is probably the most common means of storage, whether in its simplest form, with manual pick and place, or as part of a sophisticated crane and high racking system. Generally, shelving is a low-cost option which offers many of the advantages associated with storage binning, but with the added benefits of improved visibility and improved presentation of parts and materials, thereby enabling easier accessibility.

Live racking

A derivative of simple shelving, a live racking system involves taking items from the front of the racks whilst replenishments are fed in from the back, thus offering the advantage of providing a FIFO environment.

Mobile racking

Mobile racking can range from simple manual systems, where materials can be moved on the storage shelving to more sophisticated automated systems such as vertical (paternoster) and horizontal (carousel) systems. In some cases the latter may be equipped with robotic pick and place devices. These can offer a number of benefits including security, utilization of space and building volume and improved efficiency in terms of the rapid procurement of parts and materials.

Application 10.3 Advanced automated vertical storage systems

Occupying minimal floor space and making full use of available headroom, Linvar's automated vertical storage systems function on the principle of the carousel. Stock is held in containers located on shelving forming part of horizontal carriers, which rotate vertically within an outer casing. When items are selected by part number or storage location, the shelving always moves to the picking station by the shortest direction, so that they are accessed with the minimum of delay. Average retrieval time for any one of up to 3400 individual storage compartments is just 10 seconds. Apertures through which objects are reached can be at ground level, first floor level, or both.

Computer controls are operated via a keypad, to which passcodes can be introduced to prevent use by unauthorized personnel. Information is displayed on a VDU. Interface ports allow the connection of barcode readers, ticket printers, weighing scales or another computer.

Bespoke programs can be written for these automated systems if required. Alternatively, a range of standard software packages of different levels of complexity is available. The Linvar PC stock control system is capable of commanding a number of units, and can display continuously up-dated stock totals. User group meetings are arranged by Linvar on a regular basis, at which customers can discuss with company personnel how best to utilize this system in the light of their specific needs.

When not in use the unit can be closed and locked to make it secure. This also protects its contents from dust and environmental deterioration. Operator safety is of the highest standard – light beams and trip bars stop the machine immediately if something is amiss.

Four types of vertical storage system are available from Linvar, all being 3.15 metres wide while ranging in height from 2.34 to 7.5 metres and in depth from 1.25 to 1.9 metres. A maximum of 62 shelves can be accommodated (in the largest model), each carrying up to 400 kg, and fitted with a wide range of containers and other accessories.

As the result of a distributor agreement, units can also be equipped with the computerized tool management system (CTMS), developed by Amazon Computers. CTMS consists of a number of modules, the core of which controls the issue and return of all tool items while automatically up-dating data on their status and movements. The addition of further modules provides increasing degrees of sophistication in tool management and monitoring.

Source: by kind permission of Linvar Limited.

Transportation (utility level)

Design options for transportation systems at the factory and module levels concerned the basic choice between automated and manual systems (see Figures 10.4 and 10.7). At the utility level detailed decisions on specific materials handling equipment have to be made and the design options offered by DRAMA are shown in Figure 10.11.

Manual trolleys

The use of manual trolleys has a number of advantages, not least the low cost of installation and the inherent flexibility and adaptability afforded by a human operator. However, in many instances the use of manual trolleys for transportation may not be as efficient and cost-effective in operation as other forms of materials handling. Less control, especially computerized control, can be exercised over manual systems and they will always be prone to the inevitable errors that humans will make from time to time. Despite these

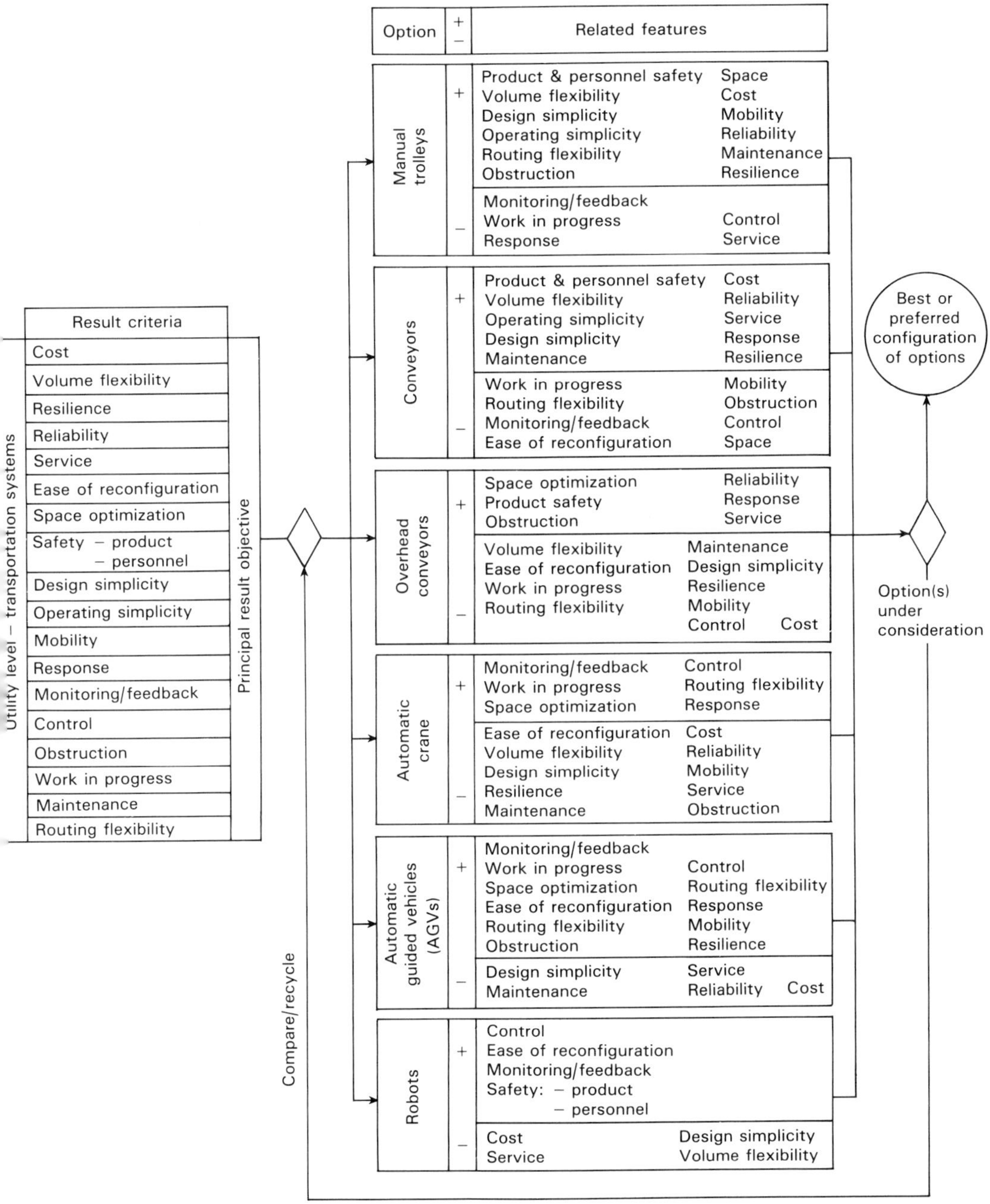

Figure 10.11 Design option guide for transportation (utility level).

drawbacks, manual transportation systems are most appropriate where development costs are heavily constrained and where the organization is more concerned with risk avoidance.

Conveyors

Conveyors can provide an efficient method of parts and materials transportation and their technology is now relatively well advanced. The choice of a conveyorized system is therefore a relatively safe option. However, conveyors do have significant shortcomings including a lack of routing flexibility, difficulties in reconfiguration, the tendency to have work-in-progress build-up on them, the space they require on the shopfloor and the physical obstruction they represent within the factory for personnel and materials movement. Given these caveats the conveyor is widely acknowledged as most appropriate where production volumes are high and materials movement standardized and unidirectional, hence their frequent association with flowline facilities.

Overhead conveyors

These serve a similar purpose to conveyors, but with certain advantages in particular situations. They do not obstruct the factory floor in the way that floor conveyors do, nor do they have the same demands on space. In certain situations they provide physical advantages in terms of being able to drop and lift materials into and out of workstations and processes (e.g. in and out of vats and furnaces).

Automatic cranes

Automatic cranes are becoming increasingly popular for transporting materials within and between work areas, as well as for high level storage 'pick and place' activities. They make more efficient use of space than conveyors and also facilitate high levels of computer control. However, they have a fixed, usually linear, routing and are difficult to reconfigure when modifying existing facilities.

Automated guided vehicles (AGVs)

AGVs are robotic transportation carts which function under direct computer control which transmits movement instructions to the vehicle. AGVs represent an increasingly attractive option for the systems designers as the technology continues to evolve. They offer flexibility in routing, especially where infra-red signals, radio control or painted lines, etc., are used as the media instead of the longer-established buried wire. The use of AGVs also eliminates the obstruction incurred by other materials-handling systems, such as conveyors and cranes, so do not physically constrain manual movement within the workplace. In addition, the use of AGVs can be used to limit the amount of work-in-progress within the factory and can be flexible to cope with volume changes, although this may well involve the purchase of additional vehicles and therefore significant further investment. Although, in comparison to conveyors, AGVs are relatively flexible in terms of materials routing, flexibility is nevertheless constrained to the layout of the track design. However, they offer the benefit of reconfigurability and are less obstructive to the physical environment and operation of the factory. The main problem of AGVs is the cost involved in buying and then developing the hardware and control software, and the consequent risk involved with this substantial investment. (A comprehensive review of AGV specification, development and implementation can be found in Castleberry, 1992.)

Application 10.4 AGVs for warehouse applications

The AGV installation is in a warehouse in the UK that is owned and operated by a leading third-party warehousing and distribution company. Operating 24 hours a day, seven days a week, it receives, stores and distributes canned products. Pallet loads are received in mixed batches of 18 pallets on curtain-sided trailers fitted with conveyor beds. Pallets are stored within the warehouse in double-deep racking. There are 44 rows of this racking with 22 aisles. At each aisle there is a gravity roller input station and a separate output station, each with the capacity to hold two pallets. Pallets are put away and retrieved by narrow-aisle trucks fitted with double-reach forks.

Why AGVs and not conveyors?

The operational requirements of the installation were studied and closely specified by consultants. Four suppliers were invited to tender for a conveyor solution with the option of AGVs. Two suppliers were shortlisted: one offered a single 'carousel' conveyor with powered input/output stations, the other offered AGVs with gravity input/output stations. The decision to opt for AGVs was taken following the discussion of the following factors:

1. The initial capital cost was virtually the same for both solutions.
2. While both provided an automatic link between incoming trailers and the aisle input/output stations, only the AGVs provided automatic transfer to the marshalling lanes. This link could have been provided by conveyors, but at additional cost and complexity.
3. AGVs left the floor open, which allowed unobstructed access to all parts of the warehouse; this was particularly relevant in the marshalling area.
4. AGVs eliminated all fork-truck handling other than dispatch vehicle loading at the front of the docks.

5. The AGV solution dramatically reduced the number of sensors in the system, and sensors are still the main cause of system stoppage. While sensors can normally be repaired quickly, the frequency with which they fail means that constant maintenance cover is needed.
6. Other than those on the AGVs themselves, the AGV solution eliminated all drive motors and gearboxes, lift and lower tables, divertors and turntables. This significantly extended the mean time between failures and avoided critical conveyor component failures, which could stop the flow of both inputs and outputs.
7. When an AGV fails, it can be taken out of the system while the remaining AGVs continue to operate normally, which means that immediate maintenance work is not necessary. Failure of a conveyor system, on the other hand, requires immediate attention and therefore constant maintenance cover. This is a major on-going cost in a warehouse which otherwise does not require such attendance.
8. In the event of the catastrophic failure of the AGV system, forktrucks can be drafted in to transport pallets to and from the aisle input/output stations.

Drawbacks of AGVs

Despite the benefits the AGV system has over a conveyor system, it is vulnerable in three main areas: (1) if there is a loss in the communication link with the warehouse computer; (2) where there is a failure of the AGV communication loop; and (3) when the AGV controller fails.

If the link with the main computer goes down, it can be overcome, with some loss of capacity, by manually entering data into the AGV controller into the warehouse computer via a VDU terminal. However, with the AGV controller and the main computer being located in the same building, loss of communications between them should be infrequent; it should also be noted that such a loss in communications would affect a conveyor system equally. There is no standby provided if the AGV communications loop fails, but as these are historically reliable a standby is not necessary.

Failure of the AGV controller is most likely to be caused by a circuit-board failure, so plug-in spares are held on site and diagnostics are built into the system. Occurrences of this sort should be rare after the first few weeks and will only involve an emergency call out if personnel on site cannot cope. A spare control unit is held on site in case of a major failure.

The cost of adding more AGVs into the system as its throughput increases is offset by the fact that constant maintenance is not required, fewer forktrucks are required because of lower power costs. It should be noted that the conveyor solution offered an inherent capacity beyond that provided by the maximum of 15 AGVs. This comes about because of the response time to pallets on the aisle output stations provided by a conveyor solution. The time taken for an AGV to respond will depend on where the AGV is relative to the aisle in which the output is waiting. While this has been partially offset by the provision of AGV homestations along the length of the AGV route, response times of between two and four minutes will be normal for an installation with over 50 AGV destinations. Although conveyors will consistently clear output stations within 30 seconds, this was not considered to be sufficient to outweigh the advantages of the AGVs.

The weight given to each factor when deciding whether to go for an AGV or a conveyor solution will vary between installations. For example, if the installation is in a factory where existing maintenance personnel are on call, the difference in maintenance costs between the two solutions will probably carry little weight. This stresses the importance of an informed and detailed study of the requirement and conditions before a choice is made.

Source: excerpts from Gordon Shimmings (1990). AGVs – warehousing on the right track. *Manufacturing Engineer*, **April**, 20–21.

Robots

A robot is a reprogrammable device able to manipulate and move materials and tools between machines and operators (a definition based upon that of the Robot Institute of America). Given this definition, one could argue that automatic cranes and AGVs are types of robots, and indeed they do belong to the wider family of reprogrammable devices. However, the term 'robot' is normally associated with more specific materials or tool-handling apparatus.

In this respect is has been said that a robot is a device having three principal components (Hunt, 1983):

1. One or more arms, usually situated on a fixed base, that can move in several directions.
2. A manipulator, the 'hand' of the device which holds the part to be worked or moved.
3. A controller that gives detailed movement instructions.

Robots can be seen as offering the advantages of computer control, ease of reconfiguration and safety in terms of being able to work in hazardous areas or moving awkward items which would otherwise have to be done by people. However, robotics are a major investment for any company so the questions of cost of purchase, development and maintenance must be balanced against the positive features of the technology (for further discussion on robots see Bennett, 1986).

■ 10.4 Generalized methodology

The generalized methodology flowchart summarizing the design process for physical system design is shown in Figure 10.12. As one can appreciate from

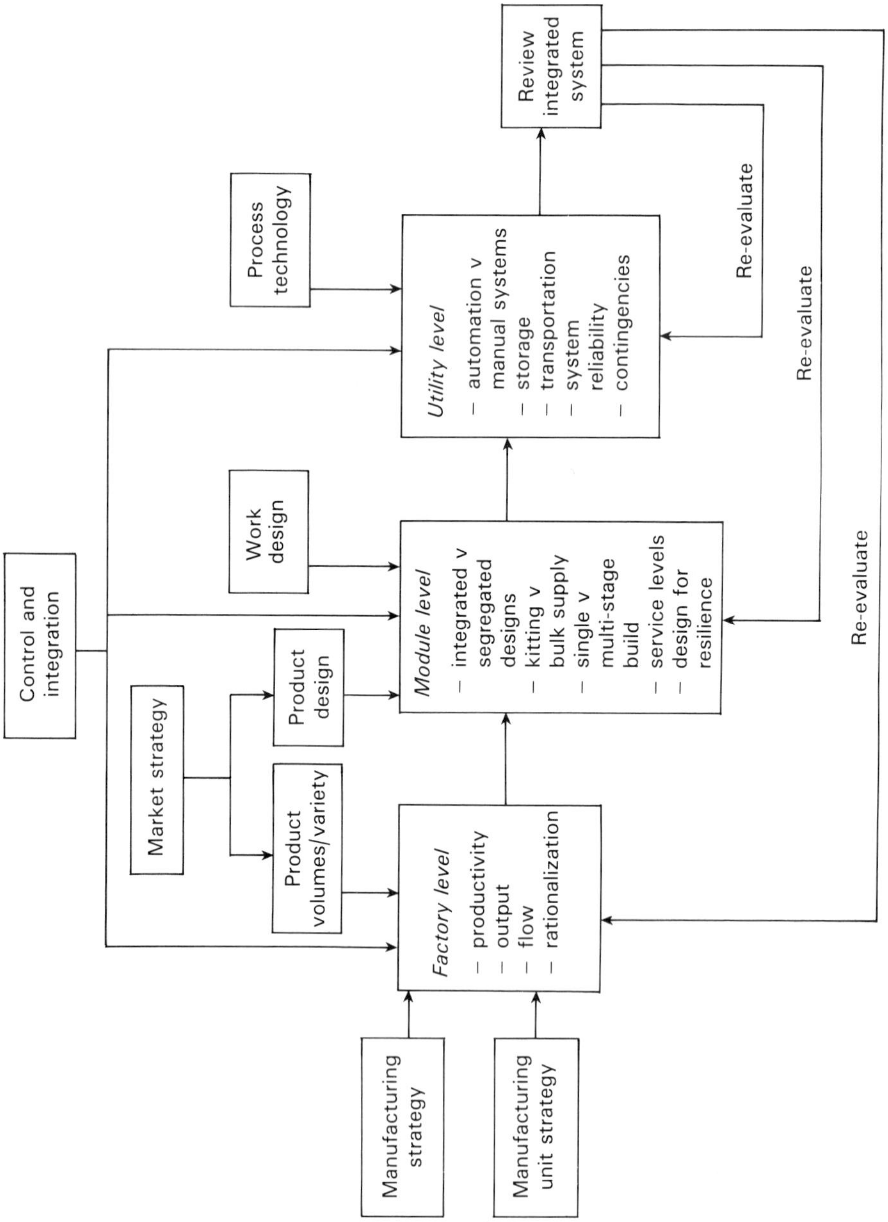

Figure 10.12 Generalized design methodology for physical system design.

the large number of design options presented in this chapter, design of the physical system involves a great many decisions at the utility level of the system. The methodology flowchart, therefore, does not go down to a very fine level of detail, but instead provides four basic steps for the system designer.

1. **Develop 'factory' level system design.** Use the DOGs in Figures 10.2 to 10.4 to develop factory-wide principles and guidelines for layout, storage and transportation.
2. **Develop 'module' level system design.** Decisions are then made for individual modules within the overall manufacturing system. For each module, decisions should be made concerning the layout, storage facilities, transportation and materials movement, physical integration, tooling and work instructions (see Figures 10.5 to 10.9).
3. **Design and incorporate facilities at the 'utility' level.** In addition to choices on process machinery and work station design, which tend to be unique to individual industries, the DOGs in Figures 10.10 and 10.11 can be used to assist in the choice of storage and materials-handling hardware.
4. **Review and evaluate integrated physical system design.** The design principles and physical system operation at the factory, module and utility levels should be assessed periodically and amended where necessary to satisfy requirements of the system.

The development of physical system design must be conducted in parallel with control and integration and work design in order to maximize the potential for creating a flexible infrastructure and to ensure the emergence of a market-oriented organization. Ways of enabling the development of an effective design include the rationalization of material flow and the alignment of separate modules in the manufacturing system along coherent routes of flow. Finally, the design of large systems and the automation of these should be monitored continuously to assess the appropriateness of the existing system towards the objectives of creating an effective, but flexible manufacturing infrastructure.

■ 10.5 Conclusions

This chapter has reviewed the literature and practice in the areas of layout, automation, flexible manufacturing and the process of physical systems design. It was suggested that a key concern in the development of modern manufacturing systems is achieving an appropriate balance between productivity of the system in terms of costs and efficiency and the flexibility offered to enable it to respond to changing markets, customer requirements and product designs. Design options for the physical system designer were then presented. A valid way in managing the factory wide transition from being technology-led to market-focused is to analyze the design of the physical system at the three levels of factory, module and utility.

The primary objective at the factory level is to ensure that output, productivity and service levels are maintained. Systems designers therefore need to establish appropriate designs and arrangements of modules that will promote manufacturing flexibility and responsiveness to market demand. The design of a module within the factory will obviously be greatly influenced by the product design and their associated processing requirements. Thus the variety of products catered for by the module will dictate its flexibility requirements through the combined configuration and design of storage, workstation and materials-handling systems. At the utility level, detailed and specific decisions are made concerning the hardware and processing kit for workstation design, storage and materials-handling systems. The options at this level are multitudinous and vary considerably from industry to industry. However, some fundamental options and considerations are presented in the DRAMA methodology and some of the more common options were evaluated in the chapter.

■ 10.6 Further reading

Critchlow, A. J. (1985) *Introduction to Robotics*, Macmillan, New York.

Groover, M. P. (1987) *Automation, Production Systems and Computer Integrated Manufacturing*, Prentice Hall, Englewood Cliffs, New Jersey.

Teale, D. C. (ed.) (1987) *New Frontiers in Manufacturing*, Springer-Verlag, London.

Tompkins, J. A. and J. A. White (1990) *Facilities Planning*, Wiley, New York.

11 Control and Integration

Objectives of the chapter

On completion of this chapter you should:

1. Appreciate the fundamental principles of push and pull inventory control systems.
2. Understand the role of management information and control systems in the running of production operations.
3. Have available a series of design options and a process methodology to guide you through the design of control and integration systems for manufacturing.

Chapter outline

This chapter considers issues of operations control and the integration of the various subsystems or modules of the overall production system through computer software and physical control procedures. The various forms of inventory control systems and scheduling techniques, including statistically based techniques, material requirements planning, just-in-time and optimized production technology are described and analyzed in detail. It is suggested how, more than merely controlling stock and sequencing job orders, the type of inventory and planning policy adopted has implications for the operation and effectiveness of the production system as a whole. Another important concept in control and integration is that of information, and the theory and practice relating to managing information within the organization is reviewed. After identifying the key parameters upon the process of developing control and integration for production, the options available for the system designer are categorized into those concerning intermodule material control, materials distribution, data collection and analysis, and production planning and control. The design process as a whole is finally summarized in a generalized methodology flowchart.

11.1 Theory and current practice

Control and integration is considered under two main areas in this chapter: first, production planning and control and, second, information control. The production planning and control topic is divided into four main headings: (1) statistical stock control; (2) material requirements planning; (3) just-in-time (JIT) management; and (4) optimized production technology. These will now be discussed in turn.

Statistical stock control

One of the major challenges facing the manufacturing manager is that of controlling materials and inventories. The general aim is to ensure that just enough stock is on hand at any point in time to satisfy requirements, but that these holdings should not be excessive, tying up large amounts of working capital and incurring high stockholding costs.

This section deals with so-called 'classical' inventory control. Resulting largely from the work of operational researchers in their quest to optimize stock holding, much of the content within this section centres upon so-called 'economic batch quantity' theory. This theory is, in effect, an attempt to balance the cost of holding inventory with the cost of ordering. Consideration may also be given to the consequences of running out of stock by evaluating the amount of safety stock required to maintain particular 'service levels'.

In recent years economic batch quantity (EBQ) theory has received many criticisms, particularly regarding its theoretical basis as an optimizing technique. The use of policies based upon EBQ has also been demonstrated to have shortcomings in practice, particularly in the face of 'requirements planning systems' and Japanese inspired 'JIT' control systems. The final part of this section therefore offers a critique of statistical inventory control and highlights some of the problems in its use. A summarized view of statistical stock control theory and practice will now be given. A more detailed description of statistical stock control can be found in Lewis (Lewis, 1981) and an alternative overview in Muhlemann *et al.* 1992.

The design of any stock control system is shaped by a number of parameters, which result in a balancing of (often conflicting) requirements. These parameters are described below:

1. **Demand and replenishment.** The design of any stock control or purchasing system needs to take into account the following aspects of demand:

   (a) the average demand per time period (A);
   (b) the variability of demand per period, usually expressed as a standard deviation (σ);
   (c) the 'lead-time', order to receipt, of material (L).

2. **Ordering costs (Co).** There are fixed costs involved each time a separate order is raised. If orders are placed too frequently the total cost of ordering becomes excessive.

3. **Storage costs (*i*Cm).** The costs of holding material in stock, including direct costs of storage, deterioration, obsolescence, insurance and opportunity costs. Expressed as a percentage (i) of the prime cost or price paid for materials (Cm).

4. **Service levels.** A measure of the efficiency of a stock control system may be indicated by the level of service it provides for its customers (internal and external) in satisfying demand ex-stock. In designing a stock system, the company usually chooses the service levels it feels are appropriate to satisfy the customer without resulting in excessive costs of holding. The most common vendor service levels adopted in practice are 95 per cent and 99 per cent (i.e. a 5 per cent and 1 per cent risk, respectively, of not being able to supply from stock).

These parameters, with their mathematical notations, are used in the design of statistical inventory control systems. There are two basic inventory control policies commonly used, 'reorder level' (ROL) and 'reorder cycle' (ROC).

Reorder level policy

The reorder level policy involves placing a fixed order, usually an economic batch quantity, to replenish inventory when the stock level for a certain item falls below a fixed value, the 'reorder level' (M). It requires the calculation of two values, the order quantity (EBQ) and the reorder level, which allows sufficient safety stock to provide a predetermined service level. The formulae for these are as follows:

$$\text{Economic batch quantity, EBQ} = \sqrt{\frac{2A\text{Co}}{i\text{Cm}}} \qquad (11.1)$$

$$\text{Reorder level, } M = AL + k\sigma\sqrt{L} \qquad (11.2)$$

where demand variation is assumed to follow the normal distribution and: A, average demand per period; L, lead-time in time periods; k, number of standard normal deviates to provide a chosen service level and σ, standard deviation of demand. (NB: $k = 1$ where vendor service level (VSL) = 84.1 per cent; $k = 2$ where VSL = 97.7 per cent and $k = 3$, where VSL = 99.9 per cent).
So the reorder level policy incorporates:

1. A fixed order quantity (EBQ).
2. A variable order cycle (time between orders), as determined by the rate of usage between orders and triggered by depletion below M.

Figure 11.1 illustrates the operation of a reorder level policy stock system.

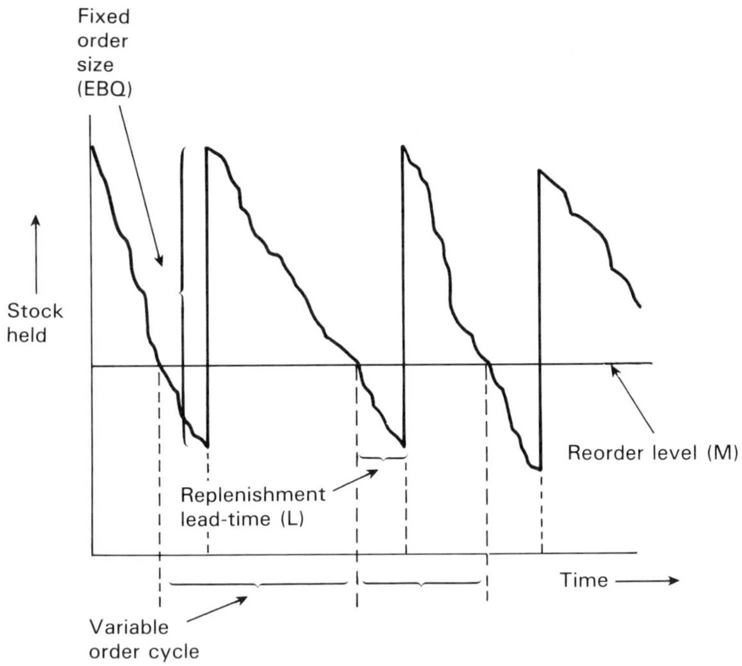

Figure 11.1 Stock control using a reorder level policy.

Reorder cycle policy

Under the reorder cycle policy (ROC) a replenishment order is placed at predetermined times with a fixed period between orders. At each periodic review the stock on hand is compared to a calculated 'maximum stock level' (S) and the difference between the two becomes the order size. So, contrary to the reorder level policy, reorder cycle has:

1. A variable order quantity.
2. A fixed order cycle, usually an economic review period.

The reorder cycle policy requires calculation of the review period (R) and the maximum stock level (S). The formulae for these are as follows:

Number of reviews per year

$$r = \sqrt{\frac{AiCm}{2Co}}; \qquad \text{therefore } R = 52/r \text{ (weeks) or } 12/r \text{ (months)} \qquad (11.3)$$

$$S = A(R + L) + k\sigma\sqrt{(R + L)} \qquad (11.4)$$

The operation of a reorder cycle policy system is shown in Figure 11.2. The reorder level and reorder cycle policies have a number of relative advantages

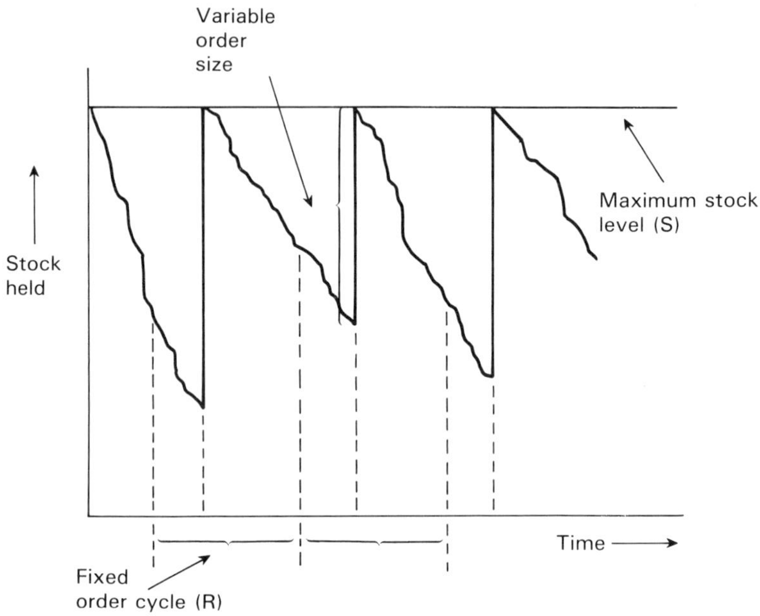

Figure 11.2 Stock control using a reorder cycle policy.

and disadvantages associated with them in practice. The ROL policy has the advantage that, in comparison with the ROC, it tends to operate with lower levels of average stockholding and can be implemented as a 'two-bin' system for physically small, less expensive, stock items. However, under ROL, it must be possible to raise replenishments at any time, and it can hinder the practice of placing multiple orders on one supplier. The relative advantages of ROC over ROL are that orders are raised at specific intervals, leading to a more even purchasing load and multiple ordering from single suppliers. Regular reviews also mean that obsolete stocks do not build up. However, ROC is a riskier policy with a higher chance of stock-outs. Therefore higher safety stocks tend to be held to compensate. ROC can also cause complexities for the supplier, with its variable order sizes for replenishment.

As well as the advantages described above, both forms of statistical stock control policy have been found to have further limitations. To illustrate these, consider the distinction between products which have 'independent' and 'dependent' demand patterns. Items are said to have independent demand where orders for their replenishment does not depend upon the demand for any other item (e.g. finished products and spares). Where demand does depend upon higher level items, such as that for components, subassemblies and raw materials, they are said to have dependent demand. Figure 11.3 shows that some form of requirements planning is more effective for the purchasing and stock control of dependent demand items. Requirements planning, as used for items with dependent demand, is now described.

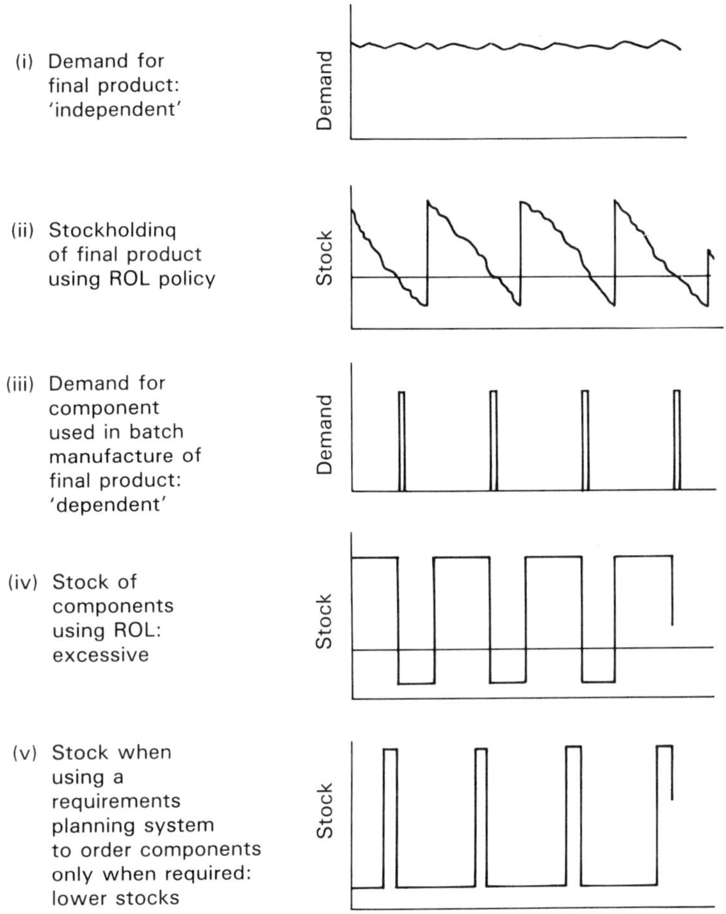

Figure 11.3 Demand patterns and stock levels for independent and dependent demand situations.

Material requirements planning

Material requirements planning (MRP) is a production planning and control technique that has become widely used over the last twenty years. Its development is closely linked to the evolution of computers and information processing. The increasing availability of low-cost computational power has made viable the MRP approach, which involves a great number of calculations in its operation. Indeed, MRP now forms the central module of most commercially available production control software packages.

MRP looks at the scheduled demand for finished products and explodes these down into net order quantities for all subassemblies, components and raw materials (made in-house or purchased), which, together, comprise the finished product.

MRP is suitable for the control of those items whose usages depend upon demand for the finished product. These are said to have 'dependent demand'. Items which have 'independent demand' include the final product itself and spare parts. These would require some other form of control such as a Reorder Level or Reorder Cycle policy.

An MRP system comprises a number of separate modules which integrate to form the overall planning and control package. Figure 11.4 illustrates the basic interaction between the components of an MRP system. Each will now be described.

The master production schedule (MPS)

The MPS is a management commitment to produce certain volumes of finished products in particular time periods into the future. This should not be confused with a sales forecast.

The bill of material (BOM)

The BOM is, in effect, the parts breakdown or 'recipe' for a finished product. It usually consists of the 'Product Structure File', which holds details of parts and materials that make up each finished product, and the 'Product Master File', holding details concerning parts numbers, unit of measurement, lead-time for procurement/manufacture, etc.

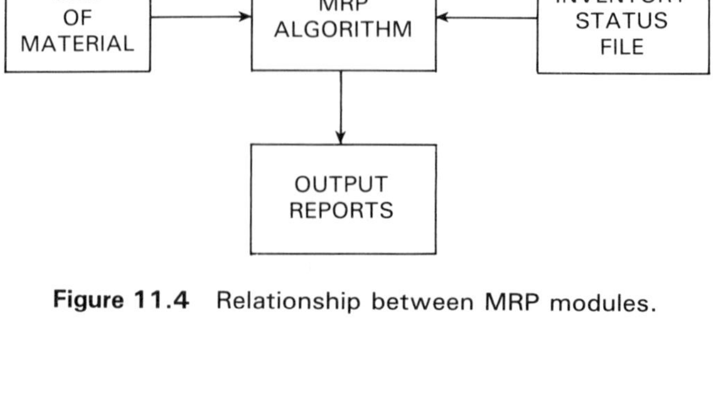

Figure 11.4 Relationship between MRP modules.

The inventory status file

This is the 'stock file', which keeps a record of all stock balances and transactions. Thus it holds information on the quantity, location and condition of materials.

The MRP algorithm

This 'calculation' module uses information contained in the MPS, BOM and inventory status files. It converts these information output reports, which detail the planned times for ordering materials or for initiating their manufacture to satisfy the production schedule for the finished product. The MRP algorithm operates by taking the MPS, relating this to the BOM file and thereby generating gross requirements for each component or raw material. These requirements are then compared with the outstanding stock by referring to the inventory status file so that net requirements (to make or to order) can be determined. Finally, these requirements are offset by the relevant lead-times, used to determine the planned start times at which to initiate manufacture or purchase of materials. This operation is illustrated schematically in Figure 11.5.

So far, we have considered MRP with the assumption that exact requirements

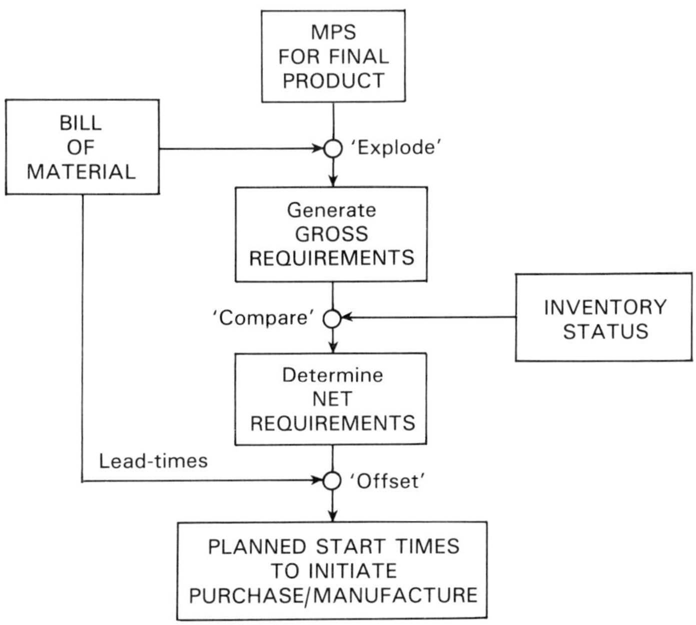

Figure 11.5 The MRP calculation process.

are ordered at distinct points in time. In practice this is over-simplistic. Ordering quantities may need to be batched, for example in tens or dozens, etc. Additionally there may be a need to take account of scrap rates and the set-up times on machines. These complications can be accommodated by making variations to the basic MRP programme. Batch sizing is possible within the operation of an MRP system through the incorporation of a series of rules. These include:

1. 'Fixed quantity' rules, where ordering is only allowed in quantities of specific increments.
2. 'Fixed period coverage' rules, where order times are specified and the material for the next order period is batched and ordered together thus reducing the cost of repetitive, small orders.
3. Complex 'dynamic' rules, for example where total least cost is the objective so a program calculates requirements and order times dynamically by balancing ordering and holding costs.

The setting of safety stocks to avoid the risk of stockouts is also viable within MRP. These can be set in terms of:

1. Fixed quantities.
2. Safety times.
3. Percentage increase in requirements (scrap rates).

The above account provides an introduction and overview of the concept and operation of MRP. More information can be found in the many texts that have been published on the subject (see, for example, Orlicky *et al.*, 1972; Bennett *et al.*, 1988; Krupp, 1984).

MRP has evolved a stage further into a concept known as manufacturing resource planning (MRPII). While MRP deals only with material ordering aspects of operations control, MRPII takes the concept further by considering the resources available. MRPII recognizes when a material requirements plan is infeasible (e.g. when the production schedules generated are greater than capacity) and suggests measures to remedy this. MRPII integrates computer control of materials with other areas of computerization within the business, including finance, marketing, sales, design and engineering. MRPII is often seen as closing the control feedback loop by constant monitoring and upgrading of the computer-based systems of the business. MRPII is therefore not merely a newer version of MRP, as MRP forms the material control module of an MRPII system while MRPII is concerned with the effective operation and control of the wider manufacturing system. (For more information on MRPII see Goodridge, 1986; Harrison, 1990; Krajewski and Ritzman, 1990).

Application 11.1 MRPII . . . and beyond, at Land Rover

Back in 1983, Land Rover (then a relatively independent member of the Rover

Group) was fighting for survival. Many of the 'traditional' high volume export markets, predominantly in Africa, had collapsed. The product range was very dated, leading to decreasing sales in 'development' European and British markets. Its manufacturing operations were spread between one site in South Wales and 14 sites in the West Midlands, each of which operated to optimize its own performance, rather than that of the whole company. Computer support systems had their origins in the late 1960s. Dramatic changes were needed.

Early moves were to upgrade the product range: Range Rover four-door models were introduced, followed by upgrading Land Rover with the introduction of the 110 and 90 models. A full review of commercial and manufacturing strategy was commenced.

During 1984, key decisions came from these reviews:

1. From Commercial, to move Range Rover up market and to enter the USA market, exploit the flexibility of, and continue upgrading, Land Rover products, and to bring to the business a focus of customer needs.
2. From Manufacturing, to consolidate all activities, properly co-ordinated onto the single site at Solihull.

To support these strategies, a new fully integrated suite of systems was seen as a requirement.

By 1985 the commercial strategy had been fleshed out, and site rationalization was progressing well with plans to complete by the end of the year.

In parallel with these activities, a group of senior managers from all disciplines began an in-depth investigation of the systems business needs to support these strategies.

It very quickly became clear that early developments should concentrate on a much improved and more responsive suite of planning systems. Shop-floor control would be achieved by use of simple, robust 'kanban'-type processes. It was further decided that unless absolutely unavoidable, a 'package' solution would be sought for the system elements, rather than bespoke development (this was because of the greater certainty of a working system being delivered within controllable cost and time-scales).

The culmination of these investigations was that elements of an MRPII package would meet most of the business needs determined, but would also demand a change in management style and organization to be fully effective. Adoption of this as a strategy was put to the Directors, with outline costs, benefits and timing, and was agreed. It was recognized that the extent of the implementations planned would not allow this to be achieved because of the lack of shop-floor planning modules.

Amongst the primary benefits targeted were:

1. Reduction from 14 'official' and numerous 'unofficial' bills of material down to one company bill of material and product definition system.
2. Reduction of inventories, both components and unfinished vehicles including obsolescence.

3. Ability to make customer delivery promises, and achieve them.
4. Ability to move away from crisis management, multiple plans and expedite lists to a controlled business environment with a single plan cascading through increasing levels of detail to material and manufacturing plans.

Implementation was achieved through a multidisciplined full-time project team, working with departmental groups and specially trained 'tutors' together with external consultants. Much emphasis was given throughout the period that it was not a 'computer system' but a set of planning and control processes, with people as the critical element in making it work.

Education of a large part of the workforce was seen as essential for success, together with specific and general training for those who would need to use the information within the system. All board members, project team members and a number of key functional and specialist staff underwent external training with MRP Limited and, subsequently, detailed practical training on Land Rovers' specific implementation was conducted by the project team. These individuals were then responsible for cascading the education and training through their own functions.

Implementation of the final modules of the main phase was completed by mid-1987 and the remaining 'old' systems were 'switched off' in November of that year. The hard work of getting acceptable reliability, accuracy of data, credibility of the process and delivery of the benefits still lay ahead.

Development and use of systems and procedures started to deliver positive benefits from late 1988, with substantial progress being made in 1989. During this time, Land Rover also adopted the 'total quality' philosophy. Initial training was again cascaded from the top and proved to be very supportive of the customer focus. Measurement, as a means of identifying improvement areas and thus focusing and directing plans, assisted overall progress. The 'eliminate waste' message of total quality also gave renewed impetus to the JIT activities, which, although progressing, were not as advanced as originally planned, and certainly had not passed the 'critical mass' to become self-sustaining.

Towards the end of 1989, we became aware that the criteria for 'A' class status (as used by MRP Limited) had changed, and now recognized that detailed shop-floor control (using MRP) was not necessary, provided effective control and prioritization of shop-floor activities was demonstrated. At last there was a real possibility of Land Rover achieving 'A' class status.

An internal review was conducted in early 1990 and, after some detailed actions were implemented, MRP Limited were called in during May 1990 to make a preliminary assessment. We worked with MRP Limited during the next few months on the 'weaker' points of our assessment and were awarded 'A' class status in October 1990.

The results achieved to date include:

70 per cent reduction in inventories.

300 per cent improvement in stock turn.

50 per cent reduction in unfinished vehicles.

$>$80 per cent of vehicles passed to sales as promised.

$>$95 per cent stores record accuracy.

$>$90 per cent vendor schedule accuracy.

$<$1-day vendor schedule arrears.

40 per cent reduction in material planning and control staff.

80 per cent increase in overall volume.

All the above includes the effect of introduction of the new vehicle, Discovery, introduced in the last quarter of 1989.

Land Rover recognizes, however, that 'A'-class MRP status is but a milestone on the way we have chosen to achieve the status of a World Class Manufacture. MRPII gives us an effective planning and control base. We need to complement this with the further development of total quality and just-in-time philosophies, until these too become a part of 'business as usual'.

By early 1991, our progress on a number of fundamental material management issues had stagnated. We had come a long way using 'progressive tightening of the screw' on our MRP implementation, but the effort necessary to continue the improvements was becoming untenable. The realization came, albeit somewhat late, that the missing link was a simple and robust execution system – JIT.

By second quarter of 1991, the 100 inventory turn programme had been launched. This programme in effect requires all material to be delivered daily, in the correct quantities. Not only will this programme take us to approaching world class in inventory turns (best automotive currently Toyota in Japan at 130, best automotive outside Japan currently about 100), but the processes necessary to make this happen are also addressing in excess of 40 other problem areas previously identified for action.

In conclusion, MRPII has provided Land Rover with a sound planning and control base. On its own it was not enough to deliver world class performance. When complemented with total quality programme and 100 inventory programme, we believe it will move us into the top league, provided we move forward at a much faster rate than world class is moving forward. We are now establishing benchmarks on a number of key business activities to measure both the movement in world class and our progress towards it. We are mindful of the lesson of the UK motorcycle industry. More recently, the lessons of the American automotive industry has been that, as the Japanese manufacturers moved in, an equivalent amount of domestic capacity has been closed. In the UK, we now have Nissan, Toyota and Honda establishing capacity to produce half a million vehicles per year. We at Land Rover do not intend to be part of the closures which will inevitably follow this. We will be there, competing with all comers on world class terms.

Source: by kind permission of Land Rover Limited and MRP Limited.

Just-in-time management

In recent years, more attention has been paid to alternative methods of production planning and control than to the more conventional approaches of statistical inventory control and requirements planning. The most widely publicized technique, developed in Japan and widely adopted by Japanese companies, is that of 'just-in-time' (JIT) management.

This section will outline both the philosophy and techniques of JIT manufacturing management. First, the methods of 'pull' logistics management, of which JIT is one form, will be compared with the 'push' systems as used by the majority of Western manufacturers. Then, following an introduction to the basic JIT concept, techniques for internal manufacture and external supply will be outlined. The links between JIT and contemporary approaches to quality through continuous improvement are then presented, and the section concludes with a discussion on the important issue of implementing JIT.

Statistical stock control (reorder level/reorder cycle) and material requirements planning are used for systems which operate in 'push' mode. In such purchasing and inventory control systems, order times and anticipated start times are usually planned and production executed in expectation of future demand. This can create difficulties in production, particularly in the case of MRP, where data accuracy is paramount and there can be discrepancies between schedules and quantities produced. This can result in over-stocking or, conversely, shortages of materials.

In contrast, there are now a number of inventory- (or usage-) based systems being developed which have been termed 'pull' systems. In these, orders are placed when parts are required at the succeeding stage of production. Thus, a triggering effect occurs. The adoption of systems which pull their requirements from preceding activities only when materials are actually needed is now viewed as an effective way of ensuring that manufacturing operations are responsive to market demands. JIT systems are the most common form of pull system in operation today.

'Push' control systems are liable to assume certain values when setting parameters for their operation including costs of ordering, lead-times/set up times and holding costs. JIT, by contrast, actively attacks these assumptions and any other problems which prevent synchronization of the operating system and thereby encourage manufacture or purchase in large batch sizes.

The Japanese philosophy of JIT is based around simple concepts of simplicity of control and operation, ensuring that material is available in the right quantity and quality at the right time and in the right place (Schonberger, 1982, 1986; Burns, 1987). Its simplicity of implementation, together with the advantages of operating with minimal inventory, has contributed to its growing application in industry (Nelleman and Smith, 1982; Celley *et al.*, 1986; Simpson, 1986; Voss and Robinson, 1987; Beal, 1988; Parnaby, 1988).

JIT is not merely a stock control technique and is not a package that can be bought 'off-the-shelf'. Rather it is a wide-ranging philosophy of waste removal on a continuous path towards 'manufacturing excellence'. It concentrates on improvement as opposed to conventional Western approaches which strive for optimization (e.g. of cost). One of the wastes removed is stock. Inventory is not seen as an asset but as an excess, which serves to hide the causes of root problems. JIT originated in Japan and demands high quality, on-time delivery and cost reductions. It is, therefore, indicative of many of the currently prevailing ideas on the role of manufacturing.

A stock control policy based on JIT suggests the supply of materials to each stage of production when, and only when, required (not before or after). Use of materials at any stage creates a need for materials from the previous stage. This implies that no inventories are held in advance of requirements or in expectation of future demand, but in practice there will always be an element of safety in JIT or any other system. When implemented, JIT results in a 'pull' mode of operation of the type described above.

The use of JIT in a company is usually linked to programmes of quality management through continual improvement. In JIT, inventory in a system is regarded as being indicative of fundamental problems and inefficiencies. The inventory is seen as hiding the problems which exist.

It is useful here to refer to the commonly used 'pond' analogy (Figure 11.6), which illustrates the issue of excessive stockholding and how JIT can be used to identify and tackle key problems. In the analogy, inventory is seen as being the water and, the greater the stockholding, the higher the water level. The inventory, however, covers a whole host of problems illustrated as 'rocks' in the analogy. By gradually lowering the 'water' level through inventory reductions (levels 1 to 2) the problems are uncovered. The largest problems become apparent first in the same way that the larger rocks are the first to be exposed. When a problem is discovered, the organization must then direct its attention to solving it or reducing its magnitude. When this is done the inventory level can be reduced again (levels 2 to 3) until the next problem is identified and solved.

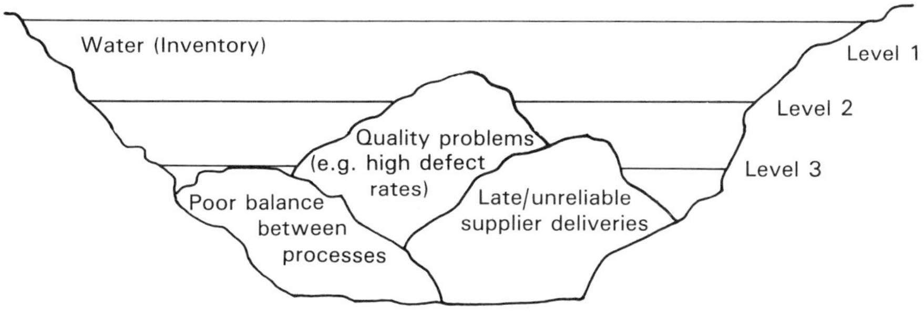

Figure 11.6 The JIT pond analogy.

This objective of JIT fits well with the modern ideas of managing quality through continual improvement. Therefore, JIT and quality management should be viewed as complementary in the development and improvement of production systems.

In practice, there is a great deal of confusion over the use of the term JIT and what it should encompass. It may be best considered as functioning in two discrete, but related, areas: internal inventory management or 'JIT production' (Sugimori, 1986; Ward, 1986; Rajput and Bennett, 1989; Womack *et al.*, 1990) and the management of supplier deliveries or 'JIT purchasing' (see O'Neal, 1987; Ansari and Modaress, 1988; and Womack *et al.*, 1990). The joint application of JIT purchasing and production enables control to be achieved throughout the materials logistics chain from the supply of raw materials to the production of finished goods. Specific approaches to each of these facets of JIT will now be presented.

The art of JIT supply management is in ensuring that just enough material arrives from suppliers at precisely the right time; not too early, as this results in high stocks, and not too late, as this would disrupt, and possibly stop, production. Typically, JIT supply management involves giving tight and rigid schedules to suppliers, but with a very short time horizon (e.g. requirements for the next day rather than the next month). The supplier is then expected to deliver in small batches on a sufficiently frequent basis to ensure just-in-time and 'just enough' material deliveries are received.

Organizations, particularly those in the West, have experienced problems when attempting to implement JIT. The following are just a few of the problems encountered and which need serious consideration when implementing JIT. First, the practice of bulk purchasing and safety stocks continues to have great attractions for price reductions and contingency management, but runs contrary to JIT philosophy. Second, lengthy changeovers serve to justify the continuance of batch production, so JIT has more commonly been implemented for assembly operations and not for preceding operations. Finally, materials from outside must be available at short notice. This causes severe problems unless suppliers and vendors are extremely well co-ordinated and able to communicate and respond efficiently. As a result, JIT suppliers tend to be local suppliers who are forced to hold stocks for the vendor to draw off. The question arises, therefore, of whether in some cases the stockholding is just being pushed back down the supply chain.

Having considered issues such as those above, there are a number of actions that should be taken when implementing JIT. For internal JIT production these include:

1. Improvement in product quality and reductions in the number of defects.
2. Reductions in set-up times on machines to allow for less expensive, more frequent machine changeovers.
3. Reductions in batch sizes when manufacturing or assembling products so as to reduce work-in-progress.

4. Location of inventories in identifiable areas in order to facilitate their reduction.

When adopting JIT for supplier deliveries the following may be used:

1. Purchase from fewer outlets and reducing the supplier base.
2. Reductions in the delivery sizes of incoming batches.
3. Increases in the frequency of deliveries to compensate for 2.
4. Establishment of long-term relationships with suppliers; working with them, not against them, for the purpose of continuous improvement.

The most well known and publicized operations control system in existence today is the Toyota 'kanban' system for JIT production (Monden, 1981, 1986; Burbidge, 1982; Shingo, 1986). Kanban has been used within Toyota's production departments since the late 1950s. A kanban is, literally, a card which must accompany parts through a particular stage of the production process. When parts are used, the kanban is returned to the previous process and acts as a signal or trigger for the supply of more material. The number of kanbans in circulation and the number of parts each kanban orders directly controls the amount of stock held. Two types of kanban cards are used by Toyota, the 'production' kanban and the 'conveyance' kanban. Production kanbans are used within an area to initiate a manufacturing operation, whereas conveyance kanbans are used between areas to authorize the movement of materials from one place to another. More recent applications of the system have seen the card replaced by the materials containers themselves acting as triggers (an empty container means it needs to be filled), or simply by 'kanban squares' painted on the factory floor.

In some companies, most notably in the car industry, a number of final assembly plants have integrated supplier operations into their own internal kanban systems. Thus, a card returned from the customer or the receipt of an empty container acts as a trigger for the supplier to manufacture and/or ship more material. The use of JIT links supposes a very close association between vendor and supplier and such supply chains are therefore characterized by geographical closeness, long-term relationships and a commitment from both parties to work together in the improvement of operations. Thus a picture of localized 'JIT networks' begins to emerge.

MRP operates in 'push' mode, whilst JIT can be described as a 'pull' type system (see Figures 1.5 and 1.6). But this does not mean that the two control approaches are incompatible. Their combined application, in the form of 'hybrid' systems has indeed been advocated as a means of capitalizing on the positive features offered by both. This enables manufacturing operations to function in a cost-effective and efficient manner (Belt, 1987; Rao and Scheraga, 1987). The configuration which appears to best promote their compatibility is where MRP operates at the supplier and customer interfaces for materials planning, purchasing and scheduling, whilst JIT is applied to the ongoing control of material flow at the operational level.

Optimized production technology

Based upon the ideas of Eli Goldratt (Goldratt and Cox, 1989), 'optimized production technology' (OPT) is an approach to production scheduling which can be described in simple terms as the 'management of bottlenecks'. The basic premise underpinning OPT is that it is pointless concentrating on increasing the throughput at all processes within the total operation, because the output rate of the system as a whole is constrained by the capacity of the bottlenecks in the system. OPT, therefore, is aimed fundamentally at scheduling operations around the bottlenecks in production and is simple in concept. However, like MRP, its full application is dependent on sophisticated computer software packages for the necessary processing of complex manufacturing information.

The idea behind OPT is that scheduling and process improvement activities should concentrate on 100 per cent loading of these bottlenecks, then trying to increase output from them through continuous improvement. The rest of the system, the 'non-critical activities', need only be loaded to support the rate of production at these bottlenecks (the 'critical' activities). Obviously, improvement of throughput rates on the critical activities may cause the bottlenecks to shift to other parts of the system, making these the critical ones and increasing the output of the manufacturing process as a whole. (For a fuller description of OPT see Harrison, 1985.)

Information control

Computer technology has become an increasingly cost-effective means of controlling manufacturing information and processes over the last twenty years or so (Warnecke, 1988). A key issue in the design of production systems today is how to integrate the various 'islands' of computer-based systems and processes (Ranky, 1990; Forrester and Hassard, 1992a). The term 'computer integrated manufacturing' (CIM) has therefore emerged. CIM, however, should not be thought of as being merely confined to production processes. It also incorporates the full range of support activities which provide and require manufacturing information, including finance, marketing, production management, etc. This necessarily requires the development of a common database for use across the whole business which, in turn, demands data accuracy and integrity across the full range of organizational functions.

CIM encompasses a wide range of technological developments including computer-aided design (CAD), computer-aided engineering (CAE), computer-aided process planning (CAPP), computer-aided manufacturing (CAM) and numerical and computer numerical control systems (NC, CNC, DNC) together with the various computer-aided production management (CAPM) systems including MRP, OPT, etc. (A fuller description of these technologies is given by Harrison, 1990.)

The establishment of an electronic link between design and manufacturing enables a reduction of design and manufacturing lead-times. Herein lies the key strategic advantage of implementing a CIM infrastructure. The primary objective of CIM is to enable the business as a whole to react faster and more efficiently to the demands of markets and customers (Burgoine, 1988), thus facilitating the establishment of a flexible manufacturing infrastructure (Saul, 1985; Bititci, 1986). Given the far-reaching nature and consequences of its approach and objectives, there is also an argument to say that CIM should be treated as an organizational goal, in much the same way as in the case of the total quality or JIT philosophies (Forrester *et al.*, 1991; Forrester and Hassard, 1992a).

In addition to satisfying particular needs within the organization, the application of computer technology must also enable the company to capitalize on future business opportunities. Identification of business and market needs allows systems designers to create and establish the appropriate computer architecture that effectively translates and communicates information across different parts of the organization. The SADT, IDEF, GRAI and CIM-OSA methodologies described in Chapter 3 may be used for this purpose.

11.2 Key parameters

Identification of key parameters

Control and integration is one of the three components of operational systems design, the other two being physical system design and work design. The design of the control systems must therefore be performed in parallel with these two other components. However, the design of control and integration systems is shaped by decisions taken within the tactical and strategic domains of the business. Figure 11.7 illustrates the main influences upon design. Each of the key parameters shaping control and integration are now considered in turn.

Market strategy

Market strategy will determine the approach to be taken in dealing with customers, and this has important consequences in the control and integration of production systems. For instance, the market strategy may stress the need for the organization to make-to-order and for finished product demand to 'pull' material requirements through the production system. In this situation the control system will be designed to respond effectively and swiftly to market needs, perhaps using JIT techniques coupled with the integration of the sales

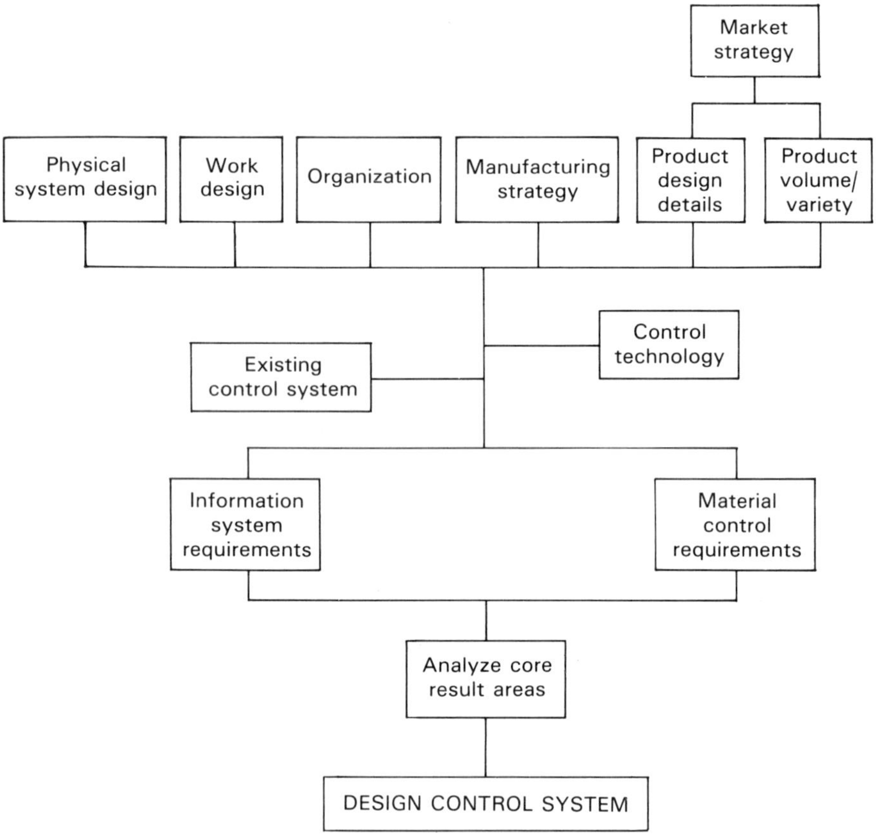

Figure 11.7 Key parameters for control and integration.

order processing, manufacturing planning and control and purchasing system. Following from the approach taken to dealing with customers, information regarding product design details are particularly important in developing an appropriate engineering and process planning system to translate basic design detail into manufactured goods. The volume and variety of products also have consequences for the complexity and sophistication of the control systems at the factory level: the greater the volume and variety, the more storage and processing space is required on the computer system.

Manufacturing strategy

The manufacturing strategy will offer a translation of corporate objectives and market strategy into objectives for manufacturing and will be more specific concerning the tasks required from control systems. In terms of the development of management information systems, the manufacturing strategy will provide a blueprint concerning the type of information required from the computer-based system, where the main priorities lie in terms of manufacturing and process control.

Organization

The structure of the organization is an important factor when developing new control systems and can determine whether the computer-based systems are centralized or decentralized, networked or stand-alone, etc. Additionally the existing and expected future state of the organization, in terms of management style and culture, has implications for whether the information should be provided on an open access basis or whether it should have restrictions on access.

Physical system design and work design

The control and integration component of the operating system must be designed, developed and implemented in parallel with the physical system and the design of work for employees at the shopfloor level. Computer-based and manual systems should be able to plan processes and allocate tasks to work-stations, control and monitor human and machine operations and then collate, process and distribute information about systems performance throughout the organization.

Control technology

Technological developments in control devices and the integration of computer-based production technology and manufacturing control continue to progress at a rapid pace. In such an environment it is essential that the organization keeps up to date, and updates its own technology wherever possible, in the light of the latest technology available on the market. For the business that wishes to remain competitive it is essential that it should scan and monitor technological developments in the environment.

Existing control system

The current state of the existing control system within an organization provides the base from which subsequent developments will take place. It is therefore essential that system designers should take into account and audit their current position before proposing new designs, control philosophies and practices. For example, the company aiming to adopt JIT principles should first seek to develop the people, organization and systems into an appropriate state in order that the new techniques and procedures will be introduced into an environment more conducive to their success.

■ 11.3 Selection and configuration of design options

Intermodule material control

The control of material within the factory is an important element in the design of control systems and their integration. Such control should be judged by the criteria specified in Figure 11.8, of which the delivery performance, inventory and logistical controls and costs would normally rate highly in the principal result objective developed by a company. The two categories of option, centralized control using a main stores or decentralized satellite

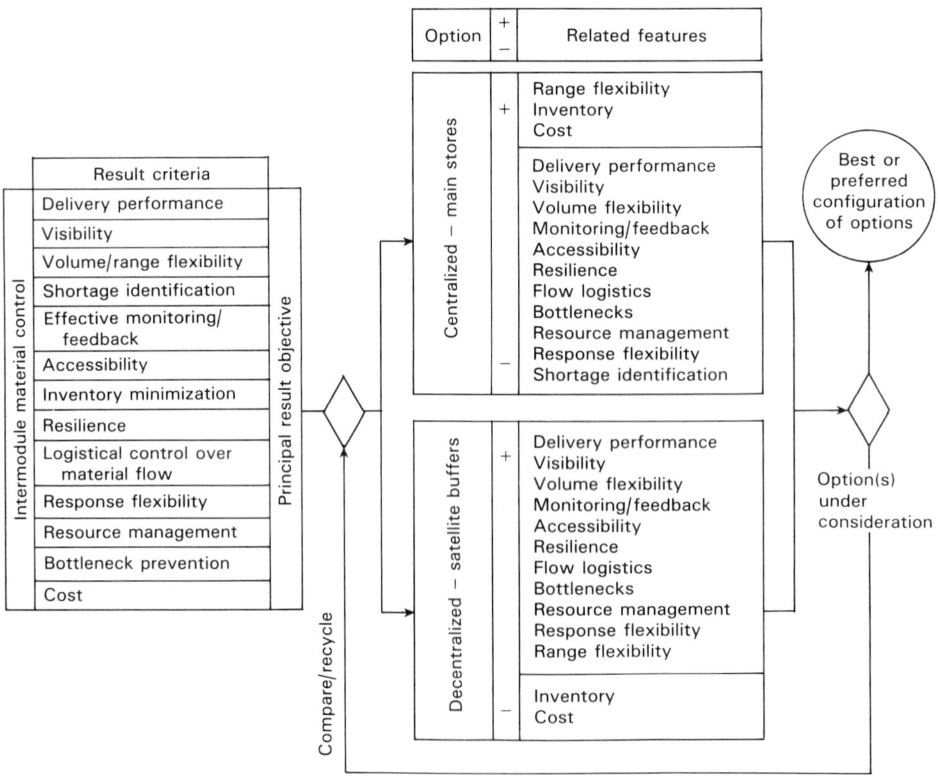

Figure 11.8 Design option guide for intermodule material control.

buffers, will now be discussed. (The issue of centralized versus decentralized stores was covered in Chapter 10 on physical system design: this section refers specifically and expands upon issues of control, not physical location and facilities, but any decision for central or distributed storage control is inevitably linked to physical facility choices.)

Centralized main stores

The centralized option is in essence the use of one or a small number of main stores within the factory. The advantage offered by centralizing storage locations is that inventory can be controlled within a confined area and so reduce holding costs. The counter argument is that use of a main stores can in fact increase stocks unless controls are exercised in the supply of materials to factory areas so that inventory is not allowed to build up on the shopfloor. Other significant drawbacks include the delay in response liable to occur in the supply of materials to functional areas resulting in a breakdown of production flow and the potential, at periods of high demand, for the stores to become a bottleneck in the overall process.

Distributed satellite stores

This alternative option could be chosen in conjunction with a smaller main stores. In some instances, it may even be seen as appropriate to eliminate the central stores area altogether. Satellite buffers enable the tactical holding of parts and materials around the factory in locations where they can quickly be drawn upon in response to demand. The objective is to hold only enough stock in these locations to satisfy the short-term requirements of a succeeding production area, and so ensure that there is no disruption to production. This is the very essence of a JIT control system using kanban, where the satellite stores could be of the input or output type presented in Chapter 10 or could merely be kanban squares. Another form of satellite store widely used in flowline assembly situations is storage at the point of use, with deliveries of incoming goods made directly to the line. The chief advantages of the satellite storage control policy are that they ensure a continuous flow of material around the factory, reduce the chance of bottlenecks and poor logistics within and between production areas, provide better service and hence improved responsiveness to production demands and are conducive to the adoption of JIT techniques as part of a manufacturing 'pull' strategy. The main caveat is that tight disciplines must be introduced and maintained over the issuing and holding of stocks in order that inventory levels and consequent costs do not proliferate.

Materials distribution

Once the type of material storage has been agreed, the method of distribution must be decided. The design option guide in Figure 11.9 shows there are two basic choices: kitting and bulk issue.

Parts kitting

Parts kitting is where the material to produce or manufacture one or a number of predetermined components or products are gathered together in the stores or a specialized kitting area and are then distributed as a 'package' to the point of use. The benefits of this approach are numerous. First, material shortages are identified in advance of manufacture. This means that production will only commence when the parts or materials required are available. Second, the kit containers can be used within a kanban control system, the number of containers in circulation at any one time being controlled and an empty container acting as a trigger for replenishment. In this way materials can be simply, but effectively controlled and inventory can be reduced by attempting to minimize the number of containers in operation. Other associated advantages are shown in Figure 11.9. The major drawback of kitting is that it can increase

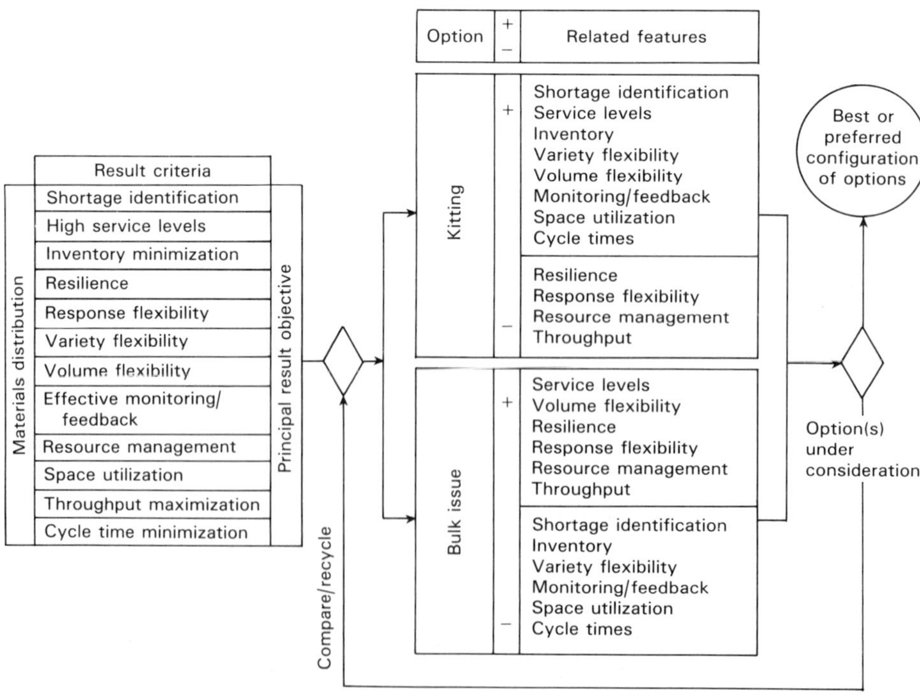

Figure 11.9 Design option guide for materials distribution.

throughput times and has been criticized as a poor use of manufacturing resources. The argument therefore is that kitting represents an additional activity which contributes nothing to the actual production of manufactured goods and so runs contrary to the principle embedded within JIT of eliminating indirect and non-value creating activities wherever possible.

Bulk issue

The other, and more commonly adopted, alternative to kitting is that of bulk issue. Here materials and parts are delivered to the manufacturing area in batches of individual items and not as part of a kit. This idea does not necessarily run contrary to JIT because the batch sizes can be very small, even as low as one, and could involve, for example, direct delivery to the point of use from external suppliers. The advantages for bulk issue include potentially faster manufacturing throughput times and it can be regarded as a better use of resources in that it does not require the additional kitting operation. The negative features include the possibility of late identification of material shortages and the consequent disruption to the flow of production and the potential for poor control over the level of inventory.

Data collection and analysis

The control of production processes is reliant upon the data that is collected from the shopfloor and the subsequent analysis of this into its usable information form. There are numerous ways in which data can be collected and four general options are illustrated in Figure 11.10.

Paper

Paper logging of operations systems is the traditional way in which data has been collected concerning output, downtime, order completion, quality, etc. Indeed, paper systems are relatively cheap to introduce and enable the collection of data in respect to a large number of performance measures, some of which it is difficult to collect using more automated procedures (for example, data concerning the detailed reasons for production downtime or quality defects rather than merely their aggregate incidences). However, paper logging can be expensive to operate in terms of the ongoing labour costs and it suffers from likely lapses through human error in data integrity and difficulties of frequent updating. It also cannot provide readily available audits without substantial time and effort (for example, stock checks) and, as a result, can result in poor operational control. A major concern is that the information system cannot ever really work in 'real-time' and will always be of a 'batch-type' providing historical information which may be out-of-date as soon as it is generated.

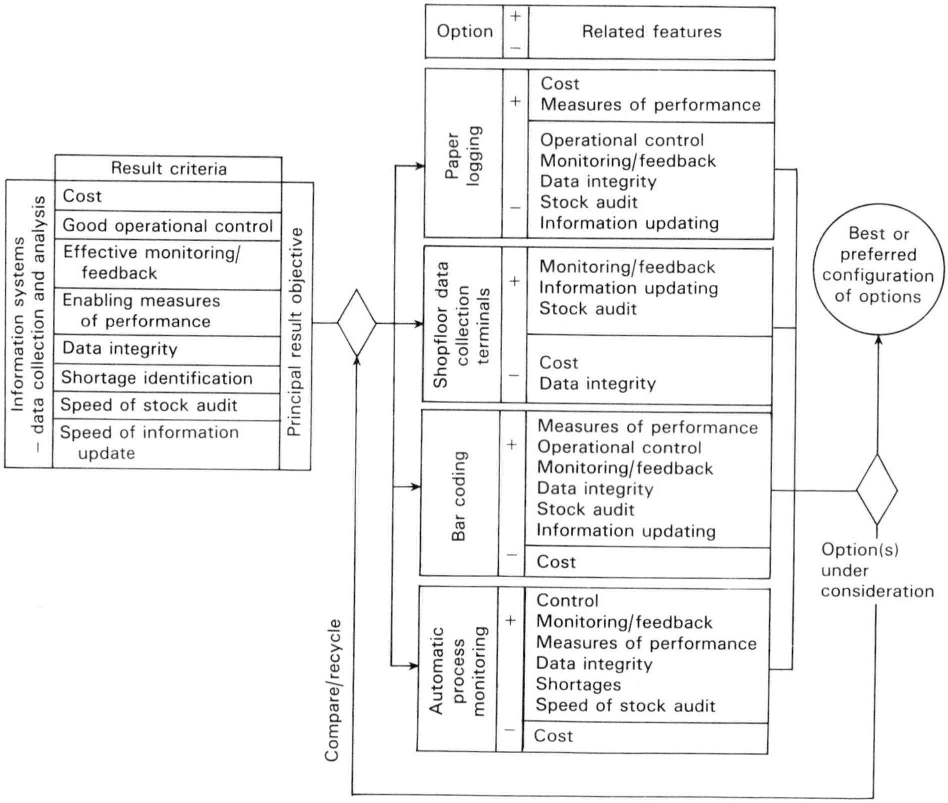

Figure 11.10 Design option guide for data collection and analysis.

Data collection terminals

A recent and widely adopted alternative to paper logging is the use of shop-floor data collection terminals which may be linked to machines and/or rely upon human input at the manufacturing process or workstation. These terminals are linked either to a local PC or a factory-wide computer system which analyzes the data and transmits the resultant information to the relevant parties. This provides instantaneous monitoring and feedback, updates information in real-time and can be used as an interrogation tool for analyzing and auditing manufacturing processes. The drawbacks are the costs, not so much of the terminals, but of the development of the computerized system. In many cases data collection terminals also still rely on human input and so may be subject to low levels of data integrity due to mistakes in the recording of data.

Barcoding

Barcoding technology has become increasingly available and attractive to a wide range of manufacturing business as experience in its use develops and the necessary hardware and software continues to reduce in price. It provides a flexible and adaptable means of recording data and triggering the issuing of instructions and commands throughout the manufacturing process. Input or system interrogation can be via a pencil-like wand or a reading gun, and increasingly in the manufacturing environment automatic barcode scanners are being used to read all incoming codes at strategic points in the production process. Such points include the entry or exit of materials to and from a stores area or when items are being moved into or out of production areas. The main drawback of automatic scanning continues to be the cost and perceived risk of unreliability on the part of many potential adopters of such a system.

Automatic process monitoring

Automatic process monitoring has been in existence for a considerable time in

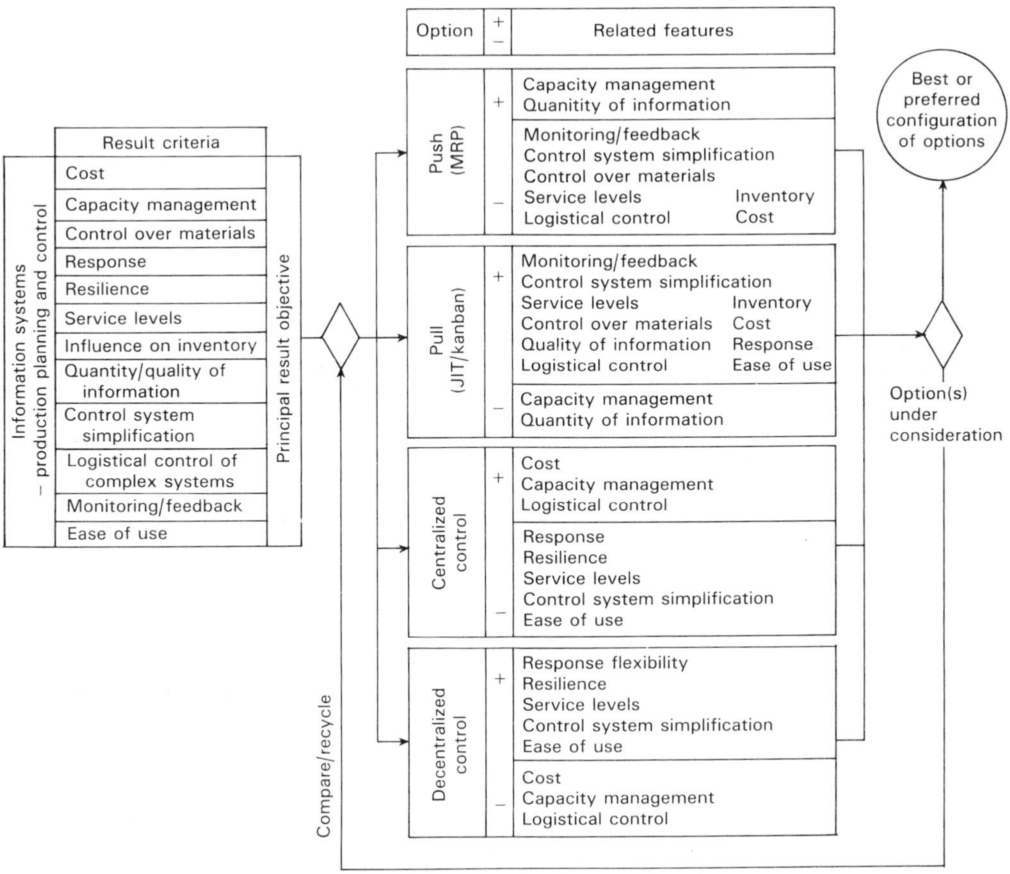

Figure 11.11 Design option guide for production planning and control.

the form of production output counters, machine speed-readers and the like. These established forms of data collection system are usually transferred to a paper logging system by means of a manual check or reading of indicators or dials. More recently, however, there has been a revolution in process monitoring technologies in the form of sensing equipment and systems linked to barcoding and 'programmable logic controllers' (PLCs), which interface directly with a computerized information system. The benefits are those of data integrity and real-time processing and information provision referred to above, but again these attractions must be balanced against the cost of purchasing, developing and operating such a system.

Production planning and control

The planning and control of production is a major concern to system designers. This is evidenced by the content of this chapter, which has outlined alternative planning and control policies that may be adopted by manufacturing companies. As has been illustrated, the issue is more than merely the planning of purchase orders, initiating dates for manufacture and stock control. The type of production planning and control policy adopted has important consequences upon the way the factory operates in terms of efficiency, cost, service and response to demand. Figure 11.11 shows the main options for control, the choice between push and pull system, and also illustrates the features associated with centralized and decentralized control.

Application 11.2 Production scheduling at Aluminium Corporation Limited

Aluminium Corporation Limited (ACL) is a specialist producer of non-standard aluminium rolled products within the British Alcan group of companies. Its production facilities at Dolgarrog in North Wales includes a casting operation, a hot mill and a wide range of cold rolling mills and finishing equipment. In addition to producing aluminium circles to customer order (for use, for example, in the manufacture of saucepans) the company produce circles for light fittings and satellite dishes, plain sheet and strip, patterned and special finished sheet, superplastic alloys and high purity products, to name just a few.

The operation is, of necessity, one of batch production, and a number of process routes through the factory could be identified. In the late 1980s ACL managers began to make a critical appraisal of the way in which orders were scheduled on to and around the shopfloor. In addition to high work-in-progress (WIP) the company faced the problem of high manufacturing throughput times as a result of the inherent queuing in the production system. The decision was taken to purchase the optimized production technology

(OPT) scheduling software and to nurture the OPT philosophy in the way in which production planning and control was conducted within the factory.

The first step was to develop a model of the factory and its manufacturing processes. On the basis of this a number of 'critical processes' (i.e. the bottlenecks) were identified, which determined the rates of output for the factory operations as a whole. In terms of scheduling, the key was to load orders to the critical resources which, in operation, would 'pull' work from preceding activities to ensure a continual flow of work and no excess build-up of WIP in front of these processes. The OPT software, in scheduling orders, would also limit the total amount of work in circulation in the factory at any one time.

The second initiative, linked in with the company's programme of quality management entitled 'Gwella' (Welsh for 'continuous improvement') was to seek to increase output and improve process reliability at the critical operations. After all, any improvement in productivity at the bottleneck would result in improved productivity for the system overall. In some cases the resultant improvements had the effect of turning critical into non-critical activities, with other processes becoming critical. In addition to the direct benefits of reductions of 50 per cent in WIP and collapsed manufacturing throughput times, ACL also realized the following as a consequence of the new approach to manufacturing management:

1. Improved product quality through quality improvement programmes, reduced handling and storage damage, etc.
2. Improved customer service in terms of order to delivery lead-times and delivery reliability.
3. More effective maintenance programmes where uptime on critical processes could be maximized and maintenance jobs prioritized.

In 1991 there was some debate within the company over whether the results achieved could have been achieved without the purchase of the software and merely from the application of the OPT philosophy. There was a consensus, however, that the investment in OPT software and packages had focused the efforts of personnel within the company towards improved manufacturing planning and control and that without the management commitment to achieve a return on their investment the changes would have been harder and taken longer to implement. However, it was becoming apparent that more recent planning and control software was becoming available and so, in 1992, ACL replaced the OPT package with 'Provisa' from AT&T Systems. Still operating under the OPT philosophy, Provisa has offered a number of further advantages:

1. It is more user-friendly and easier to operate.
2. It requires less processing time for scheduling runs.
3. Through Provisa simulation abilities, the model of the production system can be utilized for analyzing the effects of changes in production system

design or manufacturing planning and control rules without disrupting the actual system.

ACL now report that, in addition to increasing their control and providing additional facilities, the new production scheduling software has resulted in a further 30 per cent reduction in inventory levels below the platform reached through the implementation of OPT.

Source: by kind permission of Aluminium Corporation Limited.

Application 11.3 ICL: Control and integration

The modular assembly cascade is a factory-wide hierarchy of sub- and final assembly modules linked by three flows: those of materials, demand-pull and control data. The pull mode of manufacture is based upon JIT principles and is facilitated using barcode technology to track and trigger the movement of parts and kits around the factory: what has been called 'electronic kanban' within the company. The trigger for replenishments is achieved by the return of the kit-boxes (tote bins) from the assembly area, which acts as a request for more material to be supplied to the operating area (equivalent to the more usual 'conveyance kanban': a request for parts to be moved). Each module within the cascade is controlled by its own dedicated computer control system and the demand to make (i.e. the 'production kanban') comes from specific requirements for more material in the downstream operations area. In most cases this is triggered by a reduction of stock in the output level below a reorder level and a request through the computer system to replenish.

JIT kanban is used for real-time control of processes and materials. However the Ashton plant also uses MRP for long-term scheduling purposes, so the system is best described as a 'hybrid push–pull' type. MRP is used to ensure materials are ordered in time from suppliers, taking into account order to delivery lead-times. MRP is used at the factory level, that is to say 'it treats the factory as a machine', and does not attempt to schedule material internally. Thus, once material enters the factory, it moves from MRP to JIT control.

Initiating JIT between modules requires the connection of individual PC module controllers to the factory material controller. There are in effect four tiers of control as described below:

Level 1. Local device controllers (e.g. programmable logic controllers), which command the movements of automated materials-handling devices.

Level 2. PC-based systems responsible for controlling specific functions within the module including the movement of AGVs and control of stock levels and replenishment.

Level 3. PC- or mini-computer-based systems responsible for the control of the module as a whole. This includes the integration of individual controllers within the module and the capability to download long-term production plans from level 4.

Level 4. A mini- or mainframe computer-based system, which controls the interaction between different modules, interfaces with the MRP system, and provides real-time management information on the status of operations.

In summary, the key observations of ICL Ashton's control and integration components are as follows:

1. Autonomous modules within the cascade are linked by three types of flow: materials, demand–pull and control data.
2. A JIT–pull mode of operation is facilitated via the computer control system using barcodes: what has been termed 'electronic kanban'. This provides responsiveness and adaptability required in the low volume/high variety environment of the Ashton factory and enables the development of a flexible production infrastructure.
3. Each module in the cascade is controlled by its own dedicated computer control system, triggered by an input of specific requirements from downstream operations and, ultimately, direct customer requirements.
4. The control system has four levels, from local device control through to factory control.

Source: by kind permission of International Computers Limited.

'Push' versus 'pull'

A fundamental decision must be made as to whether the manufacturing system should work predominantly in pull or push mode. Push mode will normally involve the use of an MRP system, which plans the manufacturing and supplier orders for parts and materials. Alternatively, simpler systems may be developed using statistical stock control techniques of the type previously described in this chapter. Although beneficial in terms of scheduling and capacity management and providing a wealth of inventory information, MRP has the major problem of being poor at responding to short-term fluctuations and changes in demand or unforeseen events, such as machine breakdowns or cancelled orders. A pull system, based around JIT principles, will not schedule the throughput of materials in anticipation of future events. Instead, parts and materials are pulled through the system on the basis of demand at the immediately succeeding operation. This should result in lower inventory levels than MRP and increases flexibility to respond to short-term fluctuations. Although pull and push are different philosophies, it is now increasingly common to use a hybrid push and pull system to achieve the benefits of both types of production planning and control philosophy. In a hybrid approach, MRP is used for longer-term ordering of material from suppliers and for forecasting likely implications of current and expected orders on the different areas of the factory, while JIT is used for short-term control and pulling of materials into and within production areas.

Centralized versus decentralized control

MRP, by virtue of its dependence on large-scale computer-based processing, will need to be run as a centralized planning and control system. JIT and pull, however, is predominantly a distributed system of control throughout the factory, although decisions in planning JIT, such as the planned levels of inventory and number of kanban containers in circulation, will normally be taken by senior managers at the centre of the organization. Therefore decisions need to be made on the degree of centralization for production planning and control, whether push, pull or hybrid. A centralized system enables higher level control of capacity and internal inventory control, whereas decentralized control will improve response to changes in demand and provide greater responsibility at shopfloor level.

■ 11.4 Generalized methodology

The generalized methodology for control and integration within the manufacturing process is shown in Figure 11.12. The following points should be noted in respect of the various steps.

1. **Establish the overall mode of control.** By taking into account the market and manufacturing strategies which influence design decisions, and also the available control technology, the organization should seek to establish a balanced combination of push and pull for system control. This could include MRP for long-term scheduling and JIT for in-house and supplier operations control.

2. **Establish stockholding policies.** After establishing the mode of control, the organization should develop policies for stockholding within the business as a whole. These will act as a frame of reference in steps 3. and 4.

3. **Develop material control policy.** By referring to the physical system design, material control policies should be established which take into account the points shown in Figure 11.12.

4. **Develop and design information systems.** The existing information should be evolved or extended to provide information for managing the inventory and order control systems and to supply management with information they might require to assist in decision making.

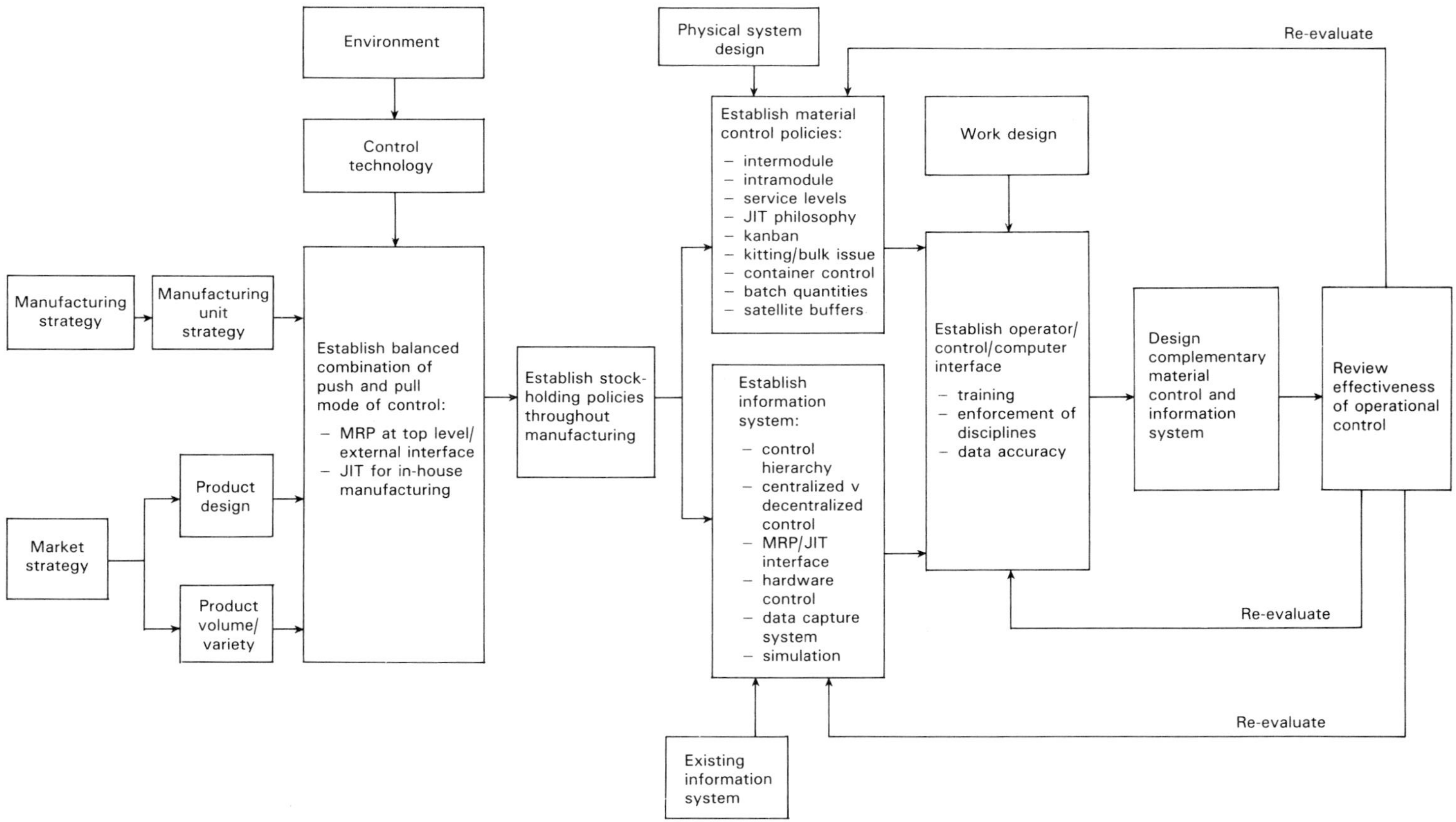

Figure 11.12 Generalized design methodology for control and integration.

5. **Establish and maintain control system interfaces.** Of particular importance here is consideration of the control system's interface with the human operator and the requisite demands for training, etc.
6. **Design the material control and information system.** This will involve integrating steps 3. to 5. in order to create the overall control system.
7. **Review the effectiveness of operational control.** In the light of changing market requirements and advances in the technological environment, re-evaluate material control policies, information systems and control interfaces in terms of their effectiveness and operation.

11.5 Conclusions

Over the last twenty years or so there has been considerable debate regarding the most effective means for planning and controlling production systems. This debate has been fundamental to the explanations given for the superior performance of Japanese manufacturing companies over their Western European and American competitors. However, there is no over-riding and conclusive agreement as to what constitutes the 'best' production control system.

It is clearly apparent that many manufacturing companies are not so effective at controlling operations as some of their counterparts. In these companies the importance of revising and developing control systems and their integration cannot be understated; it is often the key factor in moving towards a market-oriented approach to manufacture. As the technology for computer integrated manufacturing (CIM) becomes increasingly available, companies are finding it feasible to use a combination of push and pull systems and to create interfaces with other parts of the business. In this way the organization can capitalize on the benefits offered by a fully integrated system.

11.6 Further reading

Browne, J., J. Harhen and J. Shivnan (1988) *Production Management Systems*, Addison-Wesley, Wokingham.
Silver, D., (1991) *Decision Systems for Inventory Management*, Wiley, New York.
Waters, C. D. J. (1990) *Inventory Management and Control*, Wiley, New York.

12 Work Design

Objectives of the chapter

On completion of this chapter you should:

1. Be able to identify the factors influencing the design of work and the various forms that work design can take within manufacturing organizations.
2. Appreciate some of the wider labour management and sociological factors that should be taken into account when designing production systems.
3. Be able to use the DRAMA methodology to assist in the design of work organizations.

Chapter outline

This chapter focuses on the people-related issues in production system design and operation. It complements Chapter 7, which considered the higher-level organizational issues. The chapter first considers work organization in a number of production situations, namely in job, batch, flowline and group working. It then highlights the relative advantages and drawbacks of each type of system configuration for people working within it. The wider social issues of work are considered with particular reference to labour flexibility and codetermination in organizational decision making and, finally, the design of work for people is considered within the DRAMA methodology. The main influences upon the process of Work Design are identified as:

1. The prevailing sociotechnical factors in the industry within which the company operates and in the wider environment.
2. The decisions made at the tactical and strategic levels of organization.
3. The design of the other components of the operating system, i.e. physical system design, and control and integration.

The design options available are presented and then fitted within the context of the generalized methodology flowchart for work design.

12.1 Theory and current practice

Since the start of the twentieth century there has been a growing interest in the organization of work at shop-floor and plant level. Early recognition of the importance of human behaviour, social relationship and the organization of people in production came from Elton Mayo and the members of his research team from Harvard Business School in their famous Hawthorne studies at the Western Electric Company's plant near Chicago (see Roethlisberger and Dickson, 1939). This work was later developed by the 'sociotechnical' school, most notably in the work of the Tavistock Institute of Human Relations in London (Trist *et al.*, 1963). Although there are many ways of classifying production organization it is useful to apply the generally accepted classification of job, batch and flow systems, described in Chapter 2, when discussing the implications for work design (for further description of the characteristics of job, batch and flow systems, see Hill, 1991). Later in this chapter the nature of work in group systems is considered and the effects of recent approaches for enhancing labour flexibility are discussed.

Work design in job systems

A job system has traditionally been seen as appropriate where demand is low or unique ('one-offs'). It is based on layout by fixed position, where resources and materials come to a single location and products are then manufactured complete. The organization of work here is thus 'product-oriented', meaning that the job content of people within the system is centred upon the manufacture of the product in a 'make-complete' sense.

In relation to work design, job systems require people with a range of skills and ability to cope with the wide range of tasks demanded of them. Thus there is ample scope for job enlargement in work design. Additionally, the planning activities are simple, with all materials and resources coming to one point of use. However, job-type work designs demand a highly skilled workforce, so training needs and their associated costs tend to be high. Where volumes increase, the only way that job systems can cope is through duplication of facilities, with the danger of low resource utilization. Thus, despite the attractions of work design normally associated with job production, other types of organization usually need to be adopted for higher demand situations.

Work design in batch systems

The second category of system is that of batch production. A batch system normally uses a functional layout whereby similar machines or processes are grouped together and materials move between those processes during the manufacture of individual products. Batch systems are generally used in medium or intermittent demand situations, or where production rates exceed the rate of demand. The main justification for moving from a job to a batch system is the increased utilization of resources at individual facility locations.

Work organization within batch systems is best described as 'process oriented', with jobs directly reflecting the functional area within which the operator is working. In the past, a major argument for the use of batch production was that it offered a high degree of specialization, both in terms of manufacturing tasks and supervision. It was recognized as early as the eighteenth century that the development of functional specialization and expertise in restricted operational areas leads to increased output (Smith, 1776). However, more recently, the argument for functional specialization has been thrown into doubt. From an organizational viewpoint, it is regarded as leading to restricted practices, a lack of interaction between functions and political difficulties between the different functional areas (Butler, 1991). Additionally, from a strategic and marketing perspective, the division of the manufacturing process into discrete and separate areas causes problems for market focus, with people who are working upstream in the system finding difficulty associating with, or even identifying, the requirements of the end-customer.

Work design in flow systems

The final type of conventional system is flow (or line) production. Flow systems are used for situations where production volumes are high, or continuous, and the rate of production matches that of demand. Processes are normally laid-out in operation sequence and work organization is generally task-oriented, with people working in the system performing a confined task or restricted number of tasks on a repetitive cycle.

Much has been written in criticism of flow systems, but the logic underpinning the system (high specialization and the opportunity of balancing the line) enables products to be produced quickly and efficiently with a minimal number of set-ups and materials movement. In terms of work design, the benefits frequently quoted include the ability to use less skilled labour, economies of scale through a high specialization of tasks and, more recently, ease of automation of the simple and repetitive tasks performed. In addition to the criticisms of task specialization outlined above, flow systems have also been seen to experience many human problems through monotony and boredom at work. Specific difficulties have included problems associated with 'line-pacing' for workers, recruitment, high staff turnover and absenteeism, poor quality through a lack of commitment and, as a result, a high incidence of industrial relations problems (Gyllenhammar, 1977; Womack *et al.*, 1990).

Job, batch and flow can be regarded as the three conventional production system design options, and their implications for work organization and design are generally accepted. Over the last thirty years, however, there have been considerable efforts, by theorists and practitioners, to develop alternative approaches to fixed position, functional and line layouts. These include 'mixed-' and 'multimodel line assembly', 'group technology' (otherwise known as 'cellular manufacturing') and 'autonomous group working', which are now described below.

Work design in mixed- and multimodel lines

A major problem with conventional line configurations is the difficulty of achieving flexibility in terms of product variety and range. This is because, traditionally, products being manufactured on a line have needed to be highly standardized and of a restricted range in order to maintain line balance and operational efficiency. Mixed- and multimodel lines have been developed to deal with the problem of variety, and are still widely used in manufacturing companies today (Wild, 1990). A mixed-model line is one which is configured so that a number of different (but broadly similar) products are manufactured simultaneously. A multimodel line is one where batches of the same product are produced and the line is periodically reconfigured or 'set-up' each time the line changes over from one batch of products to the next (Aneke and Carrie, 1986).

Mixed- and multimodel lines, in addition to coping with product variety, have the added advantages of enabling job enlargement and reducing the monotony of repetitive work for employees. However, these alternatives to the conventional line do have considerable drawbacks. Most significantly, demand for each product variant must be known and fixed in order that the line can be properly balanced across all the product range. However, even when demand has been accurately forecasted, the line is still forced to operate at low efficiency due to line losses caused by the variations in work content associated with each model produced (Bennett and Nagarkar, 1988). In practice the use of mixed- and multimodel lines is an option restricted to situations where total demand for assemblies is high and predictable over a relatively long period and product variety is relatively low. Such configurations have been utilized in the assembly of motor vehicles where a range of optional features and variations are available on what is fundamentally a basic model. For situations where this is not the case, other process designs and work organization options need to be considered.

Work design in group technology

The concept of group technology GT (or 'cellular manufacturing') was developed as an alternative to the production systems normally used for batch manufacture (Edwards, 1974; Mosier and Tanbe, 1985; Bennett, 1986). Parts and/or products are grouped into 'families' on the basis of similarities in appearance, product type or processing requirements. Human and/or machine work centres (cells) are then established for their manufacture based on operation sequence or product layouts. In essence the items are produced on small lines of machines or operators. GT can, therefore, be seen as a hybrid of batch and flow manufacture which seeks to achieve the benefits of flowlines (e.g. low WIP, minimal materials movement, reduced setting/changeover times and shorter throughput times) in the batch production of parts or products.

A group technology cell can be regarded as a type of flexible manufacturing system (FMS). It can achieve autonomy for separate operations within the manufacturing process whilst at the same time satisfying overall production objectives (Heyer and Wemmerlov, 1984). However, in certain situations there is the problem of deciding upon the precise arrangement of facilities and the categorization of parts necessary to enable an appropriate segregation of the overall process into cells. As a consequence of this, GT has tended to be most widely used in the component processing industry where large numbers of very similar parts are produced which can be grouped into identifiable families (Burbidge, 1978).

Work design in autonomous working

Autonomous working was developed as a direct attempt to overcome the human and sociological problems associated with flowline production and short cycle tasks where little opportunity exists for job rotation and enlargement (Gyllenhammar, 1977; Aguren and Edgren, 1980; Bennett, 1986). In autonomous working, operators work independently of one another or are arranged into a number of small groups which, to achieve high volume production, may be duplicated and identical within the factory (Bowey and Connelly, 1975).

Like GT, autonomous working can be defined as a hybrid system. Whereas GT is an alternative to batch production, autonomous group working is an alternative to flow manufacture, which seeks to achieve some of the human benefits frequently linked with job production operations. Group working is based on abandoning the idea of long flowlines in favour of a 'make complete' approach. Thus, tasks are performed in independent work areas rather than at a series of positions along a line which are dependent on one another.

In autonomous working, employees are allowed greater freedom to organize their own activities. Often the only constraint attached is that a certain volume quota is expected to be achieved by the group in a set time period (usually a week) (Sandberg, 1982). This inevitably leads to opportunities for the rotation of tasks or, if the operators choose, to more radical changes in the way work is performed. Responsibility for planning and local management is devolved from functional specialists and managers to the group level, therefore affording a significant degree of job enrichment. Finally, payment schemes within autonomous group working usually take account of factors other than merely output, and a large proportion of earnings are normally calculated on the basis of group, not individual performance.

Autonomous working, however, does have its critics, including the authors of the 'lean production' thesis on the world automobile industry (Womack *et al.*, 1990). First, the productivity of organizations, such as Volvo, who have widely adopted group working is frequently questioned in comparison to the most efficient Western and Japanese producers. Second, there is an argument that job rotation, enlargement and enrichment can still be achieved in conventional and Japanese-influenced manufacturing operations. Finally, it is widely accepted by advocates of autonomous working that it does not necessarily function satisfactorily in every labour market and social context.

Application 12.1 ICL: Work design

The introduction of the modular assembly cascade at Ashton has had a significant effect on work design and the jobs of a large number of Ashton employees. Most significantly, its introduction has increased the demand for flexibility and mobility for shop floor operators within the factory. Compared to the situation prior to its introduction, the cascade has meant that operators are expected to conduct a wider range of assembly tasks and be more accountable for their work, particularly the quality of output.

The need for increased employee flexibility and individual responsibility has been highlighted both in the design and implementation of new production systems and in the new personnel policies. It is difficult, therefore, to distinguish the effects of the introduction of the cascade from the concurrent process of harmonization within the company. Indeed, the development of the single status organization has brought with it a profound change in the way employees are remunerated (from hourly rates to monthly salary) and their performance assessed (staff appraisals). However, the following conclusions have been drawn by the authors in their research:

1. The cascade has had varying effects on the work design and organization of different categories of Ashton employees. One common thread, however, is the increased need for flexibility in terms of process requirements, and this has tended to drive the evolution of increased job variety and rotation for production operators rather than any prevalent perceived sociological need.
2. The corporate policy of the process of harmonization has impacted upon the work design of all Ashton employees in the period whilst the cascade was being implemented. It has resulted in the development of single status conditions and a restriction of the divisive privileges for certain classes of employees.

3. The cascade has resulted in a considerable de-skilling in the jobs of test technicians within Ashton, as well as a reduction in the range of tasks performed. This is due to the introduction of automated test procedures. Test technicians reported a reduction in job satisfaction for the more skilled and qualified of these personnel and, indeed, many had left the job in search of more rewarding opportunities. Conversely, however, this means that the entrance requirements for new appointees are lower and so ICL are able to bring in less qualified and lower-cost labour to perform these operations.
4. The introduction of JIT–pull to replace the former push system means that production supervisors and managers needed to adapt to new concepts and disciplines of operation. Thus, for example, the practice of 'building ahead of schedule' needed to be broken whenever possible, as this runs contrary to the JIT principle of making parts and products only when they arc required.
5. Personnel involved in materials supply and control have seen their status and importance rise with the adoption of JIT and kitting operations. Stores personnel now need to be more accurate in their work and precise in terms of the kitting of parts, whilst the need for 'panic' manual progressing of materials and job orders has diminished with the increased control of assembly processes and materials movement.
6. Manufacturing and systems engineers have benefited greatly in terms of the technological and project management learning process whilst designing and developing the cascade. They now see the value of a participative approach to new systems development and the need to incorporate considerations from many disciplines and perspectives when designing systems.

Source: by kind permission of International Computers Limited.

Conventional versus cellular and group systems

Summarizing this discussion on conventional and alternative systems, it can be seen that production organization and its effect on work design has historically been a topic of intense debate and interest to systems designers and social scientists alike. The relative features of the established forms of batch and flow process designs have been questioned for some time and, as a result, alternative system designs have emerged. In terms of work design, however, there still persist doubts as to the superiority of these forms of manufacturing over conventional methods.

Cellular organization of processes has a number of important implications for work design within manufacturing. First, the use of cells enables discrete parts of the business to be dedicated to the manufacture of products serving distinctive markets or goods for individual large customers. It has been found therefore that cellular organization can significantly increase the market focus within a business and, importantly, in the people working within the cell, who can now identify and relate to their own external customers and markets. The result of this is that the introduction of cellular manufacturing is increasingly driven by a need to be more market-focused rather than merely to achieve the performance benefits through the GT engineering approach.

Second, it has been recognized that cellular and group organization can achieve numerous advantages by virtue of the fact that it breaks the manufacturing process into a series of small autonomous production units. Products or parts are made on a 'make-complete' basis instead of being routed around a functional layout or along a line. The advantages of such working are two-fold. First, it enables greater accountability across product lines within the manufacturing process and, in some cases, has resulted in 'business managers' taking responsibility for product design, marketing and production for the items produced in one or a restricted number of cells. Second, it is argued that the multiskilling requiring and the opportunity for job rotation frequently results in more varied and interesting jobs for workers. Converse to this, however, has been the counter claim that job satisfaction can fall due to the reduced variety of parts processed in a dedicated cell compared to the general functional department.

Flexibility and employment policy

In labour management theory, labour flexibility has been one of the most influential contemporary concepts. Building on the organizational concept of the 'core' and 'periphery' (see Handy, 1983), the model of the 'flexible firm' was evolved by Atkinson (1984). This model has been widely acknowledged as describing the emerging employment patterns of UK companies in response to the pressures of the 1980s, such as increased foreign competition and shrinking captive home markets. Atkinson argues that companies will seek three forms of flexibility in its employment policy:

1. 'Functional': the ability of employees to accomplish a range of tasks.
2. 'Numerical': the ability to adjust employee numbers as required by demand.
3. 'Financial': the ability to vary the amount paid for labour (which is linked to numerical flexibility).

This type of approach is articulated in the emergence of a group of core workers who, in return for providing functional flexibility, will receive stable employment and single status conditions of work. Numerical and financial flexibility, however, is seen as being achieved either through a pool of peripheral workers (for example temporary and part-time employees) who gain little by way of job security, by an increase in subcontracting work, or a combination of both. These actions result in increased labour flexibility for the organization, and the production systems in particular, but clearly they have consequences for the labour market and society in general.

Much of the labour management literature therefore takes a more critical perspective towards flexibility (see, for example, Rowlinson *et al.*, 1991). The issues of core and periphery and the model of the flexible firm have been linked to the more wide-ranging debate on flexible specialization, which envisaged a fundamental restructuring of employee relations, social relations and the work itself. Flexible specialization involves increasing the flexibility of labour by using what amounts to a revival of craft forms of production with multi-skilling being normal practice (Piore and Sabel, 1984). Flexible specialization has, therefore, become one of the cornerstones of the 'post-Fordism' movement which, because of the need for flexibility and to satisfy changing markets, considers that the era of mass manufacturing is now approaching its end (Murray, 1989).

Co-determination

In many countries, and within specific industries, employment agreements and legislation have facilitated the introduction of novel forms of work organization and working practices which otherwise would have been considered impractical (Lindholm, 1979; Bennett, 1986). In Sweden, where many of the contemporary developments in autonomous working have taken place, examples include a government Act, in 1972, to introduce mandatory labour representation on the Boards of Directors of corporations, and the 1977 Act on Co-determination. The Co-determination Act granted employees the right to be given extensive information, through their unions, on the activities of the company. In addition companies were compelled to consult the unions before making decisions involving major changes. In 1982 the legislation was supplemented by an industry-wide development agreement laying down the procedures by which the legally established principles could be upheld (SAF-LO/PTK, 1982).

A similar situation exists in Germany where, for several years, worker representatives have had participative rights. An Act of 1972 extended works councils' rights for information and consultation to include job design, work operations and the working environment. Again, as in Sweden, the legislation has been supplemented by collective agreements on co-determination and 'work humanization' (Schregle, 1987).

The Swedish and German examples provide an indication of the changes that have taken place in the industrial world in relation to industrial democracy and co-determination. They demonstrate a growing and important trend, which differs from country to country in its rate as a direct consequence of the extent to which governments have been willing to seek to influence and legislate for changes. The US and UK governments, for instance, have deliberately not sought to control directly the course of industrial democracy, preferring rather to encourage agreements at a local level. However, it is becoming increasingly the norm for workers to be involved in designing their own jobs and participating in company-wide decisions. Also, the consequences of the European Community's 'Social Charter' for workers is expected to accelerate this process in its member states into the future.

Application 12.2 Employee participation in the introduction of new technology

British workers have less say in planning for the introduction of new technology in their companies than workers in most of the other industrialized countries of the European Community.

This is the conclusion of an EC-wide study, which also reveals that layoffs and redundancies are commoner when new technology is introduced by British companies, than elsewhere in the Community.

The findings by the EC-backed, Foundation for Living and Working Conditions, are bound to fuel the growing debate about the European social charter of workers' rights, which has been vehemently opposed by the Government and many employers in Britain. Last week, a survey of member firms by the Confederation of British Industry showed that an overwhelming majority were opposed to any EC legislation which gave workers' legal rights to information and consultation by their employers.

The report by the Dublin-based foundation centres on a survey of employers and workers' representatives in almost 3000 enterprises throughout the 12 EC countries, concentrating on four sectors: retail, banking and insurance, mechanical engineering and electronics. It examined attitudes to participation in the introduction of computers in two stages: strategic planning and final implementation. Workers' representatives in half the British firms interviewed said they were offered no consultation or negotiation over the introduction of new technology, while only 10 per cent reported any information or involvement at the strategic planning stage.

This confirms that British employees have fewer rights to involvement than in advanced EC countries such as West Germany, Denmark, Holland and Belgium, but they are better off than in less industrialized member states such as Greece and Portugal.

However, the survey also reveals that professional managers are readier to accept higher levels of worker involvement in the future than is true of industry leaders at a national level. These findings are bound to be cited by the trade unions, who will call into question how representative the CBI denunciation of proposed EC Social Charter legislation is of line managers in industry.

On the other hand, British trade unions also seem less ambitious in their demands for early consultation at the strategic stage of planning than the European Confederation of Trade Unions. Only in Portugal are trade unionists less keen on following the lead of the ETUC and other Community unions in pressing for involvement in decision making at work.

There is also a significant difference between the experience in Britain and elsewhere in the Community about the employment effect of the introduction of computers. Nearly half the managers and the workers' representatives (45 per cent) reported additional employment resulting from the introduction of computerized systems, whereas only 15 per cent of employers and 17 per cent of workers' representatives spoke about layoffs and voluntary redundancies. In the British case, however, the figure for job losses rose to almost 30 per cent. At the same time it appears that British companies come under less pressure than firms in other EC countries to provide opportunities for redeployment, within the same enterprise, of workers affected by the introduction of new technology.

The general conclusions of the foundation's findings are being seen as ammunition for those who believe the future lies with the kind of social rights legislation promised by the Commission in the charter of workers' rights. However this is regarded by some trade unions in the Community as inadequate, and there has been criticism of the slowness of the Commission in defying the almost lone opposition of the British Government and employers to bring forward specific legislative proposals.

Meanwhile, flushed with the success in achieving a 35-hour working week this month, the West German metal workers' union is preparing a more ambitious strategy to give workers a much bigger say in the wider decisions taken by their employers. IG Metal wants, for example, to help decide the kind of products industry is planning to manufacture in the future, to make sure that they are environmentally benign. The West German union has also developed ideas first pioneered by British trade union shop stewards at Lucas Aerospace in the 1970s, which involve bargaining with employers over the adoption of 'human centred' forms of manufacturing technology. They believe the developing forms of human centred technologies – some of which are being financed by EC research programmes – offer the best way of preserving and developing a skill culture in industry.

Some employers in Europe and Japan now accept that too much new technology has resulted in deskilling. The trade unions know this has led to a reduction in workers' bargaining power, while professional managers suspect that the costs of 'time down' failures in some ill-considered computerized manufacturing systems has grown massively.

Source: John Palmer (1990). *The Guardian*, Monday 14th May, p. 8.

■ 12.2 Key parameters

Identification of key parameters

The key parameters for work design are shown in Figure 12.1. A major

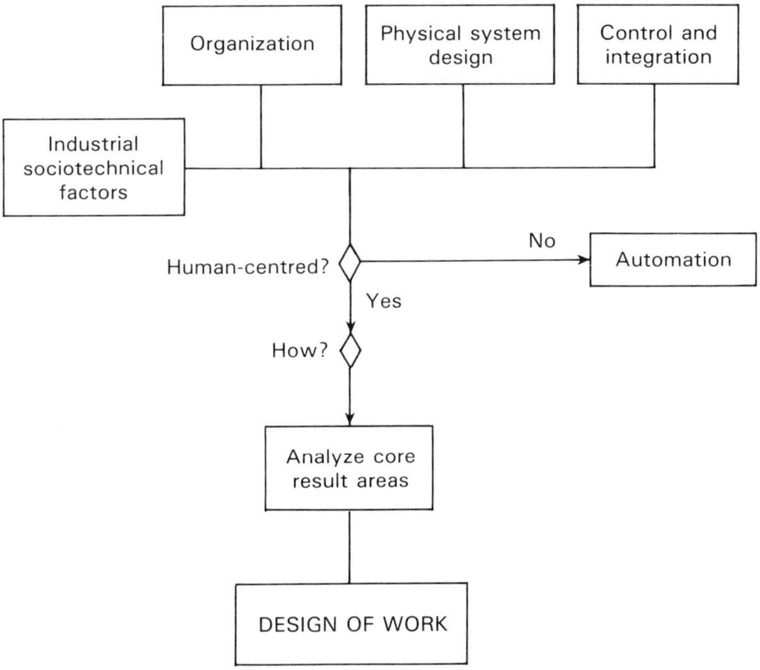

Figure 12.1 Key parameters for work design.

decision facing the organization within the physical system design and control and integration components of DRAMA is the level of automation necessary within the operation. For highly automated systems it follows that the number of people working within the factory is far less. In these systems, therefore, decisions on work design have reduced scope, although their impact may still be considerable. For other production systems, however, the human element will prevail, so the design of jobs requires significant attention if the system is to operate to its full potential. Other parameters influencing the design of work include organization and the 'sociotechnical factors' within the wider industry and environment. The key parameters are now discussed in turn.

Industrial sociotechnical factors

Any factory will have interfaces with its environment at all levels of the organization, and the shop-floor is no exception. Social interaction and working practices such as trade union activities, which span the boundaries of the organization, will result in employees developing some perception of the society and industry within which they work. In addition, managers and systems designers will try to keep up to date with the latest developments in work organization and ergonomics within other companies in order to improve the inherent potential and effectiveness of production systems. Social, technical and industry factors, therefore, have a significant bearing on the development of work design at the factory level.

Application 12.3 Volvo Uddevalla: Questions and answers about the factory

Background: The fifty years interlude: from coachbuilder to assembly-line worker to car builder

When Volvo started manufacturing cars in 1927, the men who did the work were skilled craftsmen. They were coachmakers, who had previously been building horse-drawn carriages in small workshops and forges. They had long-standing experience and a strong professional identity, all based on their tradition of craftsmanship. The coachmakers worked in teams of between five and eight people and they built their cars onto stationary bodies.

It was not until the early 1930s that Volvo adopted the production line principle according to the American model. The Torslanda plant in Gothenburg was inaugurated in 1964. At the time it was a modern and efficient plant built up around a production line with short and simple tasks. In the early 1970s, however, Volvo started to plan a new plant in Kalmar, with a new production system. This plant started in 1974 with new production principles which represented an important step away from the assembly line.

Although the Kalmar plant was still built up around the production line principle, a more flexible and efficient method of building cars was created with teamwork and longer work cycles. In 1985 the planning started for a new plant in Uddevalla. Volvo needed to increase production capacity but also to prepare for future changes at its main Torslanda plant. The planning process was unique in the way that workers and union representatives were heavily involved from the beginning. The resulting design shows a definite break with assembly-line thinking.

Assembly at the Uddevalla plant is divided into six different sections known as product plants, which are housed in two buildings that are connected, via transport passages, to a materials centre, which includes a surface treatment plant. Each plant is manned by eight autonomous assembly teams, members of which receive up to 16 months of training.

New employees are assigned to their teams as soon as they arrive at the plant. During the first four months, the employee takes part in a general programme of training, which is tailored according to his or her position. This takes place in a special training workshop. Since 1986 this workshop has served the double purpose of work training and using the training experiences to develop the new production system. Training can cover a period of up to 16 months, depending on the employee's previous knowledge of car building. One recruitment goal is that 40 per cent of the car builders should be women.

About 25 per cent are to be under the age of 25 and 25 per cent over the age of 45. The remaining 50 per cent should be between 25 and 45 years old.

Questions and answers about the factory

Researchers from the Joint Group on Work Organization and Job Design from Stockholm asked a series of questions of Volvo's management team. These, and the responses, are as follows:

Q What tasks are the assembly teams performing?
A First, of course, the assembly but also maintenance, production engineering, planning economy and personnel questions, like hiring new members.

Q Can the worker develop over the years?
A This is a learning organization. The philosophy behind it is a combination of learning by doing and learning by teaching. All members of the team are both teachers and problem-solvers, as well as car makers.

Q In Gothenburg the work cycle is around 2 minutes and at the Kalmar plant it is up to 20 minutes, but here in Uddevalla it is around 2 hours. How is it possible to memorize such a long work-cycle?
A The 1900 parts of the car are broken down in about 650 subassemblies. All small parts are packed in 90 plastic bags. These are marked in logical assembly steps.

Q What about group pressure in the teams?
A There is a strong pressure to be present in the group. If there is a high absenteeism the teams are reduced and we are put under pressure. Personally I feel difficulty in leaving my work group to work for the union. I feel pressed. (Union representative)

Q Do you regard the learning process as your normal condition here in contrast to the stable state which has characterized the thinking of the traditional factory?
A That is probably right, but it also has to do with the present development and learning process that we are undergoing in this starting up period of the factory. We have visions guiding our work here. They are rather stable and the groups have to decide themselves on how to solve their own problems in accordance with the visions.

Q What changes do you in the union want from the managers in the years to come?
A We want the managers to be smaller and smaller in influence and us to be bigger and bigger. (Union representative)

Q What is a good manager in this factory?

A One who can handle all different questions. Like in a small company. The sharing of values between the managers and the employees is also very important.

During the last fifteen years a new managerial style has grown within the Volvo group. This has been a self-selection process where managers who did not approve of changes in the spirit of Kalmar and Uddevalla have had a tough time. This has created a structural change within the Volvo leadership.

Q Why do people quit?

A There are many different reasons. People enter this factory with lots of different aspirations. We try to recruit a lot of people that are not used at all to this type of work. In some cases it works very well but not in all. Some workers also react against the system being not authoritarian enough. People must take on a lot of more responsibility for the production themselves here. Not everybody likes that. We are trying to solve this problem by moving cars that should be produced from groups where people are absent and place them in complete groups.

Q In this type of factory you have 800 quality managers – that is all the employees. Why do you then have a 10 per cent quality check of your cars?

A It is not a quality check but a quality audit. It is necessary to compare with quality goals, performance over time and to make comparisons with other Volvo factories.

As far as production and quality are concerned, we are today in the running-in phase. Our target is first to beat Gothenburg then the Kalmar factory and finally Ghent in Belgium. We want to be the best factory of them all. We think the potential in this assembly system is very big. For the moment we have a very complicated product, the 700-model. If we get a simpler product to build, like some Japanese cars, we can surely be very competitive.

Q Other factories have other strategies and stress the importance of the first-line manager. You have done away with all the first-line managers here. Why is that?

A We think our team leaders are our first-line managers. They do the important job for us and at the same time they belong to the team. We want to have a very flat organization so that the problems can be solved where they occur. For example, when we have a drop in quality we do not put in a lot of people from quality control department but we feed the figures back to the groups for themselves to store.

Q Have you met any unforeseen problems?

A Yes, we have lots of them all the time, but we are trying to solve them together between management and the teams and to sit down and discuss them in a problem solving more than a negotiating way. But we also have problems in the sense of trying to measure our productivity and quality and trying to spread our ways of measuring to other staff people within the Volvo group. We put a very high premium on quality of the product and we pay great emphasis to the learning knowledge and enthusiasm by people and that is hard to put into the traditional measurements of productivity and quality.

Source: Joint group on work organization and job design, Stockholm, Sweden. *INFO-Letter*, No. 10, Spring 1990.

Organization

Decisions taken within the strategic and tactical domains relating to the structure and state of the manufacturing organization will have direct consequences upon the design of work for employees. The corporate policy on such issues as contracts of employment, responsibility for quality and output, payment schemes and co-determination will influence designers' scope for developing jobs and will determine the role of shopfloor workers in improving quality and increasing flexibility.

Physical system design/control and integration

The two remaining DRAMA components at the level of the operating system will have direct consequences for work design. Key determinants are:

1. The level of automation.
2. The selected facilities and machines.
3. The type of control systems adopted.
4. The skills, aptitude and flexibility required to perform the full range of machine operations, accomplish production tasks and operate computer-based and other control systems.

Work design considerations should, therefore, be incorporated into the technical work involved in developing hardware and software specifications and in implementing new production systems.

12.3 Selection and configuration of design options

There is a wide range of design options for work design, and the list provided in Figure 12.2 is not exclusive. However, it does provide an illustration of the

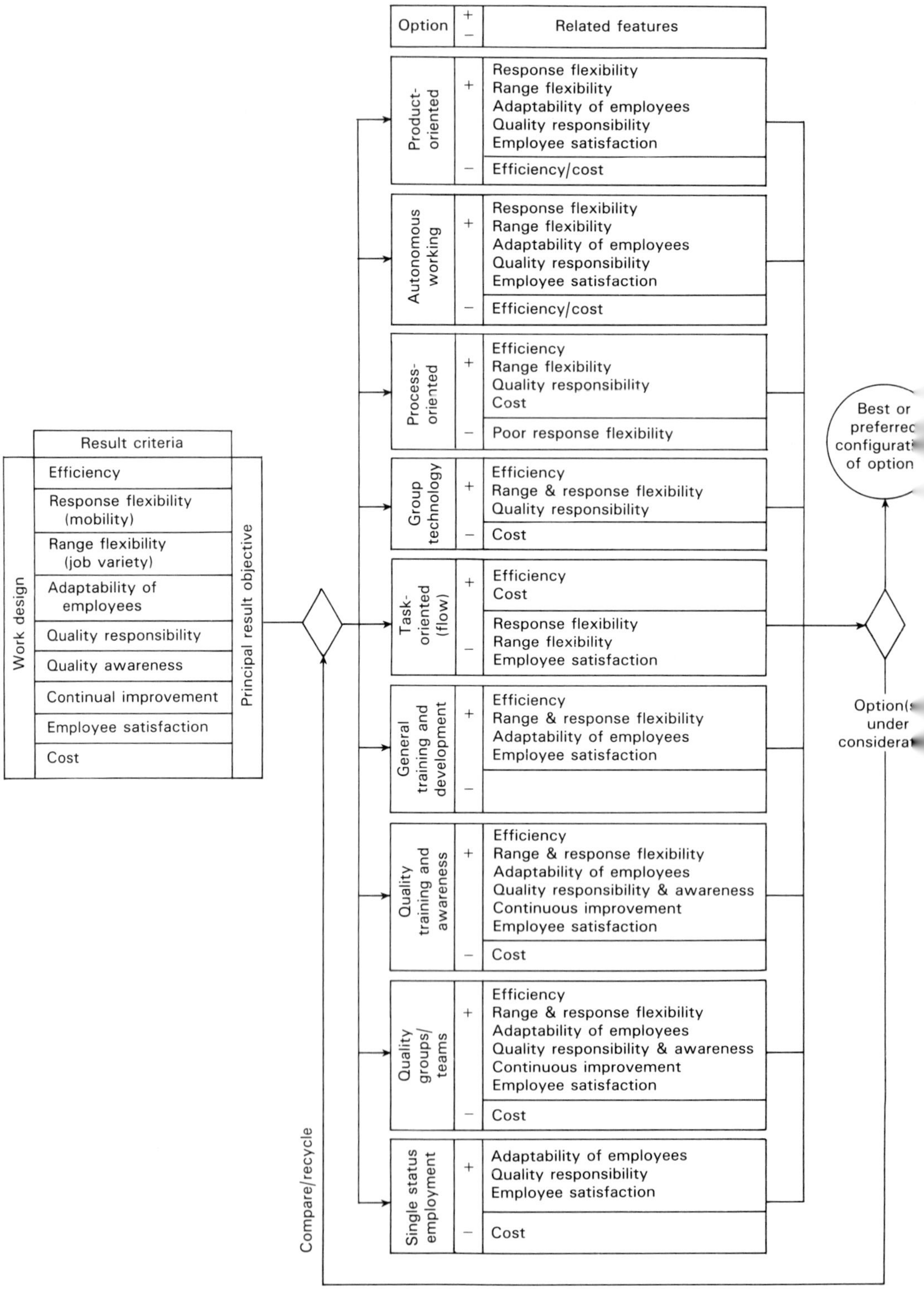

Figure 12.2 Design option guide for work design.

decisions that must be made when developing the best form of work organization. The first five options relate closely to decisions already made for the design of the physical system, the detailed implications of which have been discussed earlier in this chapter. The other four design options can be viewed as organizationally based decisions involving such factors as employee competencies and knowledge, responsibility for work carried out, and the explicit and the implicit expectations of operators and other workers.

Product-, process- or task-oriented work?

The fundamental choice between product-, process- and task-oriented work organization is dependent on, and should be taken in parallel with, the choice of physical system configurations discussed in Chapter 10. This choice is also invariably associated with the level and type of demand. It is widely acknowledged that product-oriented work organization facilitates increased employee satisfaction because of the range of operations done, identity with the finished item being produced and an absence of monotonous and repetitive tasks. However, it is widely agreed that the cost of designing work in this way becomes more prohibitive as the demand for highly standardized products increases.

Group technology

Work organization according to group technology principles is an attempt to introduce the efficiency benefits of task organization to the batch production of goods. Small group work and product focus also brings benefits in terms of job enrichment, the possibility for job rotation and increased responsibility for matters such as quality. However, the cost of installing and operating small production cells must be evaluated in detail, as these can frequently increase over and above those for process-oriented work organization. These must therefore be offset against the advantages listed above.

Autonomous working

Autonomous working involves changing from flowlines to small group work, where a major objective is to gain some of the advantages of job production and product-oriented organization. Benefits frequently cited include increased employee satisfaction, improved range and response flexibility, more responsibility for quality and output and greater adaptability of operators in the long term. There are still questions, however, regarding the efficiency of this form of organization compared with the other forms outlined above. The overall cost of operating such a system is also the subject of some controversy.

Training and development

Training and development is generally seen as a positive activity within the majority of manufacturing organizations. Not only can it equip company employees with skills, competencies and knowledge, but it can also be a direct and indirect way in forming or changing employee opinions. This may increase the management and workers' understanding of the objectives, operations and issues within the business. Cost and time are the major constraints on the level of general training given, although cost is not considered as a negative feature in the design option guide in Figure 12.2 because such expenditure, if spent wisely, should be seen as a future investment in the business as a whole.

Quality training and awareness

In recent years many companies have made significant investments in quality training and awareness because the issues of quality and customer service are seen as areas of common interest within the organization. Company-wide quality programmes will involve all employees and frequently focus upon customer needs and quality requirements. The main benefit of such programmes is frequently reported as the creation of a factory environment concerned with continuous improvement of products, processes and customer response. Despite their attractions, however, such programmes can be expensive to implement both financially and in terms of employee time, so organization designers must be clear as to the objectives of the programme and its scope.

Quality groups and teams

Many organizations place considerable value on the use of quality, or more correctly, continuous improvement groups and teams within their operations. Such programmes range from quality circles, where membership is voluntary and groups select their own project areas, to kaizen teams which focus on continuous process improvement. The direct benefits of such activities include cost savings and improvements in efficiency, whilst identified indirect benefits are increased organizational, systems and people flexibility and improving customer satisfaction.

Single status employment

The elimination of separate status for employees within an organization can yield benefits such as improved satisfaction, increased responsibility for quality and enhanced flexibility on the shop-floor. However, there can often be objections on the part of technical and salaried staff to their erosion of status and the costs of promoting all employees to the level of salaried staff must be evaluated in detail.

Application 12.4 Human resource management at Venture Pressings Limited

Beginnings

Jaguar used to source its pressings from the Austin Rover's Swindon plant. Ideally, the company would have liked to take the pressing operation in-house, but the cost would have been too prohibitive. The other option was to come to some arrangement with GKN. Getting the panels from GKN itself was a possibility, but the two companies chose a different option and Venture Pressings Ltd was born.

The choice was made to ensure Jaguar got the high quality pressings it required. As Derrick Dean, company general manager, puts it, 'the flavour of today with Jaguar and its new owners is quality, quality, quality'.

But the standard yardstick of a poor quality pressing operation tended to be a plant that was dirty, smelly and disorganized, characterized by poor materials-handling procedures – remote toolrooms and satellite die stores. This had a complementary effect on worker motivation – labour turnover would be high and job satisfaction low. Venture Pressings had to find a way out of this dead end.

The company started life at GKN's former tractor cap shop in Telford and it took two years for the initial vision of an advanced plant to become a reality in the newly refurbished factory. One major advantage of setting up on a greenfield site was that the plant itself could be organized properly. Cleanliness and order were the aims, and the layout of die lines and toolrooms meant that the movement of materials and dies was minimized. More importantly, however, the company viewed lack of motivation and job satisfaction as the most crucial factor in undermining any quality programme. In grandiose American corporate fashion, a 'mission statement' was drawn up to form the foundation of the company's quality management strategy. It says the usual kind of thing you would expect to find in such a document – including the standard 'we recognize that people are our most valued resource' – but the commitment to employee involvement and reward seems to go beyond PR hype.

Work environment

The 'vision' of Venture Pressings as a pressing company that would rank alongside the leading manufacturers of automotive body pressings world-wide required the company, and more particularly its human resources manager, to create a dynamic work environment which promoted teamwork and job mobility within the workforce. But this isn't a code for removing workers' rights and lowering standards and conditions of employment.

Employees are represented by the GMB and ATU in a single-union agreement which offers common terms and conditions to all workers, and which takes as a central point the recognition that all employees have a valued part to play in the success of the company. Traditional measures, such as clocking-in, have been removed and the cellular manufacturing system which has been set up relies on so-called 'zone teams' with team leaders and cell managers. All members of the zone teams play an active role in developing communications and deciding approaches to operational issues. To assist with this, only core responsibilities have been defined; there are no job descriptions and no limitations except for capability and safety. Employees set up their own presses, carry out minor maintenance where necessary and operate the materials handling equipment to feed and unload press lines. Employee communication is seen as vital for quality and customer service, and workers are expected to come up with efficiency procedures.

Derrick Dean elaborates on the qualities needed for the new breed of worker:

> We had to take away all the things that would normally hamstring a person. Teamwork does not just happen – you have to work very hard at it. For that reason we only want employees who share our beliefs. At the end of the day, it's the employees who will keep the quality end up.

To achieve this, of course, has also meant a rethink of the way in which the company recruits and trains its workforce. Each one of the 200 staff has to face a selection procedure that some might find overly rigorous. Prior to a personal interview, each candidate undergoes a three-hour aptitude and psychometric assessment, where loners and people who lack flexibility or who wouldn't make good team players are weeded out.

The tests are based on Saville and Holsworth aptitude tests, originally used for the US armed forces, and whose validity has been proved statistically in the relationship between test results and subsequent work performance.

Those in search of supervisory appointments must also take part in a day-long assessment, comprising a series of individual and group exercises simulating work-place responsibilities. Subsequent training includes a four-week induction programme (six weeks for management and supervisory staff) and there is a personal development plan for every employee. The aim is to stimulate initiative and team building skills, and to produce a workforce that can fulfil its potential.

Expenditure on training is high. The proportion of payroll spent on training in 1990 was a hefty 5 per cent and, while this will fall once the plant is up and running fully, it is still expected to exceed the rather paltry UK base of 1 per cent by a significant margin. Added to this, eight full days per employee are planned to be spent on training in 1992.

This effort seems to have paid off. As Dean admits:

> Working in a press shop is a fairly uninteresting occupation, to be honest, so we have to work very hard on the human resources front,

but I think we have the ingredients which most manufacturing companies will not acquire in my working life.

Source: *Industrial Manager*, May 1990, 23–24.

12.4 Generalized methodology

Figure 12.3 shows the generalized methodology flowchart for the work design component of DRAMA. The four stages in this process are now described.

1. **Establish rationale for work design.** The primary inputs for work design are the company's overall organization design (from Chapter 7) and the general social factors within the wider environmont. An analysis of these two factors will enable the company to establish the overall rationale for work design as it relates to different categories of employee within the organization. Consideration will be given to flexibility requirements, responsibilities for quality and the micro-organization design in each operational area.
2. **Develop work specifications for individual operators.** The overall rationale should then be compared with the physical system design and control and integration elements (Chapters 10 and 11) as well as with the existing work design factors. The system designer can then develop detailed work specifications for all employee types. Considerations here include the exact tasks that need to be accomplished by individuals, the degree and nature of the interface between people and machines or computers, and the communication links and interaction between employees.
3. **Translate and develop into individual work designs.** After step 2. individual work designs are developed, but these then need to be put into practice, and this may entail considerable amounts of consultation and training.
4. **Review work design.** The organization should conduct a periodic review of its work designs regardless of whether systems are themselves being redesigned. There should be an assessment of whether the new work designs are operating in a manner which is in keeping with the overall rationale developed (in step 1) and in the way described by the work design specifications (in step 2).

12.5 Conclusions

Work design has been given considerable attention in the literature on manufacturing systems, particularly since the early 1960s, but its importance

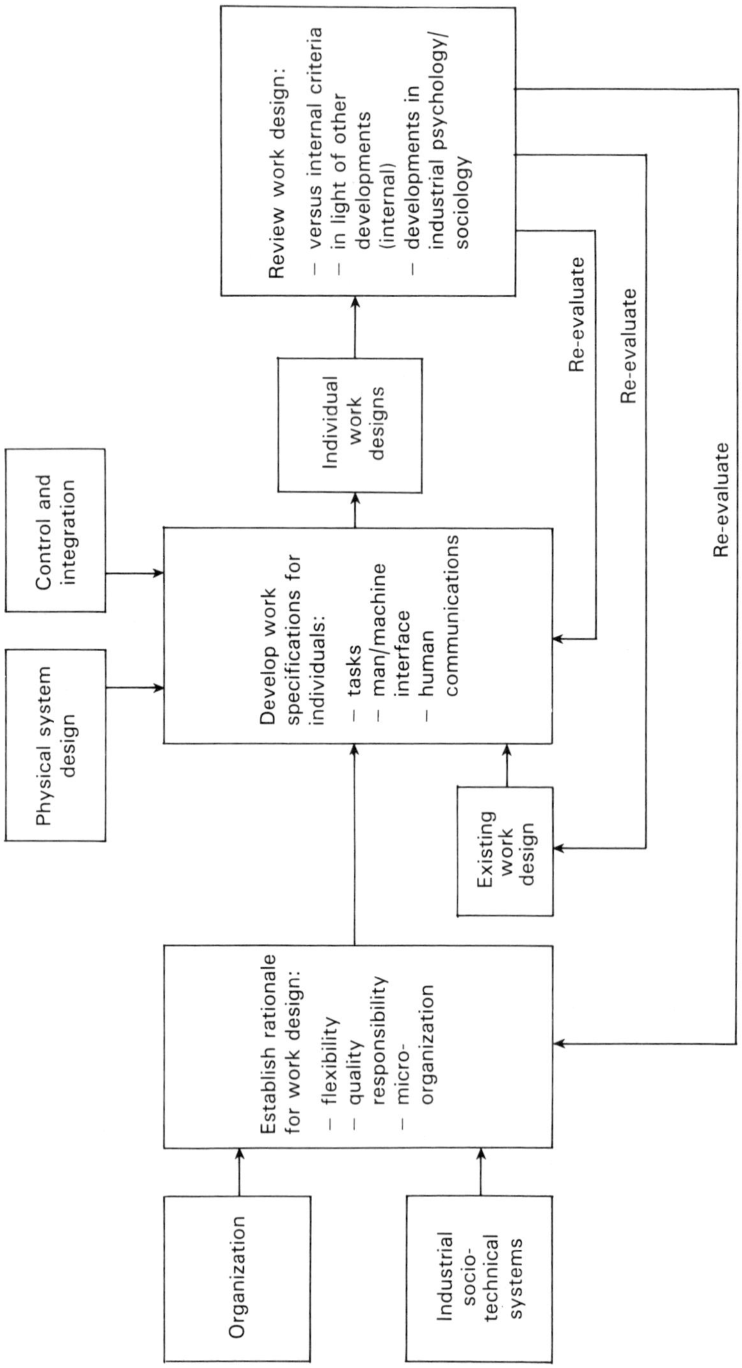

Figure 12.3 Generalized design methodology for work design.

is frequently overlooked by system designers when introducing new production systems. Many engineers and managers think of production systems as comprising the hardware and software of 'physical system design' and 'control and integration' and ignore the vital role played by direct and indirect personnel. No system, however, even if automated, can run uninterrupted without some human intervention, and organizations which acknowledge and account for this fact in the design of work will operate more smoothly and efficiently than those that do not.

A number of forms of work design have been developed over recent years as alternatives to conventional batch and flowline working. Most notably, group technology and autonomous working have evolved for the purpose of improving flexibility, increasing market responsiveness and improving the jobs and working conditions for people within the factory. In many companies, the fundamental ideas of line production and functional layouts have been largely retained. However, there is a general trend towards less rigid plant formats and facilities, where increased operator flexibility and increased participation for employees in the day-to-day decision-making processes of the organization are becoming the norm.

The main conclusion is that work design must be given the same or, in some cases, greater importance than the design of the physical and control systems in order to develop the most effective infrastructure for manufacturing. This chapter has described the process of designing jobs in parallel with developing the design for other components of the operating system. An overall rationale for work design is established through the translation of organization design in the strategic and tactical domains down to the level of the operating system. By considering the prevailing labour market conditions and the hardware and software systems, either in operation or to be introduced, specific details of appropriate work organization can be determined and implemented. The development of work designs must be viewed as an ongoing and iterative process. It should be reviewed and evolved in response to changes in the organization structure, developments in the labour market and social environment or when new physical and control systems are introduced.

12.6 Further reading

Bailey, J. (1983) *Job Design and Work Organization*, Prentice Hall, Hemel Hempstead.

McLoughlin, I. and J. Clark (1988) *Technological Change at Work*, Open University Press, Milton Keynes.

13 Implementation

Objectives of the chapter

On completion of this chapter you should:

1. Appreciate the key practical issues regarding the implementation of new production systems.
2. Understand how the barriers to successful implementation may be overcome.
3. Be able to use the methodology to assist in the process of implementation.

■ Chapter outline

This chapter considers the important topic of implementation. It begins by reviewing the literature on this subject before discussing some of the issues involved in the implementation process. Despite the importance of the subject, there are relatively few texts which assist the production system designer in converting plans into practical systems. The main parameters influencing the implementation stage are identified as justification, project management and the three components of physical system design, control and integration and work design, which together comprise the operating system. The various options available to companies for the implementation of new systems are presented and put into context with the use of a generalized methodology flowchart, which summarizes the overall process.

13.1 Theory and current practice

Literature on implementation

With the increasing rate of technological change in production systems design and the high costs of investment involved, it is reasonable to expect there to be a wealth of literature concerning the implementation phase. However, only a small number of texts appear to have explored this subject, with most authors concentrating on the design and development of production processes rather than examining how new systems are installed and put into operation.

The book edited by Rhodes and Wield is unusual in that it seeks to address the neglected area of the implementation of systems and facilities by providing case study and analytical material on the introduction of processes resulting from large-scale projects (Rhodes and Wield, 1985). Asch also considers the financial implications of the implementation stage but only from a financial perspective (Asch, 1987). However, despite the insights it provides, a major criticism of most of this literature on implementation processes is that it tends to be linear in nature and does not capture the iterative character of implementation as happens in practice.

Hygen has identified three dominant and equally important aspects of implementing production systems: people, work procedure and software (Hygen, 1984). He argues that these have traditionally been handled quite separately and, increasingly, companies are falling into the trap of concentrating upon the technical development of computerized planning and control systems to the detriment of other factors. Past experience clearly demonstrates the drawbacks of ignoring the organizational and training factors in implementation and justifies the need for end-user involvement in design in order to smooth the way for subsequent implementation.

Whipp and Clark address the criticism of linearity and over-simplicity by considering implementation as one of four stages in the design process, following design 'conception' and 'development' but preceding 'operation' (Whipp and Clark, 1986). They add a further dimension to their framework by linking the stages of development with the Utterbeck and Abernathy classification of product, process and work organization (Utterbeck and Abernathy, 1975). The result of this is a matrix on which can be mapped the relationships between different elements within the stages of design, including implementation. Whipp and Clark go on to apply the framework to the analysis of a major design case study, that of the Rover SDI automobile project (1968–76), which resulted in the introduction of the Rover 3500, 2600, 2300 and 2000 models in 1976. They argue that implementation is generally a phased progression, but that a number of factors and complications can, and invariably do, arise causing a delineation of implementation plans. These delineations include feedback loops to earlier design stages, time delays and accelerations, and the influence of organizational politics. All of this creates a picture of implementation as being far more dynamic than is commonly suggested by more general guides and descriptions.

Installing new systems and working practices has always been recognized as an inherent part of 'work study' procedure (Bennett, 1986). Here, much depends on the personal qualities of the person or people responsible for implementation. Active support is required from management and trade unions alike so the qualities required of such a person are the ability to explain clearly and simply what should be done, coupled with a gift for getting along with other people and winning their trust (ILO, 1979).

Implementation in practice

Most theory is not as informative to the practitioner as might be hoped and there is considerable scope for research aimed at analyzing the implementation of complex system design projects with a view to offering guidelines for the improvement of this process. It is to the practice of implementation, therefore, that we must turn to establish any general guidelines. In industry a phased approach to implementation is now frequently recognized as having advantages over implementation of the whole system as a single project. Obviously, the degree to which implementation can be phased is largely dependent upon whether the physical system proposed is of a modular design, or whether its functional control is designed to allow easy separation into discrete segments for operation (e.g. installation under manual control, implementation of automation and computer control systems, followed by integration of design elements).

It has been argued that installation of a new production system can be divided into five stages (Bennett, 1986), namely:

1. Gaining acceptance of the change by departmental supervision.
2. Gaining approval of the change by the management.
3. Gaining acceptance of the change by the employees involved and their representatives.
4. Retraining employees to operate the new methods.
5. Maintaining close contact with the progress of the implementation until satisfied that the new system is running as intended.

The desirability of gaining the acceptance of people who will need to operate and work in the new system becomes stronger where changes are proposed which affect the number of direct and indirect operators. This is often the case if the changes are aimed at improving manufacturing efficiency. In some organizations such consultation is routine where it has been discovered that a failure to involve the workforce can cripple a new production system through disputes and grievances.

There are very few guidelines for implementers of new manufacturing processes. However, on the basis of research conducted by the authors into the implementation of production systems, a number of issues have emerged which enable some factors in the success or failure of the implementation stage to be proposed. These are:

1. Phasing of the design and development process in the manner outlined above offers the potential for implementing solutions in incremental stages. This enables resources to be distributed more evenly over the time period for installation and allows new facilities to be introduced and accepted by users in small, less radical increments. If planned effectively, this approach can also reduce disruption to manufacturing where new systems are replacing established operations.
2. The successful implementation of major operational changes requires that their implications are recognized early and an accurate evaluation should be carried out into their full impact upon all parts of, and people within, the organization. Where this is not done, costly time delays will probably be incurred and the confidence of system users together with their enthusiasm for the approach taken and the final objective will be reduced.
3. A balance must be drawn between the time taken to develop a solution by user-consensus and the extent of 'design imposition' on the part of the system designer. The former will usually result in a more acceptable decision for all parties, but the time taken to obtain a consensus may cause the project to over-run or, if constrained by rigid completion time-scales, will reduce the time available for system implementation. Design time may be reduced where the design is imposed on the user with less consultation, but this may cause significant difficulties at the implementation stage with the emergence of unforeseen problems. Somewhere on this continuum between consultation and imposition is an optimal point for the situation in question where total project time is sufficiently short and an effective and generally acceptable solution is accomplished (Figure 13.1).
4. A frequent occurrence in the design and implementation of successful production systems is the emergence of one or more 'project champions'. A champion is a person who promotes the project and its design from an early stage and follows it through to implementation. Project champions frequently exert a heavy influence upon the project and expedite it towards a conclusion. Organizations should, therefore, not only consider carefully the appointment of a project leader or manager, but should also realize that their choice of leader is critical in ensuring good progress towards implementation.

Summarizing this account of the issues in implementation, it is apparent that few publications have focused on the nature of the implementation process. This manifests itself by the fact that the various design and decision models and methodologies described in Chapter 3 disregard the detailed question of implementation and, as a consequence, tend to treat it as a 'black-box' area. It is to industrial practice that we must turn to find guidelines and these have been listed as the need to phase implementation, to conduct an analysis of the

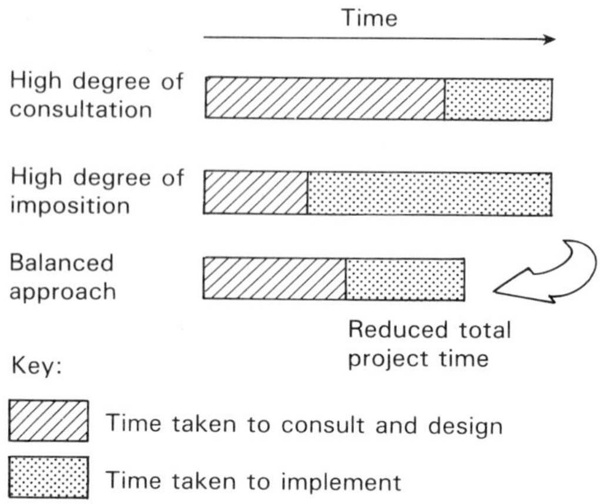

Figure 13.1 Achieving the balance between design consultation and imposition.

impact of the new system within the organization, to maintain a balance between imposition and consultation, and the identification of potential project champions to support the new system from its conception to introduction.

Application 13.1 Implementing new production facilities at Black and Decker

Since the opening of its plant at Spennymoor, in the North-East of England, in 1965, Black and Decker have gained considerable repute for their skill in implementing new production facilities. The company see people as being the key asset and the various human resource policies and teamwork approaches adopted by Black and Decker are seen as being important in smoothing the way for implementation. Early involvement of people in the design process is seen as critical so as to identify and solve any potential problems and to communicate to people within the plant the company's intentions and plans at an early stage.

Two factors explaining Black and Decker's success in implementation were identified by George Eldrington, the Manufacturing Engineering Manager at the Spennymoor plant:

1. Engineers at Spennymoor now have many years of experience in the introduction of automated systems, know of many potential pitfalls and realize the importance of user involvement when attempting to translate design into operational systems. For many, the introduction of automated new production systems is seen as a 'way of life'.

2. Design is not coloured by one single view. A committee of development personnel is responsible for design and the proposed systems are reviewed by other user groups, including manufacturing, prior to installation and 'before any buttons are pushed on the line'.

A key part of Black and Decker's strategy for implementation is the use of turnkey subcontractors for the installation of new systems in order to take advantage of specialist knowledge and to help the drain on internal resources at the installation stage. One turnkey supplier, ADM Automation of Washington, Tyne and Wear, has just completed the installation of a new flexible assembly and test line for the production of a range of electrical hand power tools at the Spennymoor plant. Based upon their relationship with Black and Decker during this project and also earlier projects at the plant, Chris Smith, a Director of ADM, identified two reasons explaining the successful track record at the Spennymoor plant in implementation:

1. Product design is performed within the plant and this combination of product and process knowledge on the part of personnel within the company enables problems to be identified and avoided before reaching the implementation stage.
2. People at the plant are used to (and not afraid of) change. New products and processes are readily accepted by all employees and, within such a climate, implementation is facilitated.

Source: by kind permission of Black and Decker Limited and ADM Automation.

13.2 Key parameters

Identification of key parameters

Implementation comprises a number of activities for translating designs into functioning systems and processes. It can be seen as an internalized process which brings together the results of previous DRAMA components, most of which involve considerable amounts of interdisciplinary and multifunctional work, and forms appropriate new or modified production systems. The key parameters for implementation shown in Figure 13.2 are therefore outputs from preceding design components. A major decision needs to be made before starting the implementation phase of whether to conduct implementation in-house or, if there are inadequate resources, to enter into a 'turnkey' contract with an external agent that has the skills, resources and expertise to introduce new systems for third parties. The key parameters are shown in Figure 13.2 and are now described in turn.

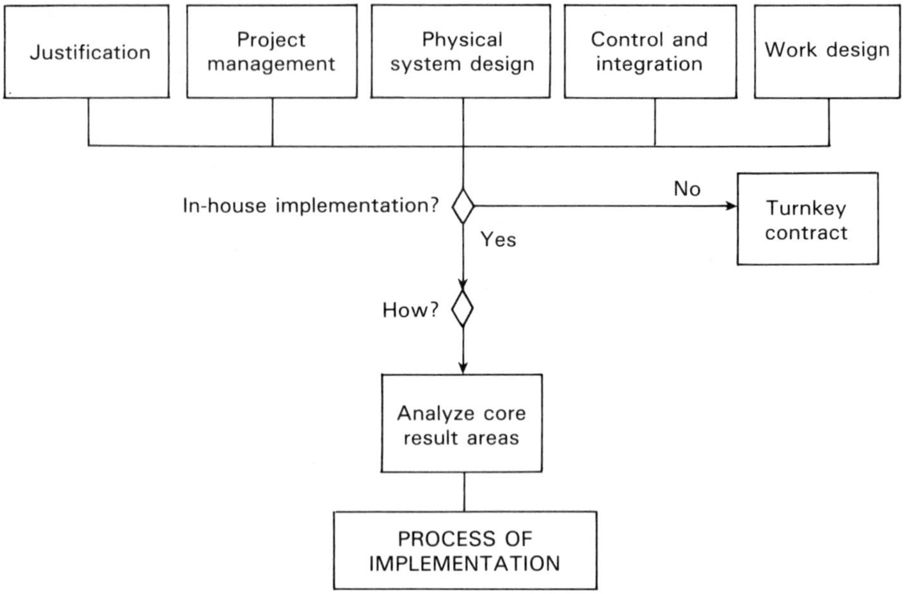

Figure 13.2 Key parameters for implementation.

Justification

The process of justification (see Chapter 8) involves consideration of the market and environment and the strategies developed for the company to respond to the factors identified. Thus the output from the justification process, usually some form of 'capital paper', provides a blueprint for development in terms of limits to expenditure, time-scale to allow successful implementation, expected performance of the system and the anticipated benefits deriving from the new system.

Project management

Implementation will start the design of the physical system, control systems and working practices finish. In many organizations the project team and the implementation team comprise the same group of people who may have been given the task of designing and implementing the proposed system at the outset or was later seen as being the most appropriate group to put the design into operation. In some companies the two groups are separate, with the project team comprising specialists with detailed technical or organizational expertise and the implementers being people who will operate the system; often called the 'user-led' implementation team. This arrangement would be based on the argument that there is an increased chance of full acceptance of new systems.

Design of the operating system components

The DRAMA components of physical system design, control and integration and work design involve developing an emergent design for the production system. Implementation is the process that translates these designs into operational systems.

13.3 Selection and configuration of design options

The main decisions to be made at the implementation stage relate to the question of who should be involved in the process and their degree of involvement. Figure 13.3 provides a list of design options for implementation which will now be considered in turn.

In-house or turnkey implementation

The organization needs to determine whether or not it has the capacity and resources required to develop systems designs into fully operational facilities. Conducting the implementation project in-house will allow full control over its development, but can cause an extensive drain on the resources of the organization. Many small companies have found that conducting ambitious programmes of implementation can create enormous pressure on the business and are increasingly looking towards the services of outside agents. These are often equipment and software vendors who follow up their involvement in design with a development contract. The use of turnkey contractors can supplement the internal resources and expertise within the company and can build upon the vendor relations created during the design stage. However, the cost of turnkey implementation can often be high and there can be reservations within the organization regarding the potential loss of management control.

Phased versus single project approach

Implementation of designs can be conducted in one large project, with all components of the system installed together, or, more commonly, via a phased approach. The single stage implementation of process designs in one project can be seen as cutting down on time taken, but can cause an immense drain

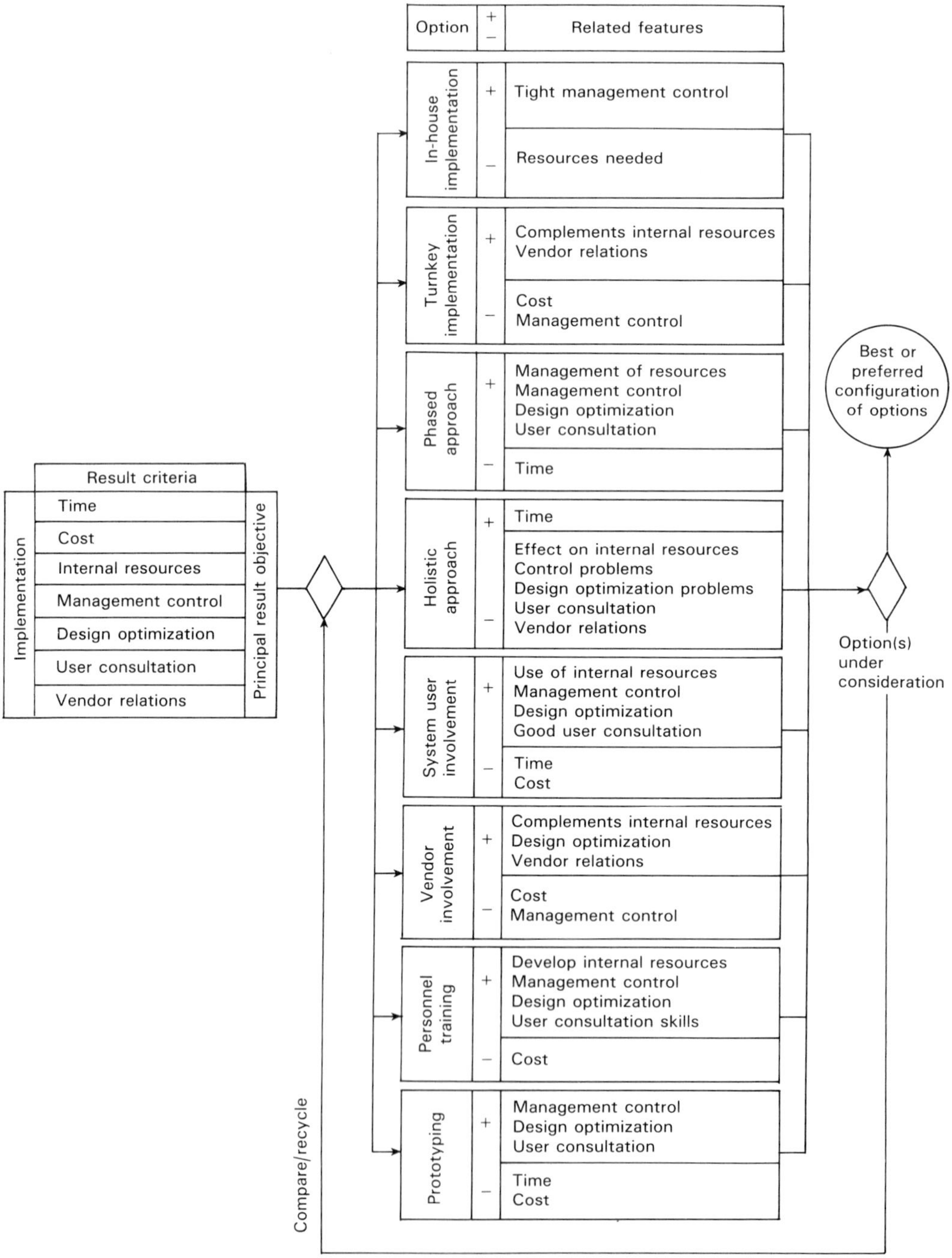

Figure 13.3 Design option guide for implementation.

on corporate resources during implementation. Implementers therefore normally look for ways to implement designs in a step-wise manner, therefore smoothing out demands upon resources and investment funds.

System user involvement

System users should always have a part to play in implementation. After all it is they who have a major influence on the success or failure of the design when put into operation. They also have a wealth of knowledge to bring to the development programme. The question is one of the degree of user involvement there should be. Canvassing the views of all those affected by design changes may be desirable in theory but, in practice, may unduly increase the time for introduction and the cost of implementation. Organizations must therefore attempt to locate an optimal point on the user involvement continuum.

Vendor involvement

Vendors of equipment and computer software will usually have valuable experience in the implementation of the products they are selling. It appears to make good sense, therefore, to include system vendors in implementation programmes for the introduction of new systems to complement internal competencies and optimize the systems potential. The cost of doing this must be evaluated in detail.

Personnel training

An often forgotten or neglected factor in production systems design is the training given to operating personnel. This can frequently be critical to the success of the design project. During the implementation stage attention should be given to explaining the function and operation of new systems to those who are going to operate them, although this process is often constrained by the expenditure available.

Prototyping

One method of optimizing the performance of new systems is to develop a prototype of the design and to experiment with modifications and adjustments. This not only solves many of the potential start-up problems when the system is introduced to full operation, but also provides an opportunity to involve system users in the continuous improvement of the evolving design.

The costs and time involved in prototyping production systems, however, can be high and so the organization must be sure that the benefits likely to accrue from the prototype will be recovered.

Application 13.2 ICL: Implementation

Implementation of the modular assembly cascade at ICL was conducted over a three-year period between 1985 and 1988. It adopted a phased approach, with separate modules being installed, commencing production and being integrated with other modules through the materials handling and control systems in a step-wise progression. The key observations in the implementation process at Ashton were as follows:

1. A phased approach to implementation at Ashton was seen as necessary for a number of reasons, including constraints upon human and technical resources given the size of the plant, the need to test, validate and debug separate modules before introducing further elements of the cascade, and the need to take into account the continual learning process of system designers and operators alike.
2. Despite the benefits of a phased approach and the desire to leave periods of reflection between each phase of module development, the pressure to complete the factory-wide system meant that such time was not present in practice.
3. From the situation where ICL relied upon the turnkey implementation of the first module, Mercury, by a third party vendor, ICL staff progressed along the learning curve for the introduction of advanced production technology to the point where implementation is now managed in-house.
4. In early module developments ICL found that inadequate or very loose functional specifications to equipment vendors at the tendering stage caused misunderstandings later on in implementation. In the light of these experiences, ICL now pay great attention to accuracy and detail in setting out their systems requirements.
5. The involvement of users and operating personnel early on in the design process and throughout systems development is considered highly desirable by ICL engineers in smoothing the path for efficient implementation. The degree of user involvement in system design, therefore, increased during the development of the cascade.
6. The development of the cascade has highlighted the importance of substantial personnel training at the implementation stage in operating and managing new systems. This is especially critical where new philosophies of manufacture (e.g. just-in-time) and new demands for flexibility and quality responsibility are being placed upon employees.
7. System operation was monitored closely immediately following the start-up of new modules and adjustments made where necessary. In some cases

ICL has found it necessary to second a member of the system development team to the operating module for a period following system introduction where teething problems occur or new ways of working have to be ingrained and maintained in the module operation.

Source: by kind permission of International Computers Limited.

13.4 Generalized methodology

Figure 13.4 illustrates the generalized methodology for the Implementation component of DRAMA. A major input to this process, which governs the future activity in designing and installing systems, is the capital justification paper developed in Chapter 8, which sets out in some detail the objectives of the new system and its likely configuration. Allied to this, the organization will have some idea of its own internal capacity for managing the implementation of new, possibly complex, production systems. This feeds into the first step in the generalized methodology.

1. **Decide upon in-house or outside implementations.** By conducting an internal audit of its expertise, the company should decide whether implementation can be managed in-house or whether some form of 'turn-key' installation contract with an external party is required. Should a turn-key approach to implementation be chosen, the key parameters of project management, physical system design, control and integration and work design will all need to be communicated to the outside party by means of a system specification for installation. If an in-house programme of implementation is preferred the same key parameters should be considered, although the company must itself decide upon the implementation plan and its effective management.
2. **Manage the implementation of the new production system.** A good working relationship must exist between system designers and the end users and adequate training must be given to personnel who will be involved in the operation and management of the system. Where there is prior experience of implementing large-scale integrated projects a set of implementation practices should already be in existence, either explicitly stated or, more probably, in the form of custom and practice. These should be evaluated and used as the basis for shaping the implementation plan and its conduct.
3. **Start the operation of the system.** Once installed, the new production system will come under a period of start-up and, eventually, will reach full operation. This stage of implementation must be managed carefully, ensuring that any initial start-up problems are quickly solved.

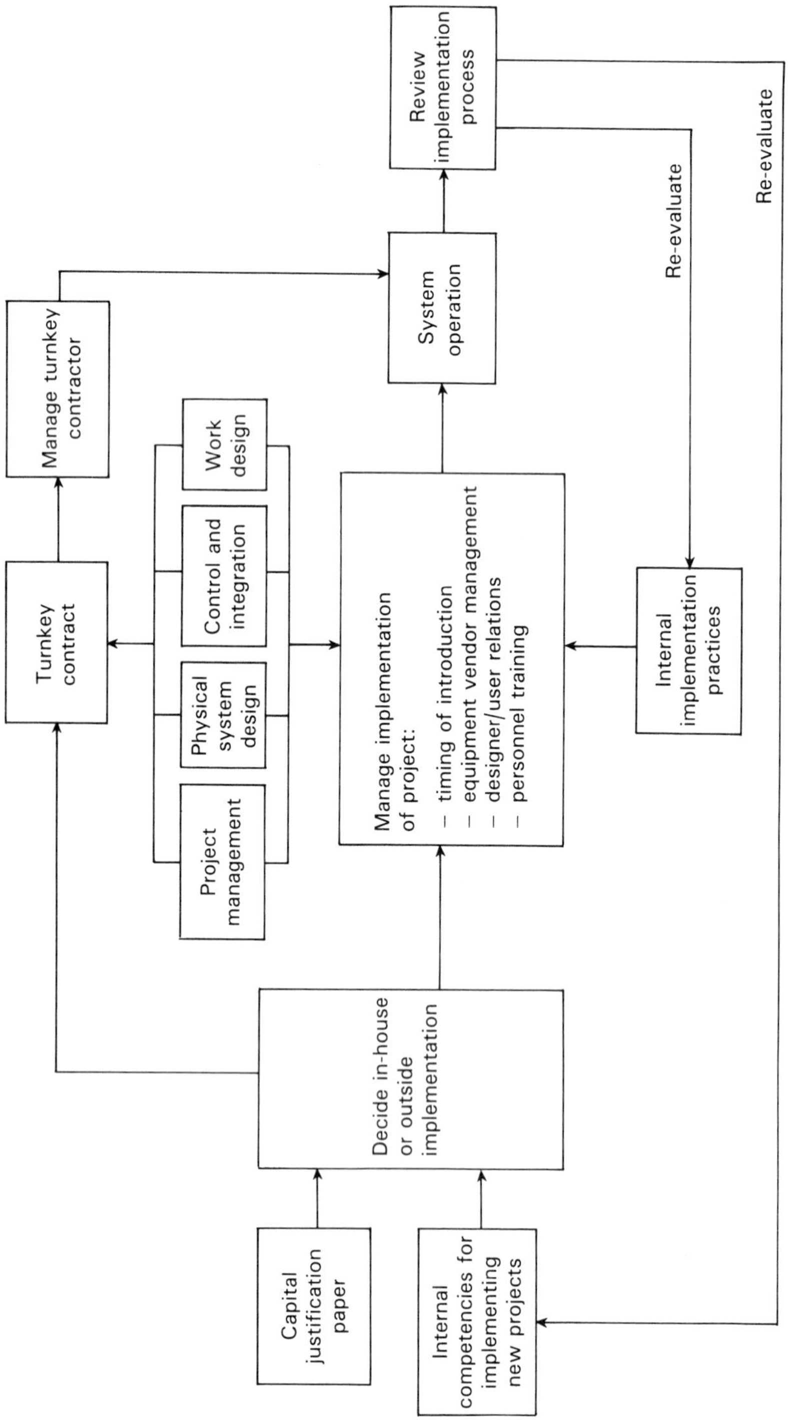

Figure 13.4 Generalized design methodology for implementation.

4. **Review the practice of implementation within the organization.** Once the project has been implemented the organization should begin to analyze its implementation procedures in an attempt to learn lessons for future projects. Processes and procedures should be reviewed and evaluated in order to develop further internal skills and practices for the future.

■ 13.5 Conclusions

The realization of the benefits to be derived from designing and installing a new production system is not just dependent on the specification and organization of the design process and the component parts of physical system, control and integration and work design. It also greatly depends on the quality of implementation. Implementation is important because the most technically sophisticated system can still fail due to unforeseen circumstances. such as a lack of employee acceptance. However, the critical nature of implementation is often overlooked by managers and engineers alike and this is compounded by a lack of guidance and theory in the available literature on this subject.

Many of the problems of implementation stem from organizational as well as individual resistance. Resistance from individuals can be answered in a number of different ways depending on the particular circumstances, and this chapter has considered some of the methods in which this can be done. They generally depend on providing greater involvement of employees early on in the system design process in order to ease the process of implementation.

Implementation entails pulling together all design work so far conducted and converting this into the installation of effective and efficient new systems, which perform to the level required by the organization. There are a number of inputs to this process, most importantly the three areas of operating system design (physical, control and work design) and these must be co-ordinated to best effect through effective implementation. This chapter presented the design options associated with implementation and provided a methodology to assist in the process. Finally, it is important that organizations learn from their successes and failures in order to improve the efficiency with which designs are translated into installed systems in the future.

■ 13.6 Further reading

Rhodes, E. and D. Wield (1985) *Implementing New Technologies: Choice, Design and Change in Manufacturing*, Blackwell, Oxford.

14 Evaluation

Objectives of the chapter

On completion of this chapter you should:

1. Understand the debate about evaluation of manufacturing performance in the operational, tactical and strategic domains.
2. Appreciate how the process of evaluation can be conducted at each of these levels in the organization.

■ Chapter outline

Evaluation of newly installed and operational production systems should occur at all levels of the business. This chapter therefore reviews the literature and practice in evaluation in relation to the evaluation of operating performance and of the production system itself within the operational and tactical domains. It then considers structured methods of strategic evaluation, most notably the post-audit of capital investment following the introduction of the system. It explains how all the preceding components of the DRAMA methodology form the key parameters for the process of evaluation, and how each of these needs to be reviewed and revised, where necessary, to fine-tune the system's effectiveness. The range of options for the evaluation of the production system and its design process are presented within a single DOG and, finally, the overall evaluation process is illustrated within the generalized methodology flowchart.

■ 14.1 Theory and current practice

An evaluation of new systems can be carried out at all levels of the manufacturing business. This ranges from a strategic analysis of performance against corporate objectives and market needs to an evaluation of actual versus forecasted performance by managers and engineers. At the operating level, evaluation comprises an assessment of day-to-day operations. Theory and practice therefore covers a wide range of topics and so, for the purposes of this book, they have been subdivided into four areas: evaluation of 'operating performance', evaluation of 'system design' and, in the strategic domain, 'post-audit of capital projects' and 'current practice'. These will now be discussed in turn.

Evaluation of operating performance

A recurrent theme of recent reports dealing with the practice of operational evaluation of manufacturing systems is the development of market-driven corporate policies and the increased customer service orientation in manufacturing (for more discussion on this see Bennett *et al.*, 1992). This has provoked a radical review of performance measures used to assess the effectiveness of operation of production systems (Bowen *et al.*, 1989). The development of appropriate control and performance measures is now seen as vital to the implementation of market-driven corporate policies at the operating level, with evaluation procedures needing to be reviewed and possibly revised if they are to reflect the flexibility, quality and delivery requirements demanded by the customer.

Means should be developed for measuring output from the operating system in terms of customer satisfaction and which replace the traditional measures based solely on factors such as cost and standard hour labour performance. Customer-oriented measures of output incorporate such factors as quality, service and delivery. Increasingly these are being adopted to provide the primary assessment of operating systems (Kenny and Dunk, 1989). This is not to say that more traditional methods of manufacturing systems assessment are not appropriate in all circumstances. Such performance measures may be suitable where resource utilization remains a key factor. In most situations, however, cost and efficiency measures used in isolation will not reveal many of the aspects of the operating system performance most in need of evaluation. System performance, therefore, should be viewed in totality; there should not be an over-dependence on a single measure and there should be regard to factors such as quality, delivery and levels of inventory to supplement values for manufacturing cost, output and productivity.

Evaluation of system design

The evaluation of a production system should take place at two stages. First, the proposal must be evaluated prior to any decision being made to proceed with the project. Second, the actual system should be evaluated after implementation to ensure that the planned benefits, as stated in the capital justification paper, are being achieved.

At the first stage – evaluating the proposal – a number of design methodologies have recently been developed whose primary aims are to assist the decision-maker in the design of production systems and to help in the choice of the arrangement and configuration of facilities. (Some of the most widely adopted of these are reviewed in Chapter 3.) These can also be viewed as conceptual frameworks for system evaluation and used to provide an assessment of existing, or newly installed production systems. For example, IDEF can be employed for an analysis of system design using a structured modelling approach and flowcharting techniques to link design considerations at various levels in the organization (see, for example, Sackett and Williams, 1988). Likewise, the GRAI conceptual model can be used as a structure to analyze hierarchical decision-making linking operating systems with information flows throughout the organization (Doumeingts *et al.*, 1986). The GRAI methodology for production control systems can also be used to provide an assessment of the current state of an organization's systems through the use of 'GRAI-nets' (see Roboam and Pun, 1989; and Chapter 3).

Computer simulation is increasingly being used as a method of analyzing the design and operation of production systems prior to their introduction. This provides a dynamic representation of the real system and its operation in the form of a purpose-built computer-based model. Simulation models can be used to test the effects of changing elements and parameters within existing systems without the need to disrupt them by conducting live experiments (for more information on simulation see Moskowitz and Wright, 1979; Moore, 1982; Pidd, 1988; Urry, 1991).

When eventually the new system is installed and, following a period of 'running-in', has reached its planned level of activity, the second evaluation can take place. The objective of this is to confirm that the system is providing the returns, both financially and operationally, that were predicted. In practice, many organizations spend considerable amounts of money on new systems yet largely neglect the post-project evaluation, relying instead on the rather limited productivity measures that were described in the previous section of operating performance (Bennett, 1986). It is true that to monitor all the relevant factors requires time and effort, and itself incurs further costs in employing staff to conduct the necessary studies. Nevertheless, these costs can often be negligible compared with the total cost of the system and the benefits of a thorough evaluation will often be more than justified. The precise list of factors which should be monitored in this post-implementation evaluation will depend on the particular type of system under consideration. However, for the market-focused organization, factors such as customer service, quality, manufacturing throughput and set-up times should receive a high priority as well as machine utilization, costs, output and labour efficiency. Such factors can be assessed using an 'audit checklist' which the organization can develop over time to specifically test the effectiveness of new systems (for further details see Skinner, 1978; Richardson and Gordon, 1980).

Application 14.1 ICL: Evaluation

Evaluation of new systems can be seen as taking place at a number of levels within the ICL organization, reflecting the three domains of decision making assumed by DRAMA.

Operational evaluation

Evaluation of the operating system is performed through the use of quality improvement groups on the shop-floor supplemented by the more traditional methods of system performance monitoring and reporting. The design process and the way in which systems are introduced were also seen as being important areas for evaluation both by engineers and users.

Tactical evaluation

Important considerations here are the assessment of facility design and operating performance not merely at an efficiency level, but also in the extent to which these systems are effective in relation to strategy and satisfying market needs. Also, project management structures and methods are reviewed by ICL Ashton factory management in an attempt to improve the effectiveness of the design and decision processes.

Strategic evaluation

An important consideration for senior managers in the operation of new systems is the need to perform investment appraisals to ensure that the expenditure is being paid back in terms of those benefits quoted in the capital paper. Factory performance is also monitored on an ongoing basis and the introduction of advanced real-time management information systems has assisted in the speed and efficiency of providing information to senior decision makers.

The key observations of evaluation practices within ICL are as follows:

1. Quality improvement groups and corrective action teams are important means of evaluating and improving performance within ICL Ashton.

2. There was a distinct shift between 1985 to 1989 away from performance measures based upon cost and efficiency towards a more balanced collection of customer-oriented indicators, including quality and service. System monitoring is now performed using performance measures derived from the five-point quantifications in manufacturing strategy, namely cost, productivity, quality, inventory and service. Evaluation of system performance is therefore monitored by managers using this information which is supplied to them via an on-line computer network.
3. Evaluation of ICL Ashton's actual performance data shows there to have been an improvement in terms of inventory holding, throughput times, service levels and cost since the introduction of the cascade. However, it is difficult to evaluate what proportion of these improvements are derived from the process design, the new products which were designed for manufacture, or as a result of increased quality awareness and continual improvement of employees within the plant.
4. Investment appraisal by senior ICL managers relies upon comparing actual system performance and effectiveness with those figures quoted in the capital paper at the justification stage.
5. ICL corporate management now adopt a long-term 'strategic' view of manufacturing and its role as opposed to the short-term 'financial control' approach so frequently used by company executives.
6. ICL Ashton exhibits signs of an organization that has progressed along the organizational learning curve for the design and introduction of new advanced manufacturing systems. Particular efforts have been made to learn from experiences in the critical area of project management so that the design process can be formalized in the future.

Source: by kind permission of International Computers Limited.

Post-audit of capital projects

In the capital planning and general accountancy literature the theory of planning and selecting large-scale projects is relatively well advanced (a review of the various techniques is given by Arnold and Hope, 1990). When one considers the post-implementation evaluation of capital projects, however, there is little to assist or guide manufacturing companies in assessing the success of implemented production systems. It has been recognized that post-auditing and evaluation of capital expenditure has, until recently, received little attention (Sackett and Williams, 1988). Even models developed in the field of organization studies, which claim to capture the realities and complexities of decision making, give scant regard to the process of evaluation following the implementation of decisions (for further discussion of this point see Bennett, 1990).

As the importance of the need for post-completion audits of design and implementation projects has gained increasingly more recognition, more serious attention has been paid to evaluation (Norgaard, 1979; Ghobadian and Smyth, 1989). The problems associated with the process of evaluation have also been explored, and conclusions have been reached upon the value or otherwise of post-audit programmes. The benefits and advantages of post-audit (from Neale and Holmes, 1988) are:

1. Improved quality of decision making.
2. Greater realism of project appraisal.
3. Improved corporate performance.
4. Identification of key variables and success factors.
5. Timely modifications of under-performing projects.
6. Increased frequency of termination for inappropriate projects.

The problems with post-audits can be listed as follows (developed from Neale and Holmes, 1988; Ghobadian and Smyth, 1989):

1. Post-audits can be resource hungry, and so difficult to justify in organizations suffering a lack of available resources.
2. There is often an absence of accurate and relevant data with which to carry out the audit.
3. There is a feeling in some organizations that conducting post-audits can interfere with the operational control systems.
4. Management and engineering personnel resist audits because of a reluctance for personal failures to be identified.

Post-audit of capital projects would appear to be essential for any organization wishing to derive the complete benefits of the high investment incurred in the installation of new production systems. However, in practice, there are a number of reasons why intensive audits are not carried out, not least of all the perceived costs of conducting the study coupled with the types of resistances against detailed capital audits outlined above.

Application 14.2 Accounting for advanced manufacturing technologies

Recently, a world-wide authority on accountancy, Professor Robert Kaplan of the Harvard Business School, titled an article in the Harvard Business Review *Yesterday's accounting undermines production*. Along with other critics of accounting methods, he believes that the conventions and systems that accountants use in business are actually hindering success in a competitive world.

Such an argument is not surprising. New technology, greater competition, the greater need for quality and the overall atmosphere of change co-habit with cost accounting systems that were developed for another era.

That such a situation has arisen, however, may seem strange. Management accountants often claim that their systems are developed for the individual business; that, unlike published accounts for shareholders, accounts for managers are designed to fit the needs of decision making and for the control of the business, the people, the products and the technology. Each system does not depend on rigid standards imposed from outside but is specific to the business for which it is created.

Together with colleague Robin Cooper, Kaplan investigated cost accounting in four companies with complex manufacturing and distribution operations. These were advanced companies, which might have been expected to have cost accounting systems to match their advanced manufacturing activities. Kaplan and Cooper found that all four companies used absorption costing based on direct labour. That is to say that they spread the cost of indirect, support activities over products on the basis of the direct labour time required to make the products.

When direct labour is the most important factor in production this is a reasonable approach. Factory overheads represent the cost of the basic manufacturing facility; labour hours are the capacity; each hour of use of capacity has to contribute a share of the cost of the facility. Technologies of the past led to a cost profile in which labour costs carried an overhead of perhaps 25 per cent. This overhead was often smaller but could sometimes climb near to 100 per cent (of labour costs). Modern manufacture has turned this on its head. It mechanizes and automates. The time for each productive operation shrinks to fractions of a second per product. Direct labour costs become a minute proportion of total cost.

Automated manufacture and advanced methods of production control have replaced direct labour with indirect staff, who are not subject to the disciplines of clock cards, time sheets and job tickets. The continued output of the factory is dependent on them but individual product performance cannot be identified with their individual efforts. There does not seem to be a measurable benefit. How does the business know that it is getting value for money? The only answer is to examine each activity to identify how it is to be measured. There can be no standard accounting system that fits every kind of support activity.

Some systems may be directly related to elements of production, to a set-up or a batch. Others will have their own outputs: quality to the total cost of quality or, rather, of poor quality; planned maintenance to the amount of lost production (in one very troubled process plant, the manager equated every minute of downtime to £500 in lost revenue); dispatch to the number of orders or the weight of orders dispatched.

Another fertile area for change is in the standard costing systems. These contain allowances for expected wastage, scrap, inefficiency or poor quality. Ideal costs are reduced by these allowances to arrive at an expected cost to use as a standard.

This standard is then used to compare with the actual costs of actual production. The variances that are disclosed are reported and when they are negative they are expected to generate corrective action. The result is that, generally, variances are relatively low unless gross inefficiencies have occurred or unless external forces create them. Management attention is not directed to small variances. Yet these mask an opportunity for major change.

Ideally, the management accountant should report variance on the basis of a comparison between what actually occurred and the ideal standards. These comparisons will disclose the true cost of scrap, waste and poor quality that the business bears and allow management attention to be directed towards production management improvements.

It is quite clear that many traditional cost accounting systems are not ideal for the modern manufacturing environment. There is, however, a considerable armoury of different approaches already available to the management accountant. Once the most appropriate have been selected and tried, new ones may still need creating.

The modern manufacturing philosophies of JIT, total quality control and so on must be underpinned and driven by the appropriate cost accounting systems.

Source: excerpts from Claret, J. (1986) 'Accounting for AMT', *Industrial Computing*, May, 22–4.

Current practice

Current practice in manufacturing companies would seem to reflect the new ideas suggested in the literature. Many organizations are moving away from operating performance measures based solely upon cost and resource inputs towards a more balanced collection of assessment and evaluation criteria which give a high priority to customer satisfaction. A strong factor determining the rate of this shift within individual organizations seems to be the nature of the historical development of the company. Those companies still advocating the need for tight financial control tend to be slower to adopt the new measures, while those that are adopting them are companies who see the value of a management style based upon strategic direction and market focus.

Structured programmes of post-implementation evaluation for new systems, where actual results are compared with those predicted, are not frequently adopted by manufacturing companies. The reason often cited is that the dynamic environments within which businesses now operate demand that more importance is attached to continuous improvement of existing facilities and the design of new systems than to retrospective analyses of past investment decisions. So, post-audits of newly installed manufacturing systems are not normally considered important enough to be carried out. This is not a view that can be supported, so the DRAMA component for evaluation provides a means for organizations to perform such an analysis.

■ 14.2 Key parameters

Identification of key parameters

The key parameters for evaluation, shown in Figure 14.1, comprise the nine preceding components of the DRAMA methodology in all three domains of decision making. These components were used in Chapters 5 to 13 to provide a set of guiding principles for manufacturing organizations wishing to design and implement market-focused production systems. This categorization can also be used as a framework for analyzing the design and performance of the operating system once introduced and, furthermore, in the evaluation of existing manufacturing systems. Each of the generalized methodology flowcharts in Chapters 5 to 13 incorporated evaluation feedback loops for assessing the design component in question. The evaluation component can thus be interpreted as being the operation of these feedback cycles for each component. This chapter brings together what has gone before in this book and thus provides an overview of the learning process through which

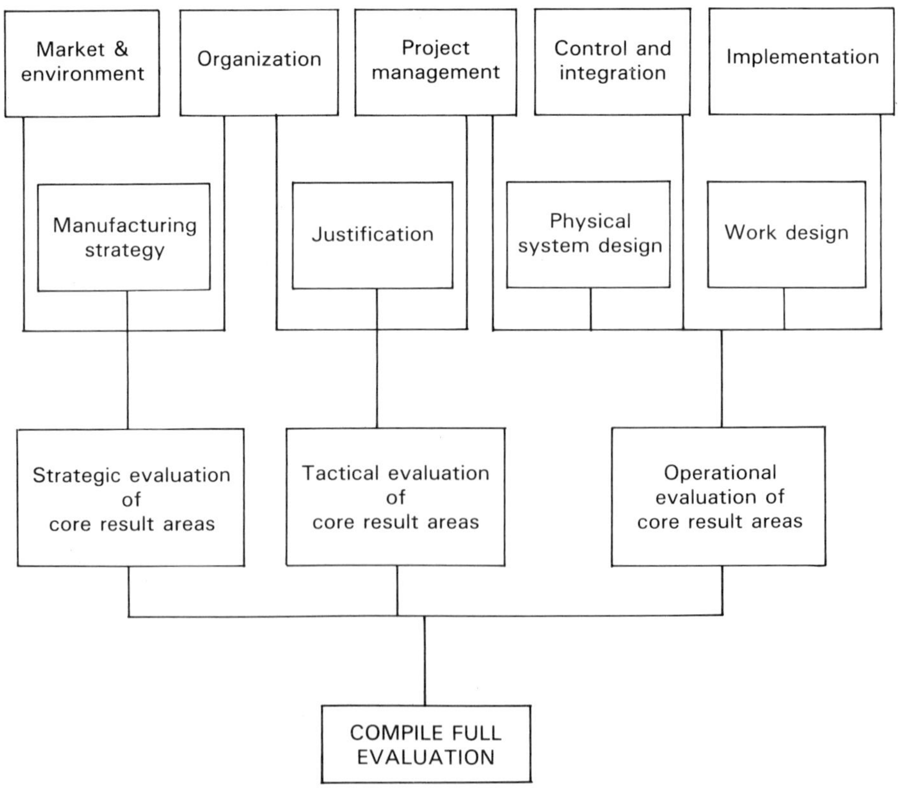

Figure 14.1 Key parameters for evaluation.

organizations need to progress when designing and refining market and customer-oriented manufacturing systems.

Strategic parameters

In the strategic domain interpretations of the state of the organization within the context of its competitive and market environment must be developed as inputs to a higher level evaluation of the company. Explicit statements of existing corporate, marketing and manufacturing strategies should accompany this, together with the current organizational structure and views on the state of the organization in relation to culture, attitudes and management style.

Tactical parameters

In the tactical domain the existing procedures and mechanisms used for justification and project management must be expressed and the organization design of the factory should be available for appraisal.

Operational parameters

Finally, at the operational level of the business, the existing design of processes, the planning and control infrastructure and employees' jobs provide the input for evaluation of the production system itself. Additionally the process of implementation currently employed by the company in question should be available for scrutiny.

14.3 Selection and configuration of design options

The previous components in the DRAMA methodology provide the main basis for evaluating existing strategies, tactical processes and production facilities. By working back through each component following implementation the methodology user can appraise the related features of the design options with their own findings and develop the design option guides further, making them more appropriate to their own needs. At the same time, the user should take actions, where appropriate, to further refine the newly installed production system with the aim of improving its overall effectiveness. In addition to this comprehensive evaluation of strategies, procedures and systems, a design option guide is provided to present and review some of the most widely adopted general evaluative procedures (Figure 14.2). Each option is now considered in turn.

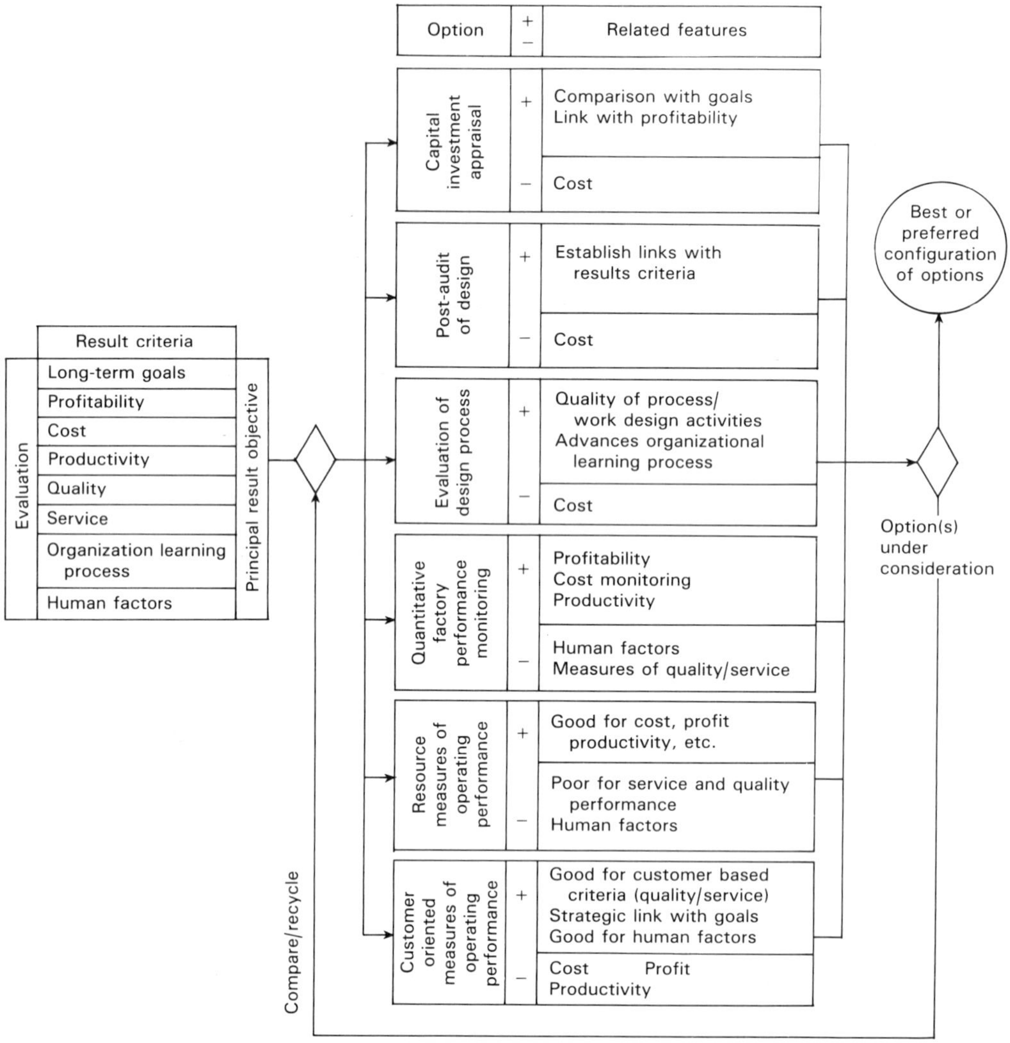

Figure 14.2 Design option guide for evaluation.

Capital investment appraisal

At the strategic and tactical levels of the organization the performance of implemented system can be compared to that performance expected when capital expenditure was sanctioned at the end of the justification process. Performing this comparison for all major investments will enable the company to develop, over time, an interpretation of areas where investment has been successful and those where it has been less successful. The positive benefits of carrying out this exercise are that implemented system performance is compared with overall corporate goals and the contribution of the new facilities to profitability can be assessed. Somewhat surprisingly, however, many organizations do not carry out a detailed capital investment appraisal exercise for fully specified manufacturing systems. The reasons given for this include the cost of performing a retrospective exercise and the fact that the market, environment and customer needs may have changed since the system was first designed, so a detailed appraisal may be of little value.

Post-audit of design

The functioning and performance of the design itself can be evaluated in terms of assessing the physical equipment, the planning and control procedures and disciplines, and the work of people within the system. In this way, recommendations for the continual improvement of the production system can be developed and the design of the system can evolve as customer and market circumstances dictate. In so doing, the performance of the system should be compared with the selected result objectives for each of the operational design components, the mix of objectives being refined or changed, as required, over time.

Evaluation of design process

In addition to evaluating strategies and systems, the process of designing and developing these should be periodically reviewed. The DRAMA methodology is one prescription for a method of design, but individual organizations may have their own variants upon the design process. Following the introduction of new systems, it can be advisable to find time to consider the ways in which the efficiency of design can be improved for future production systems projects.

Factory performance monitoring

It is normal to measure and monitor factory performance at the level of the operating system. Quantitative measures are usually used as part of the routine management control process described in Chapter 11. This provides values for assessing the contribution to profits of the unit in question, a means of controlling costs and of measuring productivity. However, quantitative measures of this sort may not be as effective in measuring criteria such as human performance and levels of quality and service, for which some element of qualitative analysis will normally be necessary.

Measuring operating performance

Linked to factory performance evaluation, the operation of individual production areas will also require an on-going assessment. Traditional measures used have been those of cost, productivity, efficiency and labour standard hours performance. Although these may be important and necessary, reliance upon these measures is increasingly open to criticism. The major objection is that they are merely measures of the input to, and not the output from, the system. Customer oriented measures of output are thus becoming more widely adopted to complement the traditional performance measures mentioned above. Examples include quality performance, measures of service provided and conformance to delivery times quoted. In this way organizations should develop a more balanced set of criteria for evaluating the performance of operating systems.

■ 14.4 Generalized methodology

Figure 14.3 presents the generalized methodology flowchart for evaluation within the market-focused organization.

1. **Evaluate any shortcoming in performance of the new system.** The starting point for evaluation following the installation of a new system is for the organization to evaluate the performance gap in its operations for each of the components. This will provide an overall view of the latest performance at the strategic, tactical and organizational levels of the business.
2. **Evaluate system performance from a strategic perspective.** Strategic evaluation involves the assessment of corporate and market strategies from the market and environment component, manufacturing strategy and corporate and departmental organization, in relation to the latest circumstances of the company. The evaluation feedback loops in the generalized methodology flowcharts for each of these three DRAMA components (Figures 5.6, 6.6 and 7.4) provide some detail about the precise nature and contents of these assessments, but generally current performance should be measured against the goals, profitability, quality and service targets laid down for the new system.
3. **Evaluate system performance at the tactical level.** The details of the evaluation for the tactical domain components of DRAMA are shown in Figures 7.4, 8.3 and 9.6. Generally, however, tactical evaluation involves measuring and assessing performance against predetermined targets for productivity, cost, quality, output, inventory, human resources and service.

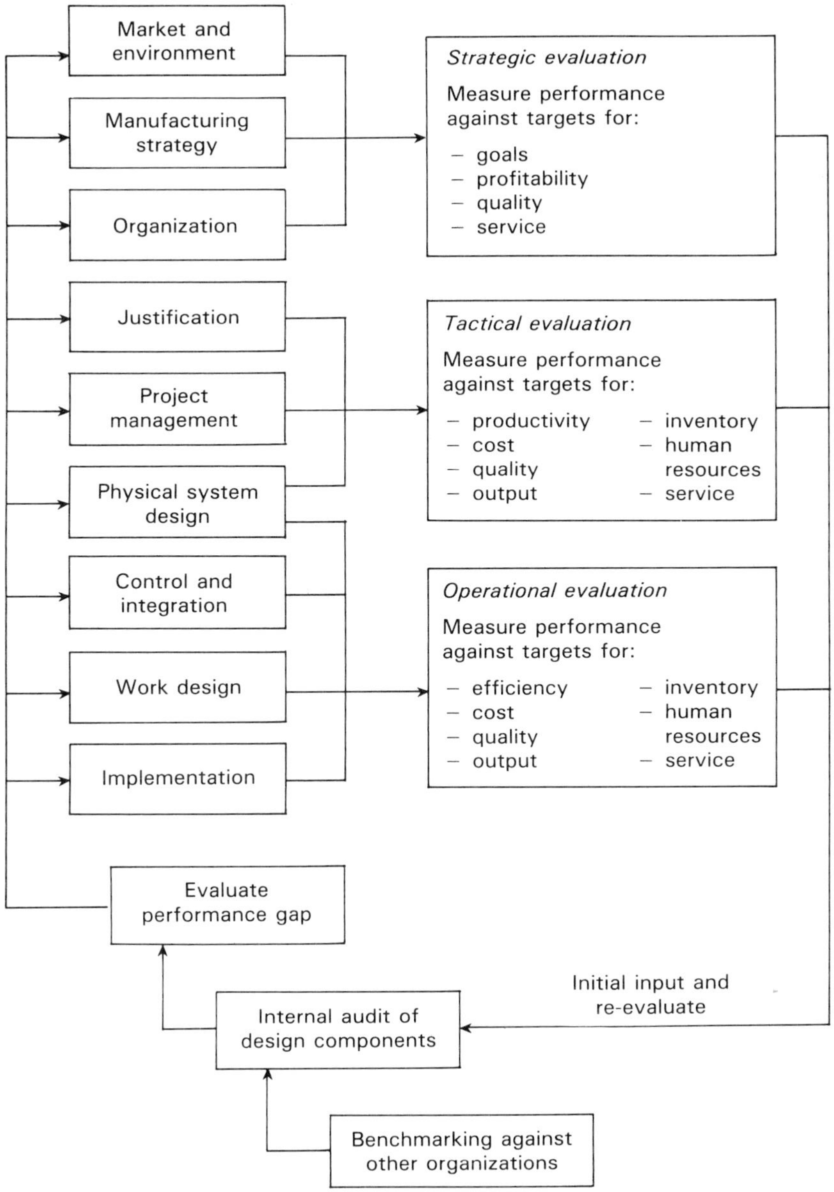

Figure 14.3 Generalized methodology for evaluation.

4. **Evaluate the operational performance of the system.** Operational evaluation involves consideration of the three elements of the operating system, physical system design, control and integration and work design, together with a review of the implementation process adopted for new systems. The precise details of the form and content of this evaluation is

provided by the feedback loops in the generalized methodology flowcharts for these components (see Figures 10.12, 11.12, 12.3 and 13.4).

5. **Audit the design process for DRAMA design components.** Finally, the results of these evaluations should be combined to give an overall assessment of the design process for new production systems within the organization together with the lessons fed back into the procedures for each DRAMA component.

14.5 Conclusions

Evaluation of production systems design is important because, if properly conducted, it can result in corrections and modifications to the various design components should any of the factors contributing to system performance not be meeting expectations. Evaluation is often overlooked or neglected by manufacturing companies, but should take place at all levels of the organization, from an audit of the investment decisions for facilities to an assessment of system operation and efficiency.

In relation to the design process, evaluation should take place both before and after implementation. When carried out beforehand it forms part of the justification process and will comprise a feasibility analysis together with a financial assessment using one or more of the techniques described in Chapter 8 and, in some cases, a simulation of the planned system. A post-project evaluation is equally important because it is necessary that the planned objectives of the new system are being achieved. It will involve detailed monitoring of the factors which contribute to overall system performance, using both quantitative and qualitative measures.

Designing production systems is not just a matter of specifying machines and equipment together with new working methods. It forms part of a total corporate, market and manufacturing strategy and must be linked closely with the plans and controls associated with all the business functions. For this reason, the DRAMA methodology for evaluation involves a retrospective examination of all the components of the production systems design methodology and the refinement of the design process and actual operating system.

14.6 Further reading

Adam, E. E., J. G. Hershauer and W. A. Ruch (1981) *Productivity and Quality: Measurement as a Basis for Improvement*, Prentice Hall, Englewood Cliffs, New Jersey.

Saaty, T. L. (1980) *The Analytic Hierarchy Process*, McGraw-Hill, New York.

Appendix I

Cellular Organization

■ I.1 Group technology

Group technology (GT) is an alternative to the conventional batch production systems based on process-oriented work organization and functional layout. Like batch production it is associated with medium (intermittent) demand.

The main drawbacks associated with conventional batch production, as identified in Chapter 2, are:

1. High working capital costs.
2. Frequent set ups.
3. High material handling costs.
4. Long delivery lead-times.

Additionally, there are problems of:

1. Extensive material movement.
2. Complex production control with regular work in progress monitoring.
3. Lack of job enrichment and product responsibility.

Group technology has three aspects:

1. Grouping *parts* into *families*, with the aim of reducing setting times.
2. Grouping *processes or machines* into *cells*, with the aim of reducing material movement.
3. Creating groups of *multiskilled operators*, with the aim of increasing group autonomy and job flexibility.

The methods of creating groups are as follows:

1. 'By eye' or 'rule of thumb' using local product knowledge.
2. Coding and classification using a universal or tailor-made system.
3. Using production flow analysis.
4. Designing products for group technology manufacture.

Grouping parts into families is the first step of GT and the following points should be noted:

1. The principle of the 'composite component' can be applied, as shown in Figure I.1.
2. It is concerned mainly with single machines.
3. It is limited largely to rotational components.

Families of components can be identified by eye or using a coding system of the type shown in Figure I.2

Grouping processes into 'cells' complements the first step of grouping parts into families and involves the following:

1. Changing from functional layouts to operation sequence layouts, as shown in Figure I.3.
2. Reducing the batch sizes.
3. Changing the approach to production planning, as illustrated in Figure I.4.

Coding and classification proves less effective at identifying groups of machines and families of parts, so the technique of 'production flow analysis' is probably more appropriate (Figure I.5). Other family and cell formation techniques include mathematical methods, such as cluster analysis and rank order clustering. Where products are not already in existence, designing them specifically for GT production may be a way of achieving the best results.

Creating groups of multiskilled operators was not one of the primary objectives of GT but the grouping of people and the requirement for wider skill and flexibility is a significant spin-off. This is often seen as a disadvantage

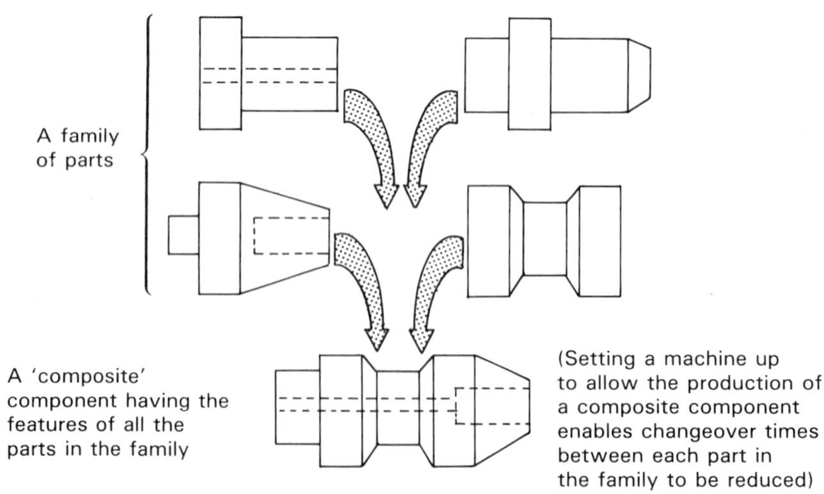

Figure I.1 A family of parts and composite component.

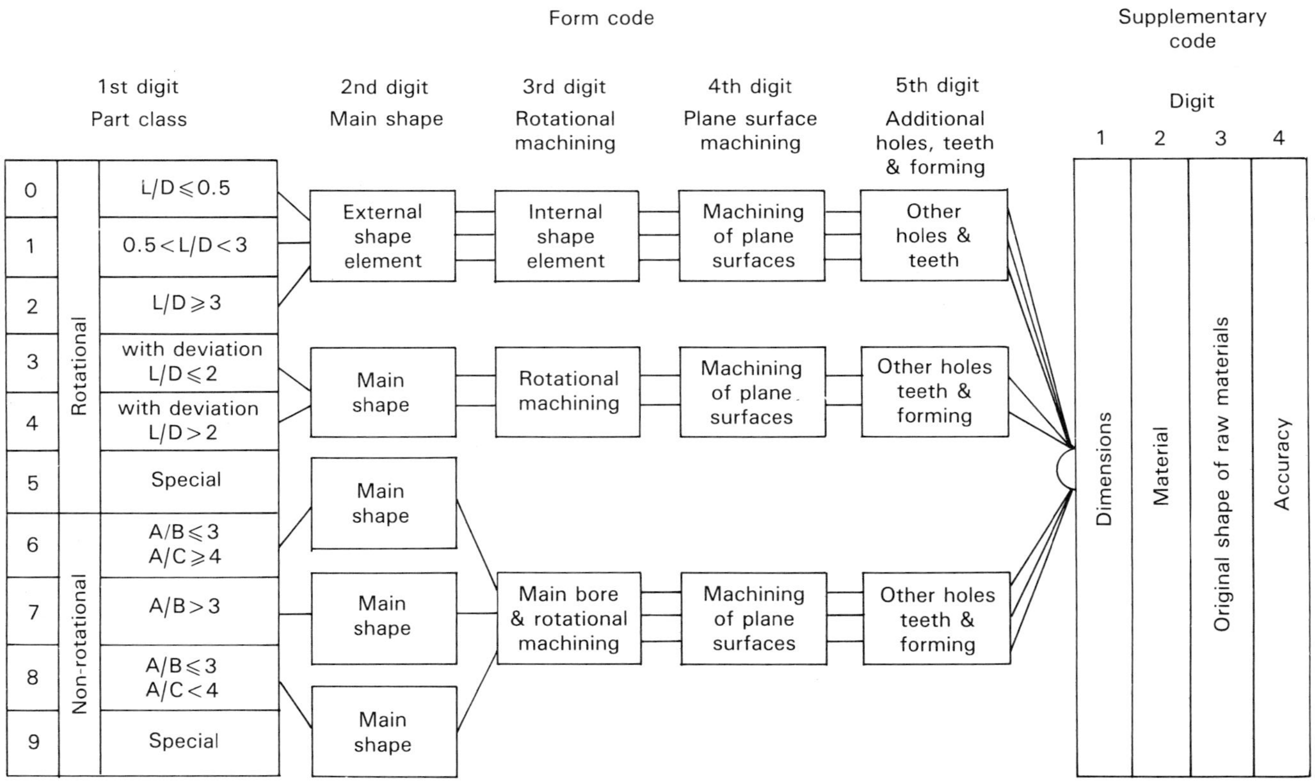

Figure I.2 Basic structure of the 'Opitz' parts coding system developed in Germany (Opitz, 1970).

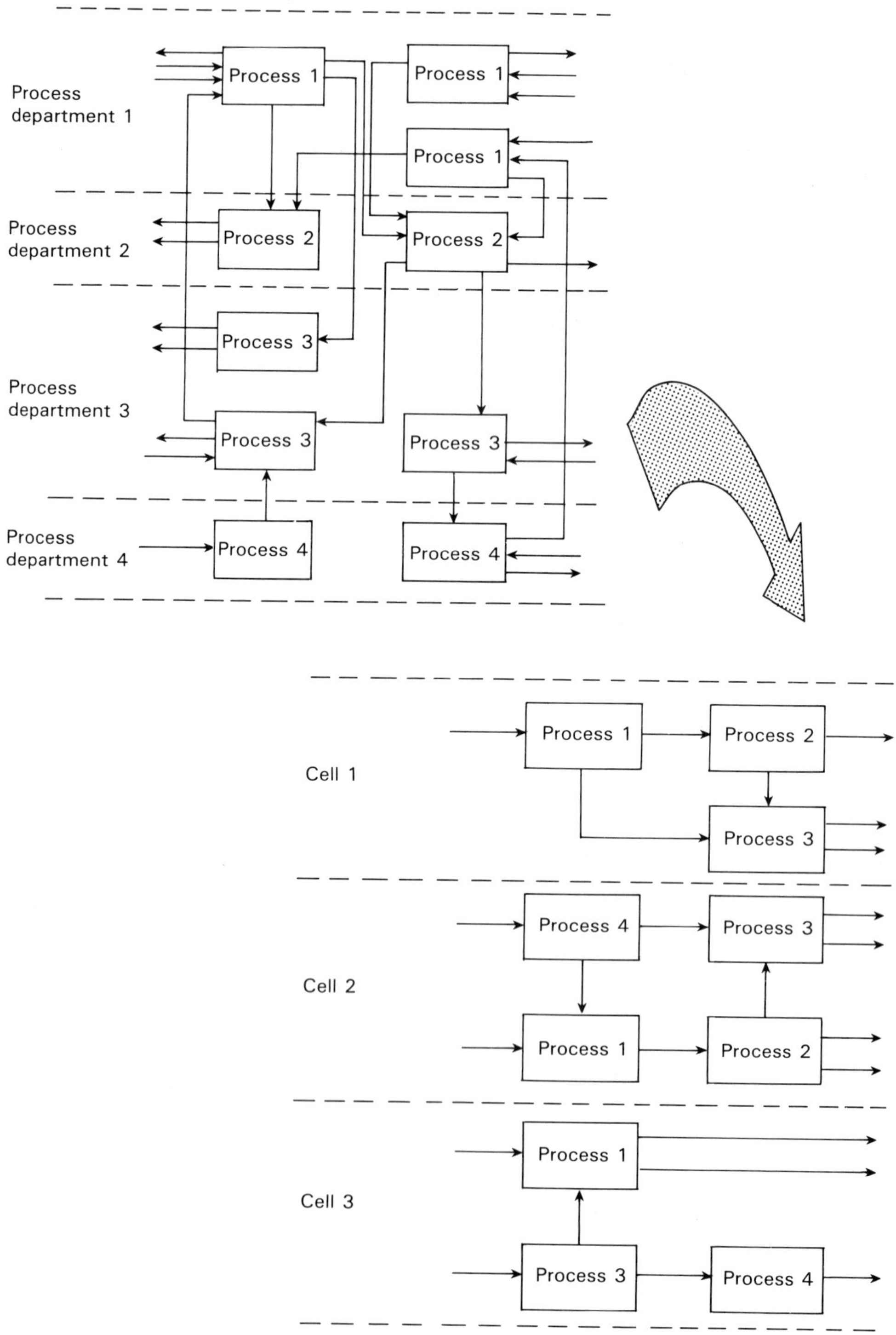

Figure I.3 The change from functional layout to group technology cells.

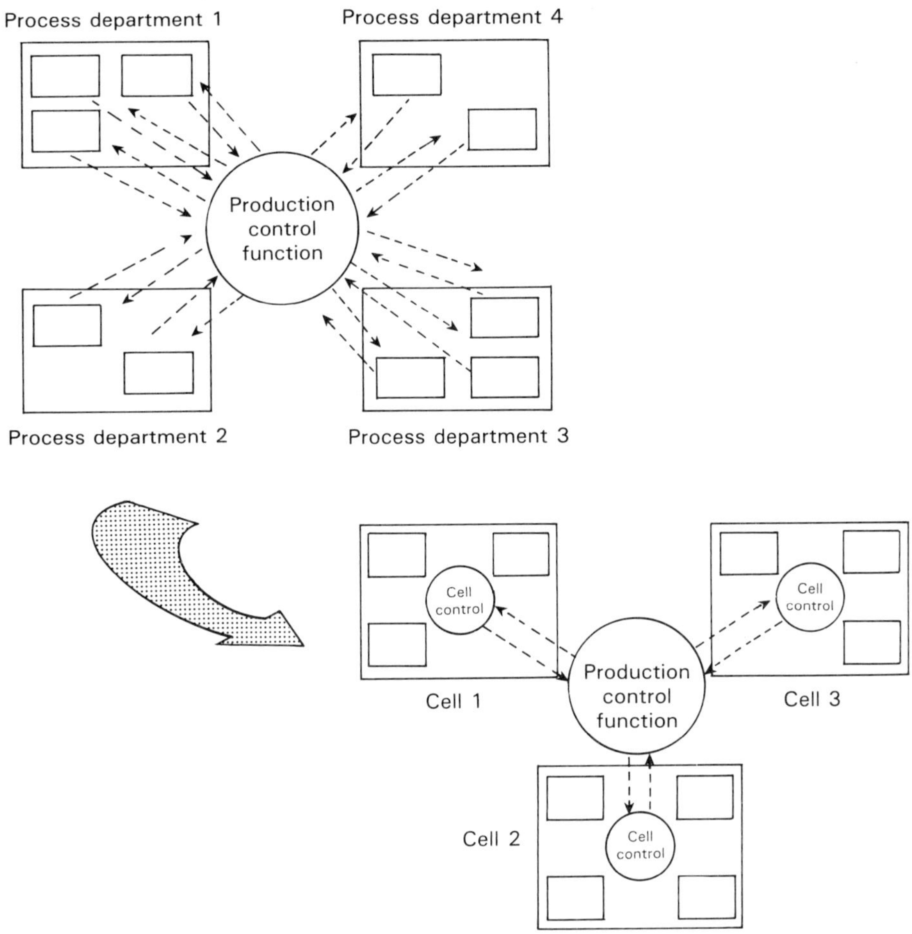

Figure I.4 Production planning with group technology.

by engineers, but behavioural scientists see it as a way of enabling job enrichment. This feature has been deliberately exploited in some countries, such as Sweden, where the idea of creating groups was later extended to include the assembly function.

The claimed benefits of GT can, therefore, be summarized as follows:

1. Setting times are reduced due to similar nature and sequence of machining operations.
2. Material handling is reduced due to shorter distances between machines.
3. Throughput times are shorter due to the rapid flow of parts through the cell.
4. Levels of work in progress are reduced because large batches of partly completed items are not waiting between machines.

Matrix of components and processes

Components

Processes	K48251A	L48388	L48267B	M44276E	M47693F	L48388M	M48195C	M44276D	E41795	E48596	E34267	E12204	E12288	K47697	E47782	E48586	K34596	E33494	M48265D	K44276C	M45691D	M45691B	M48386H	K34098A	E7392	E46384	E33295	K45199	K43390	M61592	E18694
DMT (3)		✓									✓			✓	✓	✓								✓		✓		✓	✓		✓
DM (3)			✓	✓				✓				✓		✓			✓	✓	✓					✓			✓	✓	✓	✓	
PG			✓			✓											✓	✓	✓	✓											
OXY (3)	✓	✓	✓														✓							✓				✓	✓	✓	
P&GR					✓																										
PGR													✓													✓					
PGH																															
PGG																					✓	✓					✓				
P&G							✓	✓	✓	✓	✓	✓											✓	✓			✓	✓	✓	✓	✓
RP																									✓						
PGB				✓										✓	✓						✓	✓					✓				
W&P	✓						✓																	✓				✓	✓		
WG (3)												✓																			

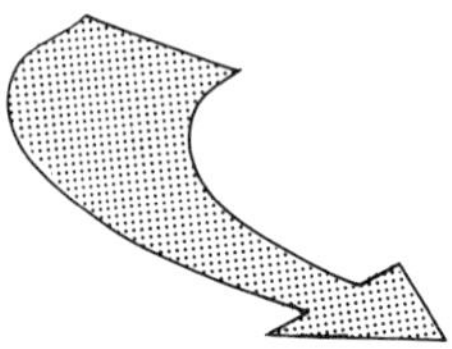

Identified component families and process groups

Components

Processes	L48267B	K34596	M48265D	E33494	K44276C	L48388M	E7392	K34098A	K45199	K43390	M61592	M48195C	M44276D	E34267	E12204	E18694	E41795	E48586	M48366H	K48251A	L48388	M45691D	M45691B	M44276E	K47697	E47782	E46384	E12288	E48596	E33295	M47693F
PG	✓	✓	✓	✓	✓	✓																									
DM	✓	✓	✓	✓																											
OXY	✓	✓																													
RP							✓																								
P&G								✓	✓	✓	✓	✓	✓	✓	✓	✓	✓	✓	✓											✓	
DMT								✓	✓	✓				✓		✓					✓										
DM								✓	✓	✓	✓		✓		✓					✓											
OXY								✓	✓	✓	✓	✓								✓	✓										
W&P								✓	✓	✓					✓																
WG																															
PGG																						✓	✓						✓		
PGB																						✓	✓	✓	✓	✓			✓		
PGR																											✓	✓			
DMT																									✓	✓	✓			✓	
DM																								✓	✓				✓		
P&GR																															✓

Group 1 (PG, DM, OXY, RP) – Family 1 (L48267B to E7392)

Group 2 (P&G, DMT, DM, OXY, W&P, WG) – Family 2 (K34098A to L48388)

Group 3 (PGG, PGB, PGR, DMT, DM, P&GR) – Family 3 (M45691D to M47693F)

One 'exception' – This component must be routed through groups 2 & 3 (circled ✓: P&G, E33295)

Figure I.5 Production flow analysis.

5. A cell is a single planning point which makes scheduling and monitoring easier.
6. Greater operator autonomy and responsibility enables job satisfaction and improved quality.

However, GT has had its critics, who have contended that:

1. Job satisfaction can fall due to reduction in component variety.
2. WIP will remain high because each machine will require its own pool of work rather than sharing a pool in a functional department.
3. Utilization will be lower because of machine duplication.
4. Production control is more complicated because of the difficulty of balancing load against capacity.
5. Cells are designed on the basis of a particular product mix, so they are vulnerable to changes.

Nevertheless, there is general agreement that GT is suitable for automation applications and it can be argued that a 'flexible manufacturing system' is an extension of the GT philosophy which uses modern automation methods.

Areas where automation can be applied to cellular manufacturing:

1. Individual machine operations.
2. Material handling on the machine.
3. Material handling between machines.
4. Material handling in and out of stores.
5. Tools changing on machines.
6. Movement of tools into and out of stores.

I.2 Autonomous work systems for assembly

Autonomous working is an alternative to conventional flow production systems based on task-oriented work organization and operation sequence layouts. Like flow production, it is associated with high (continuous) demand.

Some of the drawbacks associated with conventional flow production were noted in Chapter 2. The following is an extended list grouped into two categories previously identified:

1. **The physical drawbacks:**

   (a) it is rigid and difficult to change;
   (b) it is not conducive to producing a variety of products;
   (c) it cannot quickly adjust to demand variations;
   (d) the whole line is at risk in the event of a station stoppage;
   (e) a pool of spare operators is needed;
   (f) efficiency is adversely affected by balancing and system losses (as demonstrated in studies at Philips, Eindhoven).

2. **The 'human' drawbacks:**

   (a) recruitment difficulties;
   (b) high labour turnover (it has been as high as 50 per cent per annum in Scandinavia and Italy);
   (c) high absenteeism;
   (d) sabotage;
   (e) no quality responsibility.

Consider the design of a conventional flowline. This involves:

1. Deciding the required output rate or cycle time based on predicted demand.
2. Determining the required number of stations.
3. Deciding on the need for station replication or multiple resourcing.
4. Allocating tasks to each station.

However, the following constraints need to be considered:

1. Precedence constraints (i.e. restrictions on the order in which tasks can be done).
2. Zoning constraints (i.e. restrictions on where certain tasks or combinations of tasks should, or should not, take place).

There is also the problem of dealing with product variety on conventional flowlines. This has necessitated the following variations:

1. Multimodel lines, where the line is reorganized periodically to produce different models or variants.
2. Mixed model lines, where the line is designed to allow simultaneous production of any model or variant without reorganization.

System loss is a particular problem of flowline design. These are losses incurred due to pacing and variability of work times at stations and are particularly prevalent on multimodel lines. The result of excessive system loss is line stoppages or unfinished products. Possible ways of reducing system loss are:

1. Creating a supply of buffer stocks between adjacent work stations.
2. Allowing products to be available to operators for a longer period of time.

Autonomous work systems apply the principles of cellular organization to assembly operations. They overcome many of the problems associated with flowline production, i.e.:

1. Longer cycle times enable balance and system losses to be reduced or eliminated.
2. Different products can be produced in parallel.
3. Response and volume flexibility is greater.
4. Model changes can be effected more easily.

5. There is no work station interdependency.
6. Job enrichment and autonomy lessens the human problems associated with flowline working.
7. Responsibility for quality is more clearly assigned.

There are various ways in which autonomous working can be facilitated, being based generally on creating alternative forms of work organization, as illustrated in Figure I.6.

To implement autonomous working in a practical environment the broad options available are:

1. Individual versus group work.
2. Parallel (dock) systems versus line assembly (or flow groups).

These options are shown in Figure I.7.

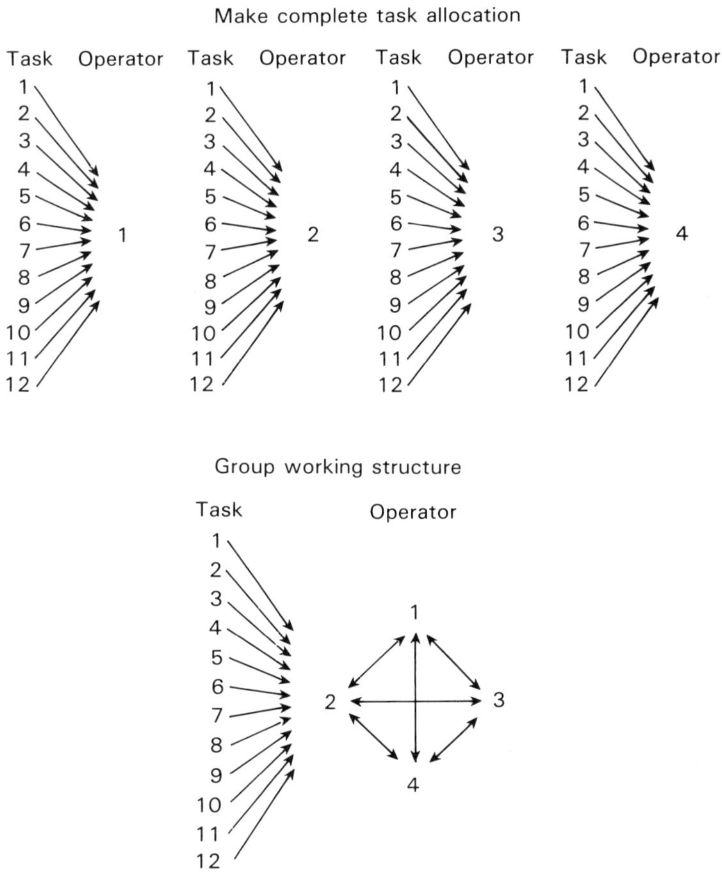

Figure I.6 Alternative work organization for autonomous working.

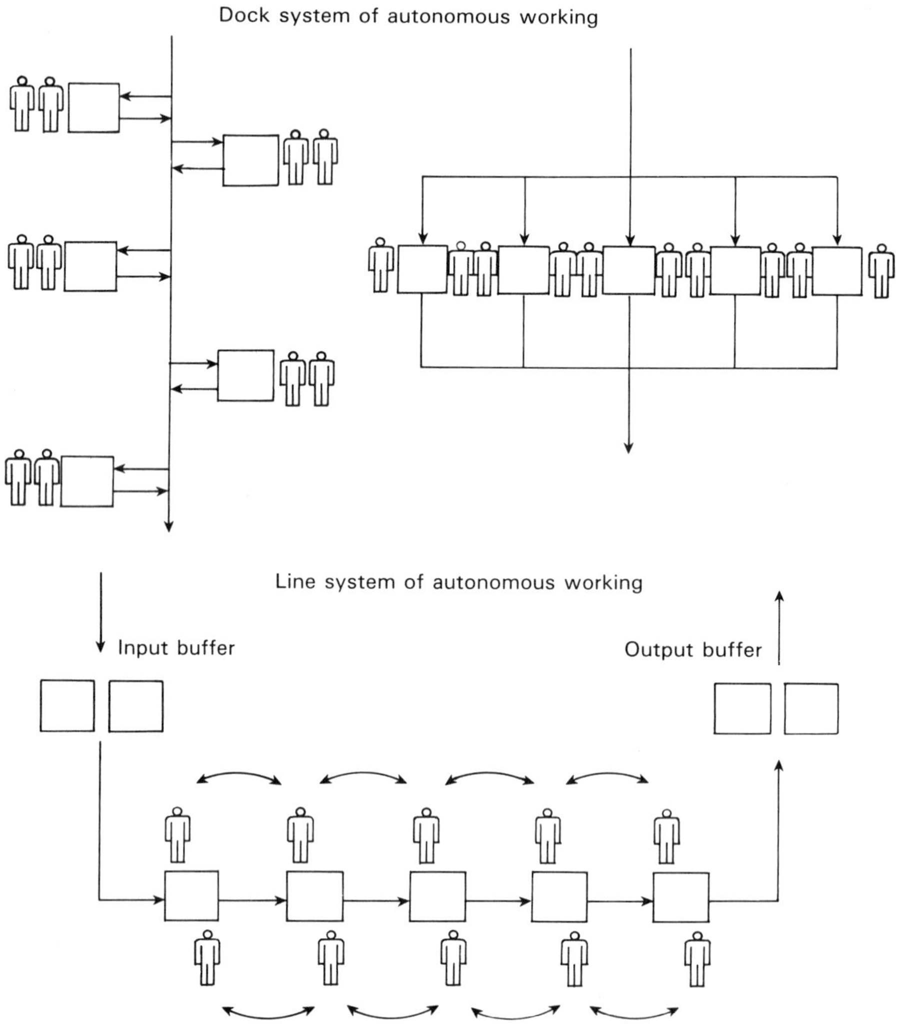

Figure I.7 Alternative layout options for autonomous working.

Implications for production control

Some of the implications for production control of adopting cellular organization have already been addressed in the previous discussion on the benefits of Group Technology. The greater complexity of autonomous work systems for assembly necessitates a control system that can monitor products and subassemblies, as well as guide them to the correct work places at the correct time.

The reduced setting times, more efficient material flow and modular nature of cellular organization facilitates 'just-in-time' production using 'kanbans'.

Material requirements planning (MRP), described in Chapter 11, can be used for longer term material purchasing.

Implications for design

One of the most important requirements for successful implementation of cellular organization is that the design of the product lends itself to the 'completeness of task' principle. With complex products, this principle may require the design of modularized products which enable the complete production of a module to be assigned to an individual or group. Product design is one of the keys to successful implementation of cellular organization. The potential benefits of both Group Technology and Autonomous working are often inhibited when products have been designed to be manufactured on conventional systems.

Implications for payment systems

Conventional payment systems, particularly those based on an individual financial incentive, are usually inappropriate for cellular organization. A more relevant payment system would probably comprise several components depending on the nature of work organization and the motivational and reward factors being emphasized.

Some of the features of a payment scheme for autonomous work groups might be as follows:

1. An individual element based on job evaluation, e.g. education/training/range of skills/extra responsibilities/timekeeping record, etc.
2. A group element based on delegated responsibilities, e.g. production planning/quality responsibility/cost accounting/administration/social responsibility (no rejection of members).
3. A result related element (possibly plant-wide), e.g. cost of factory workers/cost of staff/rejects and adjustment/operational supplies and tools/added material/store value/quality index.

Glossary

AGV: *See* Automated Guided Vehicle.
AMHS: Automated Materials Handling System.
AMT: Advanced Manufacturing Technology.
ASRS: Automated Storage and Retrieval System.
Automated Guided Vehicle: Computer-controlled driverless materials handling trucks.
Autonomous Working: A form of production organization where individuals or groups of operators work independently on the complete build of a product or component.

BCG Matrix: *See* Boston Consulting Group Matrix.
Bill of Materials: The multilevel parts breakdown for a final product and its subassemblies used in MRP, taking the form of 'product structure' and 'product master' files.
BOM: Bill of Materials.
Boston Consulting Group Matrix: An approach to product portfolio strategy analysis which categorizes products or business units according to market share and market growth.
Brainstorming: An idea generation approach to encourage creative thinking in which groups of people make suggestions within a forum where there is no criticism or disapproval.
Bulk Issue: Supply of material or parts in volume batches (as opposed to 'kitting').

CAD: Computer-Aided-Design.
CADCAM: Computer-Aided-Design/Computer-Aided-Manufacture. Used as a term to cover the computerized design to manufacturing relationship.
CAE: Computer-Aided-Engineering.
Capital Paper: The expenditure justification document raised for large-scale investment proposals.
CAPM: Computer-Aided Production Management.

CAPP: Computer-Aided Process Planning.

Cause and Effect Diagrams: Diagrams used by quality circles to analyze and trace the causes of problems. (Because of their appearance, they are sometimes known as 'fishbone diagrams'.)

Carousel: A rotating materials storage facility: can be 'horizontal' or 'vertical' (paternoster store).

Cascade: ICL Ashton's modularized flexible assembly system (modular assembly cascade).

Cell: This normally refers to a production unit within a cellular manufacturing or group technology system.

Cellular Manufacture: A form of facility layout and organization where the production process is disaggregated into a series of locally managed and operated cells.

Centres of Excellence: Manufacturing organization where volume production of specific components or final products is conducted within individual 'expert' plants.

Champion: A person who promotes an innovation or development within an organization.

CIM: Computer Integrated Manufacturing.

CIM-OSA: Computer Integrated Manufacturing – Open Systems Architecture (a methodology developed under the European Community's AMICE project).

Closed Systems: An approach to production systems design in which the external environment is not taken into account.

CNC: Computerized Numerical Control.

Co-determination: Dual management and employee participation in the decision making processes of a company.

Competitive Edge: Distinct advantage over rival organizations.

Computer Integrated Manufacturing: The integration of design engineering, production planning and control, and manufacturing automation with the wider business systems of an organization.

Concurrent Engineering: *See* Simultaneous Engineering.

Conformance to Requirements: Adherence to customer demands/expectations/requirements.

Consultation: The offering, sharing and seeking of information, especially at lower levels of the organization.

Corporate Policy: The organization's overall objectives and philosophy.

Corporate Response: The organization's reply to its market and competitive environment.

Corrective Action: Action taken to bring an outcome back to the required standard.

CPM: *See* Critical Path Method.

CRA: Core Result Area.

Critical Path Method: A network analysis technique used in project planning and control where critical activities are identified and overall completion times computed.

Customer-focus: Orientation towards market and consumer needs.

DATA: *See* Design Attribute/Time Analysis.
DCF: *See* Discounted Cash Flow.
Decision Rules for Analyzing Manufacturing Activities (DRAMA Framework): Developed from the DRAMA methodology, a model for analyzing the design of new production systems
Design Attribute/Time Analysis (DATA): A model developed by IDOM (Innovation, Design and Operations Management) for plotting the variability of design configurations against their time frame (short or long).
Design Option Guide (DOG): Schematic illustration to assist in the selection and configuration of appropriate design options.
Design Routine for Adopting Modular Assembly (DRAMA Methodology): The manual and methodology developed by Aston University's IDOM Research Unit to provide a set of guiding principles for the design and introduction of systems for high variety assembly.
Discounted Cash Flow (DCF): A financial justification technique that takes into account interest rates and the 'time value of money'.
DOG: *See* Design Option Guide.
Domain: Organizational level of decision making including 'Strategic', 'Tactical' and 'Operational'.
Downstream production: Assembly and associated operations for production of finished products.
DRAMA Framework: *See* Decision Rules for Analyzing Manufacturing Activities.
DRAMA Methodology *See* Design Routine for Adopting Modular Assembly.

EBQ: Economic Batch Quantity.
Electronic Point of Sale (EPOS): Retail computer systems usually employing barcodes.
Electronic JIT: A computer-based trigger to prompt material (or Kanban) replenishment (as opposed to a physical trigger or card).

Families (of parts): Groups of like components, or those which share similar processes.
FAS: *See* Flexible Assembly System.
FIFO: First In First Out, i.e. an inventory control system where items held in storage are removed in the order they were put in.
'Fishbone Diagrams': *See* Cause and Effect Diagrams.
Flexibility: Adaptability in terms of range, response and propensity to change.
Flexibility of Range: Ability to satisfy a wide variety of customers' requirements.
Flexibility of Response: Ability to reply to demand in a short period of time.
Flexible Assembly System (FAS): An assembly system capable of supplying a wide range of products and/or responding in a short lead-time.

Flexible Infrastructure: A basic structure or installation within an organization/process which is adaptive to change.
Flexible Manufacturing System (FMS): A production system capable of supplying a wide range of products and/or responding in a short lead-time.
Flexible Materials Handling System: A non-dedicated parts transportation system capable of supplying a wide range of process requirements.
FMS: *See* Flexible Manufacturing System.
Front-end operations: Operations which take place early on in the chain of manufacturing.
Functional Differentiation: Organization based on division into specialized skill areas.

Gantt Chart: A horizontal bar chart (named after its originator, Henry Gantt), which represents various activities that need to be accomplished against a time-scale. Otherwise known as a 'milestone plan'.
GRAI: Graphe a Resultats et Activites Interrelies: a conceptual decision model and production control system design methodology developed at the University of Bordeaux.
Group Technology (GT): A form of production system design where the total system is made up of a series of cells that manufacture families of parts.

Hard Systems: A mechanistic approach to systems thinking which concentrates on finding solutions to well-defined problems.
Harmonization: Single status employment and conditions of work.

IDEF: ICAM **DEF**inition: A methodology developed for the analysis and design of operations systems.
Industrial Democracy: Employee participation and co-determination within manufacturing operations.
Input Store: Buffer storage points holding kits of parts for assembly modules within the modular assembly cascade.
Intangibles: Items which are difficult to quantify or to which it is difficult to ascribe true value or likely value.
Integrated Design: A facility design where the utilities are physically combined.
Inventory Status File: The record of stock balances and locations as used in MRP.
Internal Rate of Return (IRR): A financial justification technique in which percentage return from an investment is based on a 'discounted' rate of repayment.
IRR: *See* Internal Rate of Return.

JIT: *See* Just-in-Time.
Job Enlargement: An approach to job design in which operators are assigned extra tasks in order to reduce the monotony associated with highly repetitive work (sometimes known as 'horizontal job loading').

Job Enrichment: An approach to job design where operators are given greater responsibilities with the aim of increasing their satisfaction and thereby their commitment to the organization (sometimes known as 'vertical job loading').
Job Rotation: Movement of workers from one task to another.
Job Variety: The range of individual tasks.
Just-In-Time (JIT): On-time delivery of parts and products, from suppliers or in-house, to satisfy demand without excessive inventory holding. Its wider philosophy can be defined as the elimination of waste.

Kaizen: A Japanese term meaning continuous improvement.
Kanban: A specific application of JIT using a system of triggers to prompt material replenishment.
Key Parameter: Important input shaping the design of individual DRAMA components.
Key Result Areas: The main objectives identified for individuals and used as a basis for assessing their performance.
Kitting: Parts supply in containers for the build of a single or predetermined quantity of assemblies: contrast with 'bulk issue'.

Lead-time: The time between initiation and completion of an activity. Examples are design lead-time (time between conception of a product/process and its completion) and order lead-time (from customer order to receipt).
Learning Curve: The (exponential) accumulation of knowledge.
Levels of Manufacturing: The degree of vertical integration.
Life-Cycle Costing: A means of calculating the cost of an investment which takes into account the expenditure incurred during every phase of its life.

Manufacturing Resource Planning (MRPII): A computerized information system that links MRP with wider business systems such as finance, marketing, design and engineering.
Manufacturing Strategy: The strategic objectives of manufacturing operations.
Market Strategy: Strategic objectives relating to the way the organization should respond to and manage its markets.
Market-driven: Governed and directed by market changes: being able to respond and adapt effectively.
Market-focused Production: A production philosophy based on the responses needed to meet the demands of customers and threat of competition.
Marketing Strategy: The strategic objectives of marketing operations.
Master Production Schedule: The plan for manufacturing finished products as used in MRP.
Material Requirements Planning (MRP): A form of production planning and control based upon projecting forecasted and known demand into requirements for subassemblies, components and materials.
Matrix Organization: An organization structure featuring dual responsibilities and channels of communication.

Mercury: The final assembly module for products up to 1 m^3 within ICL's modular assembly cascade.
Mission Statement: A brief and articulated statement of overall corporate policy and objectives.
Mixed-model Line: A flowline configuration where different models/variants are produced simultaneously on the same line.
Modular Assembly: Assembly through disaggregation of the total system into a series of individual units.
Modular Assembly Cascade: ICL's flexible manufacturing system within the Ashton factory.
Module: A subunit within the factory system.
Module (or 'cell') Controller: A computer-based control system, dedicated to a module for both hardware and materials control, which communicates with the factory's overall computerized materials control system.
MPS: Master Production Schedule.
MRP: *See* Material Requirements Planning.
MRPII: *See* Manufacturing Resource Planning.
Multimodel Line: A flowline which is reconfigured to manufacture different products.

Net Present Value (NPV): A financial appraisal technique which expresses the net value of an investment in present day monetary values.
Niche: A particular area of marketing expertise or market control.
Nichemanship: The technique of marketing through niche recognition.
NPV: *See* Net Present Value.

Off-standard: Deviation from the performance target set through work measurement.
Operational Domain: The area of decision making at the lower levels of organization.
Open Systems: An approach to production systems design in which the effect of the wider environment is taken into account. Also used to describe computer systems where the user is not limited or tied to the use of a single manufacturer's operating standards.
OPT: *See* Optimized Production Technology.
Optimized Production Technology: A production planning and control system that focuses on the bottlenecks in operations.
Order Qualifiers: Criteria which customers normally expect of products.
Order Winners: Criteria which give a product superiority, and thereby competitive advantage, in the market.
Organizational Culture: The underlying assumptions and beliefs prevailing within an organization.
Organization State: The characteristics of a corporate or divisional body.
Organization Structure: The way in which the personnel and functions of business are constructed or organized.

Output Store: Buffer storage points holding completed subassemblies/products for modules within the Modular Assembly Cascade.

Pareto Analysis: A technique for distinguishing between items which are 'important' and those which are less important (sometimes known as the ABC or 80/20 rule).
Payback Period: The time taken to recover a capital investment.
PCB: Printed Circuit Board.
Personnel Strategies: The corporate objectives for manpower management.
PERT: *See* 'Programme Evaluation and Review Technique'.
Policy: A set of (corporate) objectives detailing an organization's strategic approach.
Principal Result Objective (PRO): A particular organization's preferred mix of objectives when using DRAMA's Design Option Guide in the methodology.
PRO: *See* Principal Result Objective.
Proactive Functions: Production functions which act in anticipation of future needs.
Product-oriented/Focused Structure: An organization structure which is directed towards the satisfaction of customer and product requirements.
Programme Evaluation and Review Technique (PERT): A network technique used in project planning and control where multiple time estimates are applied to each activity and the overall time probabilities are used in connection with costing and resource allocation.
Prototyping: The testing of products/processes under design by building a test system prior to 'live' implementation.
Pull: The technique of supplying materials and other items on the basis of short-term/immediate requirements of the next operation.
Push: Production/materials supply ahead of immediate requirements, based upon forecasted or anticipated future demand.

Quality Awareness Education (QAE): ICL's quality improvement training programme for production personnel.
Quality Circle: A voluntary group of employees who select and tackle quality and efficiency problems.
Quality Culture: Quality- and customer-focused assumptions and beliefs ingrained within an organization.
Quality Improvement Process: The continual development of product and process design and conformance.
Quality of Conformance: Techniques to assure that actual quality matches design specification.
Quality of Design: Quality features acquired by the product/process at the conception and design stage.

Reactive functions: Production functions which respond to needs after they occur.

Reorder Cycle: A stock control system where an order is placed at regular, predetermined, time intervals.
Reorder Level: A stock control system where fixed order quantities are placed when stock levels decrease to a predetermined point.
ROC: *See* Reorder Cycle.
Return On Investment (ROI): A financial justification technique based on the average annual repayment of a capital investment.
ROI: *See* Return on Investment.
ROL: *See* Reorder Level.

Satellite Buffer: A storage area on the shopfloor for kits and subassemblies to enable high levels of service and flow through manufacturing.
Scientific Management: Management through looking at improvements in efficiency by method study and work measurement (from the work of F. W. Taylor).
Segregated Design: A facility design where the utilities are physically separate.
Service Levels: A quantifiable measure to evaluate performance in terms of meeting delivery dates: usually taken as the proportion of items delivered to customers in the time they were promised.
Showcase Factory: A production plant where an important aspect in its design is the need to have a modern facility to exhibit to customers for publicity reasons.
Simultaneous (Design) Engineering: Parallel development of product, process and organization designs. Otherwise know as 'concurrent engineering'.
Single-status: Equality throughout the workforce in terms of employment conditions.
Soft Systems: The theory (developed by Peter Checkland at the University of Lancaster) where a problem solution is seen as a behavioural, as opposed to a mechanistic, interplay between the real world and the abstract world of systems thinking.
Strategic Domain: The highest level of organizational decision making. Strategy functional objectives which are developed with reference to corporate policy (e.g. 'marketing', 'manufacturing' and 'factory' strategies).
Surface Mount Technology: An advanced process for PCB production where components are bonded to the surface of the board as opposed to being inserted into drilled holes.

Tactical Domain: The intermediate area of decision making between 'strategic' and 'operational'.
Technology-led: Following the leading edge in product/process design.
Time to Market: The speed with which organizations are able to introduce new products on to the market
Total Quality Control (TQC): Often used as an alternative term for 'Total Quality Management' but in practice more restricted and mainly involving statistical process control techniques.

Total Quality Management (TQM): Control and conformance to quality in a holistic sense with individual responsibilities for quality, emphasis on the removal of causes of errors, and continual improvement as an integral philosophy.
Trade-offs: The idea that a benefit to one part of a system will incur a penalty in another part
Turnkey: A contract where the vendor takes on the responsibility for project management and control of subcontractors on behalf of the client so that the facility, etc., is handed over in a ready-to-operate state.

Upstream Production: Production processes which tend towards raw material conversion and component processing as opposed to final assembly and configuration.
User Representatives: Personnel representing the interests of their department within the context of design for new manufacturing systems.
Utility: The hardware subsystems incorporated within a module, i.e. workstations, transportation and storage.

Vertical Integration: Production spanning a number of levels of manufacturing from raw materials processing to final assembly, test and packaging.

Work Design: The arrangement and organization of jobs and conditions for personnel.
Working Party: A design group, containing designers and users across a range of functions, who consider the design of products/processes.
Workstation: A bench or location where assembly and production tasks are accomplished by an operator(s).
World Class Manufacturing: A popular term describing excellence within the manufacturing function of an organization.

Zero Defects (ZD): A campaign-based method of quality assurance and awareness.

References

Abernathy, W. J. (1978) *The Productivity Dilemma: Roadblock to Innovation in the Automobile Industry*, Johns Hopkins Press, New York.

Abernathy, W. J., K. B. Clark and A. M. Kantrow (1981) The new industrial competition, *Harvard Business Review*, **59**, September–October.

Ackoff, R. L. (1977) Towards flexible organizations: A multi-dimensional design approach, *OMEGA*, **5**(6).

Aguren, S. and J. Edgren (1980) *New Factories*, Swedish Employers' Federation, Stockholm.

AMICE ESPRIT (1989) *CIM-OSA Reference Architecture*, European Community Programme for Research in Information Technology, Brussels, Belgium.

Aneke, N. A. G. and A. S. Carrie (1986) A design technique for the layout of multi-product flowlines, *International Journal of Production Research*, **24**(3).

Ansari, A. and B. Modaress (1988) JIT purchasing as a quality and productivity centre, *International Journal of Production Research*, **26**(1).

Argyris, C. (1972) *The Applicability of Organizational Sociology*, Cambridge University Press, London.

Arnold, J. and A. Hope (1990) *Accounting for Management Decisions*, Prentice Hall, Hemel Hempstead.

Asch, D. (1987) The implementation of corporate decisions, *Management Accounting*, **October**.

Atkinson, J. (1984) The Flexible Firm, *Personnel Management*, **August**.

Beal, K. (1988) Integrated materials logistics and continuous flow manufacturing, *International Journal of Production Research*, **26**(3).

Beaton, D. M. (1983) Planning for FMS, in K. Rathmill (ed.) *Proceedings of 2nd International Conference on Flexible Manufacturing Systems*, IFS/North Holland Publications, The Netherlands.

Belt, B. (1987) MRP and kanban: Possible synergy?, *Production and Inventory Management*, **1st quarter**.

Bennett, D. J. (1981) Facility decisions, in C. D. Lewis (ed.) *Operations Management in Practice*, Philip Allan, London.

Bennett, D. J. (1986) *Production Systems Design*, Butterworths, London.

Bennett, D. J. (ed.) (1990) *Design Routine for Adopting Modular Assembly: A Process Design Methodology for Electronics Industries*, Innovation, Design and Operations Management Research Unit, Aston Business School.

Bennett, D. J. and P. L. Forrester (1988) System integration of modularised assembly FMS, *Proceedings of Factory 2000 Conference*, August–September, 1988, IERE Publications, Cambridge.

Bennett, D. J. and P. L. Forrester (1991) The DRAMA methodology for analysing strategy and its links with production system design, *International Journal of Technology Management*, **6**(3/4).

Bennett, D. J., C. D. Lewis and M. H. Oakley (1988) *Operations Management*, Philip Allan, Hemel Hempstead.

Bennett, D. J. and C. V. Nagarkar (1988) Flexible manufacturing system lets small manufacturer of mainframes compete with giants, *Industrial Engineering*, **20**(11).

Bennett, D. J., P. L. Forrester and J. S. Hassard (1990) An application of decision process modelling to manufacturing system design, *OMEGA*, **18**(1).

Bennett, D. J., P. L. Forrester and J. S. Hassard (1992) Market-driven strategies and the design of flexible production systems: evidence from the electronics industry, *International Journal of Operations and Production Management*, **12**(6).

Bessant, J. (1988) Pushing boxes or solving problems? Some marketing issues in the diffusion of computer integrated manufacturing innovations, *Journal of Marketing Management*, **3**(3).

Bessant, J. and B. Haywood (1986) Flexibility in manufacturing systems, *OMEGA*, **14**(6), 465–73.

von Bertalanffy, L. (1968) *General System Theory*, Braziller, New York.

Bhattacharyya, S. K. and A. M. Gibbons (1992) Aspects of efficiency in engineering manufacturing for the 1990s, *Proceedings of Industrial Automation '92 Conference*, Singapore Industrial Automation Association/Cahners Exhibition pte, Singapore. May.

Bititci, U. S. (1986) Achieving integration as well as flexibility, *FMS Magazine*, **July**.

Blau, P. M. and R. A. Schoenherr (1971) *The Structure of Organizations*, Basic Books, New York.

Blois, K. J. (1986) Marketing strategies and the new manufacturing technologies, *International Journal of Operations and Production Management*, **6**(1).

Boer, H. and W. E. During (1987) Management of process innovation – the case of FMS: A systems approach, *International Journal of Production Research*, **25**(11).

Bolwijn, P. T. and T. Kumpe (1986) Toward the Factory of the Future, *The McKinsey Quarterly*, **Spring**.

Bowen, D. E., C. Siehl and B. Scheider (1989) A framework for analysing customer service orientations in manufacturing, *Academy of Management Review*, **14**(1).

Bowencamp, R. D. and B. H. Kleiner (1987) How to be a successful project manager, *Industrial Management and Data Systems*, March–April, 3–7.

Bower, J. L. (1985) Capital budgeting as a general management problem, in E. Rhodes and D. Wield (ed.) *Implementing New Technologies: Choice, Design and Change in Manufacturing*, Blackwell, Oxford.

Bowey, A. and R. Connelly (1975) Applications of the concept of working in groups, *Management Decision*, **4**(3).

Burbidge, J. L. (1978) *Principles of Production Control*, Macdonald & Evans, London.

Burbidge, J. L. (1982) The Japanese kanban system, *Institute of Production Control*, **January–February**.

Burgoine, J. (1988) Beyond MRPII, *Production Engineer*, **January**.

Burns, A. N. (1987) JIT manufacturing, *British Production and Inventory Control Society*, **December–January**.

Burns, T. and G. M. Stalker (1961) *The Management of Innovation*, Tavistock, London.

Butler, R. (1991) *Designing Organizations: A Decision Making Perspective*, Routledge, London.

Buzzacott, J. A. (1982) The fundamental principles of flexibility in manufacturing systems, in *Proceedings of 1st International Conference on Flexible Manufacturing Systems*, IFS/North Holland Publications, The Netherlands.

Castleberry, G. A. (1992) *AGV System Specification, Procurement and Implementation Guide*, AGV Decisions Inc, Port Washington, Wisconsin.

Caulkin, S. (1987) ICL's Lazarus act, *Management Today*, **January**.

Celley, A. F., W. H. Clegg, A. W. Smith and M. A. Vonderembse (1986) Implementation of JIT in the United States, *Journal of Purchasing and Inventory Management*, **Winter**.

Chandler, A. D. (1962) *Strategy and Structure: Chapters in the History of the Industrial Enterprise*, MIT Press, Cambridge, Massachusetts.

Checkland, P. B. (1981) *Systems Thinking, Systems Practice*, Wiley, Chichester.

Checkland, P. B. and J. Scholes (1990) *Soft Systems Methodology in Practice*, Wiley, Chichester.

Child, J. (1972) Organization structure, environment and performance: The role of strategic choices, *Sociology*, **January**.

Child, J. (1984) *Organizations: A Guide to Problems and Practice*, Harper and Row, London.

Clark, P. A. (1987) *Anglo-American Innovation*, De Gruyter, London.

Clegg, S. (1979) *The Theory of Power and Organization*, Routledge, London.

Cleland, D. I. and W. R. King (1983) *Project Management Handbook*, Van Nostrand Reinhold, New York.

Cooke, P. N. C. (1986) Rethinking investment appraisal for CIM, *FMS Magazine*, **July**.

Cooke, S. and N. Slack (1991) *Making Management Decisions*, Prentice Hall, Hemel Hempstead.

Crosby, P. B. (1979) *Quality is Free*, McGraw-Hill, New York.

Cyert, R. M. and J. G. March (1963) *A Behavioral Theory of the Firm*, Prentice Hall, Englewood Cliffs, New Jersey.

Daft, R. L. (1991) *Management*, Dryden Press, Chicago.

Day, G. S. (1990) *Market Driven Strategy: Processes for Creating Value*, The Free Press, New York.

Doumeingts, G. and D. Chen (1992) Decisional systems design using GRAI method, *Proceedings of Industrial Automation 1992 Conference*, Singapore, May.

Doumeingts, G., M. C. Maisoneuve, V. Brand and C. Berard (1986) Design methodology for computer integrated manufacturing and control of manufacturing units, in U. Rembold and R. Dillman (ed.) *Computer-Aided Design and Manufacturing: Methods and Tools*, Springer-Verlag, Berlin.

DTI (1988) *Competitive Manufacturing: A Practical Guide to the Development of a Manufacturing Strategy*, IFS Publications, Bedford.

Edwards, G. A. B. (1974) Group technology: technique, concept or new manufacturing system?, *The Chartered Mechanical Engineer*, **February and March**.

Emery, F. E. and E. L. Trist (1965) The causal texture of organizational environments, *Human Relations*, **February**.

Ettlie, J. (1988) *Taking Charge of Manufacturing*, Jossey-Bass, New York.

Farish, M. (1989) Meeting the market, *Engineers' Digest*, **May**.

Forrester, P. L. and J. S. Hassard (1992a) The CAPM/CAE Interface within CIM System Approaches in Medium Sized Companies, *Computing and Control Engineering Journal*, **3**(2).

Forrester, P. L. and J. S. Hassard (1992b) Interconnecting computer-aided design engineering and production management: Cases from UK manufacturing companies, *Proceedings of the Penang Electronics and Associated Industries (PEMAS) Conference*, Penang, Malaysia, May 1992.

Forrester, P. L., J. S. Hassard and N. Tang (1991) CIM: The ultimate goal for manufacturing?, *Proceedings of 7th National Conference on Production Research*, Hatfield, U.K., September.

Foyer, P. and P. Drazon (1986) Assembly: The Cinderella of factories, *Production Engineer*, **March**.

Frazelle, E. H. (1986) Flexibility: A strategic response in changing times, *Industrial Engineer*, **March**.

Galonek, J. (1987) Subassembly area renovation improves methods and increases work flow, *Industrial Engineer*, **April**.

Ghobadian, A. and D. A. Smyth (1989) The role of post-audit in effective control of capital projects, *Proceedings of the Operations Management Association Conference*, Dunblane, U.K., January 1989.

Goldratt, E. M. and J. Cox (1989) *The Goal: Excellence in Manufacturing*, North River Press, New York.

Goldsmith, W. and D. Clutterbuck (1985) *The Winning Streak*, Weidenfeld and Nicolson, London.

Goodman, L. J. and R. N. Love (1980) *Project Planning and Management: An Integrated Approach*, Pergamon, New York.

Goodridge, I. (1986) New concepts in manufacturing, *Management Services*, **November**.

Grinyer, P. H., M. Yasai-Ardekani and S. Al-Bazzaz (1980) Strategy, structure, the environment and financial performance in 48 United Kingdom companies, *Academy of Management Journal*, **June**.

Gustavsson, S. O. (1984) Flexibility and productivity in complex production processes, *International Journal of Production Research*, **22**(5).

Gyllenhammer, P. (1977) *People at Work*, Addison-Wesley, Reading, Massachusetts.

Halpin, J. (1966) *Zero Defects*, McGraw-Hill, New York.

Handy, C. (1983) *Understanding Organizations*, Penguin, London.

Hannah, K. H. (1987) Just-in-time: meeting the competitive challenge, *Production and Inventory Management*, **28**(3).

Harrison, M. C. (1985) The concepts of optimised production technology, *BPICS Control*, **June–July**.

Harrison, M. (1990) *Advanced Manufacturing Technology Management*, Pitman, London.

Hartley, J. (1984) *FMS at Work*, IFS Publications, Bedford, U.K.

Hayes, R. H. (1981) Why Japanese Factories Work, *Harvard Business Review*, **59**, **July–August**.

Hayes, R. H. and R. Schmenner (1978) How should you organize manufacturing?, *Harvard Business Review*, **January–February**.

Hayes, R. H. and S. C. Wheelwright (1984) *Restoring Our Competitive Edge: Competing Through Manufacturing*, Wiley, New York.

Hayes, R. H., S. C. Wheelwright and K. B. Clark (1988) *Dynamic Manufacturing: Creating the Learning Organization*, Free Press, New York.

Heizer, J. and B. Render (1988) *Production and Operations Management*, Allyn and Bacon, Boston, Massachusetts.

Hemphill, J. K. and C. M. Westie (1950), The measurement of group dimensions, *Journal of Psychology*, **29**.

Heyer, N. L. and V. Wemmerlov (1984) Group technology and productivity, *Harvard Business Review*, **62**, **July–August**.

Hickson, D. J., D. S. Pugh and D. C. Pheysey (1969) Operations technology and organization structure: An empirical reappraisal, *Administrative Science Quarterly*, **September**.

Hickson, D. J., R. J. Butler, D. Cray, G. R. Mallory and D. Wilson (1985) *Top Decisions: Strategic Decision Making in Organizations*, Blackwell, Oxford.

Hill, R. (1986) Philips Over the Years: A model of the maturing multinational, *International Management*, **August**.

Hill, T. J. (1985) *Manufacturing strategy: The strategic management of the manufacturing function*, Macmillan, London.

Hill, T. J. (1991) *Production/Operations Management: Text and Cases*, Prentice Hall, London.

HMSO (1977) *Life-Cycle Costing in the Management of Assets*, Her Majesty's Stationery Office, London.

Hunt, V. D. (1983) *Industrial Robotics Handbook*, Industrial Engineering & Management Press, Norcross, USA.

Hygen, J. (1984) An evaluation of implementation strategies for production management systems, in H. Huber (ed.) *Production Management Systems: Strategies and Tools for Design*, Elsevier, Amsterdam.

ILO (1979) *Introduction to Work Study*, International Labour Office, Geneva.

Imai, M. (1986) *Kaizen: The Key to Japan's Competitive Success*, Random House, New York.

Jacques, E. (1951) *The Changing Culture of a Factory*, Tavistock, London.

Jaffe, J. (1967) The system design phase, in P. E. Rosove (ed.) *Developing Computer-based Information Systems*, Wiley, New York.

Johnson, G. and K. Scholes (1989) *Exploring Corporate Strategy: Text and Cases*, Prentice Hall International, Hemel Hempstead.

Kanter, R. M. (1984) *The Change Masters: Corporate Entrepreneurs at Work*, Unwin, London.

Kanter, R. M. (1989) *When Giants Learn to Dance*, Unwin, London.

Kantrow, A. M. (1980) The strategy–technology connection, *Harvard Business Review*, **58** (July–August).

Kaplan, R. S. (1984) Yesterday's accounting undermines production, *Harvard Business Review*, **62**, July–August.

Kaplan, R. S. (1986) Must CIM be justified by faith alone?, *Harvard Business Review*, **64**, March–April.

Karlsson, U. and D. J. Bennett (1991) Competing via alternative work organizations: The case of car assembly in Sweden, in D. J. Bennett and C. D. Lewis (ed.) *Achieving Competitive Edge: Getting Ahead Through Technology and People*, Springer-Verlag, London.

Kast, F. E. and J. E. Rosenzweig (1970) *Organization and Management: A Systems Approach*, McGraw-Hill, New York.

Katz, R. and R. Kahn (1975) *The Social Psychology of Organizations*, Wiley, New York.

Keats, B. W. and M. Hitt (1988) A causal model of linkages among environmental dimensions, macro characteristics and performance, *Academy of Management Journal*, **September**.

Kenny, G. K. and A. S. Dunk (1989) The utility of performance measures: production managers' perceptions, *IEEE Transactions on Engineering Management*, **36**(1).

Keogh, J. P. (1987) A practical approach to automation, in D. C. Teale (ed.) *New Frontiers in Manufacturing*, Springer-Verlag, London.

Kerzner, H. (1984) *Project Management for Executives*, Van Nostrand Reinhold, New York.

Knobel, L. (1988) Hewlett-Packard's culture shock, *Management Today*, **June**.

Kotler, P. (1986) *Principles of Marketing*, Prentice Hall, Englewood Cliffs, New Jersey.

Krajewski, L. J. and L. Ritzman (1990) *Operations Management: Strategy and Analysis*, Addison-Wesley, Reading, Massachusetts.

Krupp, J. A. G. (1984) Why MRP systems fail: Traps to avoid, *Production and Inventory Management*, **25**(3).

Kumpe, T. and P. T. Bolwijn (1983) Automation without tears, in K. Rathmill (ed.) *Proceedings of 2nd International Conference on Flexible Manufacturing Systems*, IFS/North Holland Publications, The Netherlands.

Kumpe, T. and P. T. Bolwijn (1988) The new case for vertical integration, *Harvard Business Review*, **66** (March–April).

Lamming, R. (1987) Strategic approaches to just-in-time purchasing and supply in the automotive and electronics industries (UK, USA, Canada), in C. A. Voss (ed.)

Just in Time Manufacturing: Proceedings of the 2nd International Conference, IFS (Publications) Ltd, Bedford.

Lawrence, P. and J. W. Lorsch (1967) *Organization and the Environment: Managing Differentiation and Integration*, Harvard Business School Press, Boston, Massachusetts.

Lewis, C. D. (ed.) (1981) *Operations Management in Practice*, Philip Allan, Hemel Hempstead.

Lim, S. H. (1986) Flexible manufacturing systems and manufacturing flexibility in the United Kingdom, *International Journal of Operations and Production Management*, **7**(6).

Lindholm, R. (1979) Towards a new world of work: Swedish development of work organizations, production engineering and co-determination, *International Journal of Production Research*, **17**(5).

Lumby, S. (1988) *Investment Appraisal and Financing Decisions*, VNR, Wokingham, U.K.

Macbeth, D. K. and N. Ferguson (1991) Strategic aspects of supply chain management, *Integrated Manufacturing Systems*, **2**(1).

McLean, T. (1988) JIT demands a new accounting system, *Works Management*, **February**.

Maji, R. K. and S. K. Banerjee (1988) Design of advanced manufacturing systems, *Proceedings of Factory 2000 Conference*, August–September 1988. IERE Publications, Cambridge, U.K.

Manley, J. H. (1975) Implementation attitudes: A model and measurement methodology, in R. Doktor, R. L. Schultz and D. P. Slevin (ed.) *Implementing Operations Research and Management Science*, North Holland Publishers, Amsterdam.

Marca, D. A. and C. L. McGowen (1988) *Structured Analysis and Design Technique*, McGraw-Hill, New York.

Marechal, G. (1989) From factory automation to computer integrated manufacturing according to the open systems architecture CIM-OSA, in F. Kimura and A. Rolstadas (ed.) *Computer Applications in Production and Engineering*, North-Holland Publishers, Amsterdam.

Marsh, P., P. Barwise, K. Thomas and R. Wensley (1988) Managing strategic investment decisions, in A. M. Pettigrew (ed.) *Competitiveness and the Management Process*, Blackwell, Oxford.

Meredith, J. R. (ed.) (1986) *Justifying New Manufacturing Technology*, Industrial Engineering and Management Press, Atlanta.

Meredith, J. R. (1987) Automating the factory: Theory versus practice, *International Journal of Production Research*, **25**(10).

Meredith, J. R. and S. J. Mantel (1985) *Project Management*, Wiley, New York.

Meyer, M. W. (1972) Size and the structure of organizations: A causal analysis, *American Sociological Review*, **August**.

Miles, R. E. and C. C. Snow (1978) *Organization Strategy, Structure and Process*, McGraw-Hill, New York.

Miller, D. (1988) Relating Porter's business strategies to environment and structure: Analysis and performance implementations, *Academy of Management Journal*, **31** (June).

Mintzberg, H., D. Raisinghani and A. Theoret (1976) The structure of 'unstructured' decision processes, *Administrative Science Quarterly*, **21**.

Moder, J. J., C. K. Phillips and E. W. Davis (1983) *Project Management with CPM and PERT*, Van Nostrand Reinhold, New York.

Moinet, M. (1987) A modular strategy for automation and CIM, in D. C. Teale (ed.) *New Frontiers in Manufacturing*, Springer-Verlag, London.

Monden, Y. (1981) What makes the Toyota production systems really tick?, *Industrial Engineering*, **January**.

Monden, Y. (1986) *Applying just in time: The American Japanese experience*, Industrial Engineering & Management Press, Norcross, USA.
Moore, P. G. (1982) *Basic Operational Research*, Pitman, London.
Morris, P. W. G. (1983) Managing project interfaces: Key points for project success, in D. I. Cleland and W. R. King (ed.) *Project Management Handbook*, Van Nostrand Reinhold, New York.
Moskowitz, H. and G. P. Wright (1979) *Operations Research Techniques for Management*, Prentice Hall, Englewood Cliffs, New Jersey.
Mosier, C. and L. Tanbe (1985) The facets of group technology and their impacts on implementation, *OMEGA*, **13**(5).
Muhlemann, A., J. Oakland and K. Lockyer (1992) *Production and Operations Management*, Pitman, London.
Murray, R. (1989) Life after Henry (Ford), *Marxism Today*, **October**.
Neale, W. and D. Holmes (1988) Post-completion audits: The costs and benefits, *Management Accounting*, **March**.
Nelleman, D. O. and C. F. Smith (1982) Just-in-time versus just-in-case production/inventory systems: Concepts borrowed back from Japan, *Production and Inventory Management*, **23**(2).
Newell, A. and H. Simon (1972) *Human Problem Solving*, Prentice Hall, Englewood Cliffs, New Jersey.
Nicholson, H. (1991) *Interconnected Manufacturing Systems: The Problems of Advanced Manufacturing*, Peter Peregrinus Ltd, London.
Norgaard, C. J. (1979) The post-completion audit of capital projects, *Cost and Management Accounting*, **January–February**.
Nutt, P. C. (1983) Implementation approaches for project planning, *Academic Management Review*, **8**.
Oakley, M. H. (1984) *Managing Product Design*, Weidenfeld and Nicolson, London.
O'Neal, C. L. (1987) The buyer–seller linkage in a just-in-time environment, *International Journal of Production Research*, **25**(1).
Opitz, H. (1970) *A Classification to Describe Workpieces*, Pergamon, Oxford.
Orlicky, J. A., G. W. Plossl and O. W. Wight (1972) Structuring the Bill of Materials, *Production and Inventory Management*, **13**(4).
Parnaby, J. (1988) A systems approach to the implementation of JIT methodologies in Lucas Industries, *International Journal of Production Research*, **26**(3).
Pascale, R. T. and A. G. Athos (1981) *The Art of Japanese Management*, Simon and Schuster, New York.
Perrow, C. (1967) A framework for the comparitive analysis of organizations, *American Sociological Review*, **April**.
Peters, T. J. (1987) *Thriving on Chaos*, Macmillan, London.
Peters, T. J. and R. H. Waterman (1982) *In Search of Excellence: Lessons from America's Best Run Companies*, Harper and Row, New York.
Pidd, M. (1988) *Computer Simulation in Management Science*, Wiley, New York.
Pinto, J. K. and D. P. Slevin (1987) Critical factors in successful project implementation, *IEEE Transactions on Engineering Management*, **EM-34**(1), February.
Piore, M. and C. Sabel (1984) *The Second Industrial Divide: Possibilities for Prosperity*, Basic Books, New York.
Piper, A. (1985) Breaking the multinational mould: How DEC snuggled up to its customers, *International Management*, **November**.
Porter, M. E. (1985) *Competitive Advantage: Creating and Maintaining Superior Performance*, Free Press, New York.
Primrose, P. L., G. D. Creamer and R. Leonard (1985) Identifying and quantifying the company-wide benefits of CAD within the structure of a comprehensive investment programme, in E. Rhodes and D. Wield (eds.) *Implementing New Technologies: Choice, Design and Change in Manufacturing*, Blackwell, Oxford.

Primrose, P. L. and R. Leonard (1986) The use of a conceptual model to evaluate financially flexible manufacturing system projects, in J. R. Meredith (ed.) *Justifying New Manufacturing Technology*, Industrial Engineering and Management Press, Norcross, USA.

Pugh, D. S., D. J. Hickson, C. R. Hinings and C. Turner (1968) Dimensions of organizational structure, *Administrative Science Quarterly*, **13**(1).

Rajput, S. K. and D. J. Bennett (1989) Modular systems design and control for flexible assembly, *International Journal of Operations and Production Management*, **9**(7).

Randhawa, S. U. and D. Bedworth (1985) Factors identified for use in comparing conventional and flexible manufacturing systems, *Industrial Engineer*, **June**.

Ranky, P. (1983) *The Design and Operation of FMS*, IFS/North Holland Publications, Bedford, U.K.

Ranky, P. G. (1990) *Total Quality Control and JIT Management in CIM*, CIMWare Ltd, Guildford.

Rao, A. and D. I. Scheraga (1987) Systems for supporting the MRPII/JIT Environment, *CIM Review*, **Summer**.

Reed, M. (1985) *Redirections in Organizational Analysis*, Tavistock, London.

Rhodes, E. and D. Wield (1985) *Implementing New Technologies: Choice, Design and Change in Manufacturing*, Blackwell, Oxford.

Richardson, P. R. and J. R. M. Gordon (1980) Measuring total manufacturing performance, *Sloan Management Review*, **Winter**.

Ridgway, K. and S. Downey (1991) A graphical technique for planning the introduction of AMT: the GRAI method, in E. H. Robson, H. M. Ryan and D. Wilcock (ed.) *Integration and Management of Technology for Management*, Peter Peregrinus Ltd/IEE, London.

Riedel, J. C. K. H. and K. S. Pawar (1991) The strategic choice of simultaneous versus sequential engineering for the introduction of new products, *International Journal of Technology Management*, **6**(3/4).

Robbins, S. P. (1990) *Organization Theory: Structure, Design and Applications*, Prentice Hall, Englewood Cliffs, New Jersey.

Roboam, M. and L. Pun (1989) Utility of design methodology for advanced manufacturing systems, *Computer Integrated Manufacturing Systems*, **2**(1).

Roboam, M., M. Zanetlin and L. Pun (1989) GRAI-MERISE-IDEF0: Integrated methodology to analyse and design manufacturing systems, *Computer Integrated Manufacturing Systems*, **2**(2).

Roethlisberger, F. J. and W. J. Dickson (1939) *Management and the Worker*, Harvard University Press, Boston, Massachusetts.

Rogers, E. (1983) *The Diffusion of Innovations*, Free Press, New York.

Ross, D. T. (1977) Structured analysis: A language for communicating ideas, *IEEE Transactions*, **SE-3**.

Rowlinson, M., J. S. Hassard and P. L. Forrester (1991) Who wants harmonisation?: Image and reality in single status working, *Personnel Review*, **20**(5).

Russell, P. J. (1987) Open systems architecture for CIM, *Proceedings of Third CIM Europe Conference*, May, 1987. Knutsford, U.K.

Sackett, P. J. and G. M. Williams (1988) An integrated structure for production scheduling and control, *Proceedings of Factory 2000 Conference*, August–September, Cambridge, U.K.

SAF-LO/PTK (1982) *Agreement on Efficiency and Participation*, Swedish Employers' Confederation, Stockholm.

Sandberg, T. (1982) *Work Organization and Autonomous Groups*, Liber Forlag, Lund, Sweden.

Saul, G. (1985) Flexible manufacturing system is CIM implemented at the shop floor level, *Industrial Engineer*, **June**.

Schonberger, R. J. (1982) *Japanese Manufacturing Techniques: Nine Hidden Lessons in Simplicity*, Free Press, New York.
Schonberger, R. J. (1986) *World Class Manufacturing: The Lessons of Simplicity Applied*, Free Press, New York.
Schonberger, R. J. (1987) *World Class Manufacturing: Implementing JIT and TQC*, Free Press, New York.
Schregle, J. (1987) Workers participation in the Federal Republic of Germany in an international perspective, *International Labour Review*, **126**(3).
Schroeder, D. M. (1990) A dynamic perspective on the impact of process innovation upon competitive strategies, *Strategic Management Journal*, **11**(1).
Shanklin, W. L. and J. K. Ryans (1981) Is the international cash cow really a prize heifer?, *Business Horizons*, **24**(1).
Shingo, S. (1986) *Zero Quality Control, Source Inspection and the Poka-Yoke System*, Productivity Press, New York.
Silverman, D. (1970) *The Theory of Organizations*, Heinemann, London.
Simpson, A. (1986) Effective just-in-time manufacturing at Hewlett-Packard, in J. Mortimer (ed.) *Just-in-Time: An Executive Briefing*, IFS Publications, Bedford, U.K.
Sizer, J. (1979) *An Insight Into Management Accounting*, Penguin, London.
Skapinker, M. (1986) Xerox searches for life beyond boxes, *International Management*, **June**.
Skinner, W. (1969) Manufacturing – missing link in corporate strategy, *Harvard Business Review*, **May–June**.
Skinner, W. (1978) *Manufacturing in the Corporate Strategy*, Wiley, New York.
Skinner, W. (1985) *Manufacturing: The Formidable Competitive Weapon*, Wiley, New York.
Slack, N. (1983) Flexibility as a manufacturing objective, *International Journal of Operations and Production Management*, **3**(3).
Slack, N. (1986) The flexibility of manufacturing systems, *International Journal of Operations and Production Management*, **7**(4).
Slagmulder, R. and W. Bruggeman (1991) Evaluating flexible manufacturing technologies: A research agenda for investment decision making, in D. J. Bennett and C. D. Lewis (ed.) *Achieving Competitive Edge: Getting Ahead Through Technology and People*, Springer-Verlag, London.
Smith, A. (1970) *The Wealth of Nations* (first published 1776), Everyman's Library, London.
Snead, C. (1989) *Group Technology – Foundation for Competitive Manufacturing*, Van Nostrand Reinhold, New York.
Storey, J. (ed.) (1991) *New Perspectives on Human Resource Management*, Routledge, London.
Sugimori, Y. (1986) Toyota production and the kanban system, in J. Mortimer (ed.) *Just-in-Time: An Executive Briefing*, IFS Publications, Bedford, U.K.
Suzaki, K. (1987) *The New Manufacturing Challenge: Techniques for Continuous Improvement*, The Free Press, New York.
Sweeney, M. (1986) Flexible manufacturing systems: Managing their integration, in C. A. Voss (ed.) *Managing Advanced Manufacturing Technology*, IFS Publications, Bedford, U.K.
Sweeney, M. (1992) How to perform simultaneous process engineering, *Integrated Manufacturing Systems*, **3**(2).
Talysum, A. T., M. Z. Hassan and J. D. Goldhar (1986) Scale versus scope considerations in the CIM/FMS factory, in A. Kusiak (ed.) *Flexible Manufacturing Systems: Methods and Studies*, Elsevier Science Publications, Amsterdam.
Thompson, J. D. (1967) *Organizations in Action*, McGraw-Hill, New York.

Toft, L. A. (1988) Project management: A personal view, *Proceedings of the Institute of Mechanical Engineers*, **202**(131).

Trist, E. L., G. W. Higgin, H. Murray and A. B. Pollock (1963) *Organizational Choice*, Tavistock, London.

Urry, S. (1991) *Introduction to Operational Research*, Longman, Harlow, U.K.

Utterbeck, J. M. and W. J. Abernathy (1975) A Dynamic Model of Process and Product Innovation, *OMEGA*, **3**(6).

Vonderembse, M. A. and G. S. Wobser (1987) Steps for implementing a flexible manufacturing system, *Industrial Engineer*, **April**.

Voss, C. A. and S. J. Robinson (1987) Application of just-in-time manufacturing techniques in the United Kingdom, *International Journal of Operations and Production Management*, **7**(4).

Vuzelov, V. (1983) A modular approach to flexible automation, in K. Rathmill (ed.) *Proceedings of 2nd International Conference on Flexible Manufacturing Systems*, IFS/North Holland Publications, Bedford, U.K.

Ward, J. (1986) JIT production: Theory into practice, in J. Mortimer (ed.) *Just-in-Time: An Executive Briefing*, IFS Publications, Bedford, U.K.

Warnecke, H. (1988) Computer integrated manufacturing: A production strategy, *Computer Integrated Manufacturing Systems*, **1**(1).

Wheaton, M. G. (1987) Real world AMT, in D. C. Teale (ed.) *New Frontiers in Manufacturing*, Springer-Verlag, London.

Wheelwright, S. C. (1981) Japan: Where operations really are strategic, *Harvard Business Review*, **59**, July–August.

Wheelwright, S. C. and R. H. Hayes (1985) Competing through manufacturing, *Harvard Business Review*, **63**, January–February, 99–109.

Whipp, R. and P. A. Clark (1986) *Innovation and the Auto Industry*, Francis Pinter, London.

Wild, R. (1980) *Operations Management: A Policy Framework*, Pergamon, London.

Wild, R. (1984) *Production and Operations Management, Principles and Techniques*, Holt, Rinehart and Winston, London.

Wild, R. (1990) *Essentials of Production and Operations Management*, Cassell, London.

Womack, J., D. Jones and D. Roos (1990) *The Machine that Changed the World*, Rawson Associates, London.

Woodward, J. (1965) *Industrial Organization: Theory and Practice*, Oxford University Press, London.

Yeomans, P. H. (1985) *Improving Quality and Productivity in Systems Modelling using the IDEF Methodologies*, MicroMatch Limited, U.K.

Zelevnovic, D. M. (1986) Flexibility: A condition for effective production systems, *International Journal of Production Research*, **24**(3).

Zelenovic, D. M., I. P. Cosic, D. N. Sormaz and Z. D. Sisarica (1987) An approach to the design of more effective production systems, *International Journal of Production Research*, **25**(1).

Index